Exposition by Emil Artin: A Selection

Michael Rosen
Editor

Exposition by Emil Artin: A Selection

Michael Rosen
Editor

American Mathematical Society
London Mathematical Society

PENN STATE BRANDYWINE

THE PENNSYLVANIA STATE UNIVERSITY
COMMONWEALTH CAMPUS LIBRARIES
DELAWARE COUNTY

Editorial Board

American Mathematical Society
Joseph W. Dauben
Peter Duren
Karen Parshall, Chair
Michael I. Rosen

London Mathematical Society
Alex D. D. Craik
Jeremy J. Gray
Robin Wilson, Chair
Steve Russ

2000 *Mathematics Subject Classification.* Primary 01A75, 11–06.

For additional information and updates on this book, visit
www.ams.org/bookpages/hmath-30

ISBN-10: 0-8218-4172-6
ISBN-13: 978-0-8218-4172-3

Copying and reprinting. Individual readers of this publication, and nonprofit libraries acting for them, are permitted to make fair use of the material, such as to copy a chapter for use in teaching or research. Permission is granted to quote brief passages from this publication in reviews, provided the customary acknowledgment of the source is given.

Republication, systematic copying, or multiple reproduction of any material in this publication is permitted only under license from the American Mathematical Society. Requests for such permission should be addressed to the Acquisitions Department, American Mathematical Society, 201 Charles Street, Providence, Rhode Island 02904-2294, USA. Requests can also be made by e-mail to reprint-permission@ams.org.

© 2007 by the American Mathematical Society. All rights reserved.
Printed in the United States of America.

The American Mathematical Society retains all rights
except those granted to the United States Government.
∞ The paper used in this book is acid-free and falls within the guidelines
established to ensure permanence and durability.
The London Mathematical Society is incorporated under Royal Charter
and is registered with the Charity Commissioners.
Visit the AMS home page at http://www.ams.org/

10 9 8 7 6 5 4 3 2 1 12 11 10 09 08 07

Contents

Photo by Natascha Artin Brunswick. Published courtesy of the Estate of Natascha Artin Brunswick.

Emil Artin standing in front of the main building at the University of Hamburg, early to mid-1930s.

Photo by Natascha Artin Brunswick. Published courtesy of the Estate of Natascha Artin Brunswick.

Emil Artin in the garden outside of the Artin's Hamburg apartment, early to mid-1930s.

Photo by Natascha Artin Brunswick. Published courtesy of the Estate of Natascha Artin Brunswick.

Emil Artin and a younger colleague, Erich Kähler, taking a coffee break at a garden café outside the center of Hamburg, early to mid-1930s. Kähler assumed Artin's chair at the University of Hamburg in 1964, following Artin's death.

Credits and Acknowledgments

The American Mathematical Society gratefully acknowledges the kindness of these individuals/institutions in granting the following permissions:

American Scientist

"The theory of braids", American Scientist **38** (1950), pp. 112–119 (pp. 291–298 in this volume). Reprinted by permission of American Scientist, magazine of Sigma Xi, The Scientific Research Society.

Annals of Mathematics, Princeton University

"Theory of braids", Ann. of Math. **48** (1947), pp. 121–126 (pp. 299–324 in this volume).

Michael Artin

"A proof of the Krein-Milman Theorem", originally appearing in Max Zorn's informally published *Piccayune Sentinel*, 1950, pp. 335–337 in this volume.

Tom Artin

Front cover photo of Emil Artin, inside the Artin's Hamburg apartment, early to mid-1930s. Taken by Natascha Artin Brunswick, published courtesy of the Estate of Natascha Artin Brunswick.

Back cover photo of Emil Artin, Emmy Noether, and an unidentified man, in either Hamburg or Göttingen, sometime before 1933. Of this photo, Tom Artin writes, "...this picture resonates especially for me because it depicts how—as a child—I so often witnessed mathematics being done. To the end of his life, my father was an inveterate walker. I have vivid memories of trailing behind him and his colleagues, hearing their incomprehensible conversations in the mysterious language of mathematics." Taken by Natascha Artin Brunswick, published courtesy of the Estate of Natascha Artin Brunswick.

Three photos of Emil Artin, pp. vi–vii. Taken by Natascha Artin Brunswick, published courtesy of the Estate of Natascha Artin Brunswick.

Mathematic Seminary, University of Hamburg

English translations of:

"A characterization of the field of real algebraic numbers (Kennzeichnung der Körper der reellen algebraischen Zahlen)", Hamb. Abh. **3** (1924), pp. 319–323 (pp. 269–272 in this volume).

"Algebraic construction of real fields (Algebraische Konstruction reeller Körper (with O. Schreier))", Hamb. Abh. **5** (1926), pp. 85–99 (pp. 273–283 in this volume).

"A characterization of real closed fields" (Kennzeichnung der reell abgeschlossen Körper (with O. Schreier))", Hamb. Abh. **5** (1927), pp. 225–231 (pp. 285–290 in this volume).

Thomson Learning

The Gamma Function, pp. 13–59 in this volume. From *The Gamma Function* 1st edition by ARTIN E / Translated by Michael Butler. ©1964. Reprinted with permission of Brooks/Cole, a division of Thomson Learning: www.thomsonrights.com. Fax 800 730-2215.

Public Domain

Galois Theory, pp. 61–107 in this volume, is a work in the public domain.

Theory of Algebraic Numbers, pp. 109–237 in this volume, is a work in the public domain.

"On the theory of complex functions", pp. 325–334 in this volume, is a work in the public domain.

All other works in this volume were originally published by the American Mathematical Society.

Introduction

Michael Rosen

Emil Artin was one of my intellectual heroes. As an undergraduate math major I studied, in detail, his well-known booklet *Galois Theory* and spent a summer mastering his wonderful book *Geometric Algebra*. Moreover, in class I learned algebra from the famous textbook *Modern Algebra* of B. L. van der Waerden, which was based, at least in part, on lectures of Emil Artin and Emmy Noether. Artin's influence seemed to be everywhere, and not just in these expository works. He was a renowned research mathematician, one of the most important algebraists and number theorists of the twentieth century. Already during this early stage of my mathematical education, I had heard of the Artin reciprocity law, Artin L-functions, Artinian rings, and the famous Artin conjectures. For all these reasons, I formed the ambition of attending Princeton University as a graduate student and studying under his direction. Unfortunately, when I arrived at Princeton in the fall of 1959 he was no longer on the faculty. He had accepted a position at Hamburg University in Germany.

Luckily, even without the author, the printed word remained. I continued to learn from Artin's expository writings. His works have a gem-like quality, brilliant in form and content. Unfortunately, many of them are no longer readily available. This small volume represents a modest attempt to correct this situation. The reader will find assembled here a selection of Artin's books and articles, all of which retain their power to both instruct and delight.

Part 1

Emil Artin was born in Vienna on March 3, 1898. His father was an art dealer and his mother an opera singer. Art and music were important to him throughout his life. When he was only four years old, his father died and he was sent to live with his grandmother. Two years later his mother remarried and he returned to live with her and his new stepfather, who was the owner of a cloth factory in Reichenberg, Bohemia. With the exception of a happy year of study in Paris at the age of fourteen, Artin lived in Reichenberg until his graduation from high school. After only one semester at the University of Vienna, he was drafted into the Austrian army where he served until the end of the First World War.

In January of 1919 he entered the University of Leipzig where he studied mathematics under the supervision of Gustav Herglotz. His relationship with Herglotz was excellent. According to Serge Lang and John Tate, his future Ph.D. students, Herglotz was the only person that Artin recognized as his "teacher". He made rapid strides. Within two years he was awarded a doctorate on the basis of an outstanding thesis, a seminal work that influenced many later developments in number theory and algebraic geometry. After graduating from the University of Leipzig, Artin went to Göttingen, the center of the mathematical world at that time. Although he stayed there only a year, this period turned out to be very important for his future development, in particular because it was there that he became aware of the great paper of Teiji Takagi on class field theory. In an attempt to better understand Takagi's results Artin was led to formulate what is now known as the Artin reciprocity law, undoubtedly his crowning mathematical achievement, a pinnacle of algebraic number theory. The search for a proof and the successful completion of that search did not take place at Göttingen. Artin moved in 1922 to the relatively new University of Hamburg, which soon became a great center of mathematical activity. It was here that he discovered the proof of his reciprocity law. Artin moved up rapidly through the academic ranks, becoming a Full Professor in 1926 at the age of 28. Thereafter he, together with Erich Hecke and Wilhelm Blaschke, directed the activities of the Mathematical Seminar of Hamburg. To get some idea of the rich mathematical life of that time and place, one only has to leaf through issues of the *Abhandlungen der Mathematisches Seminar Hamburg* of that period.

The years 1921–1931 were amazingly productive for Artin. As Richard Brauer expresses it in his valuable article on Artin's career [20], "The ten year period 1921–1931 of Artin's life had seen an activity not often equalled in the life of a mathematician". In his thesis Artin developed the theory of hyperelliptic curves over finite fields on the model of the theory of quadratic number fields. In so doing he introduced new types of zeta and L-functions and proved in special cases a new type of Riemann hypothesis connected with these functions. He discovered and proved the Artin reciprocity law (generalizing in one blow all previous reciprocity laws). He invented Artin L-functions and proved many of their most important properties. Moreover, he made a conjecture about them that remains to this day one of the great unsolved problems of number theory. All this work was in number theory. In topology he invented the braid group and proved a series of important results about them. Braid groups turn out to be of great importance in topology, algebraic geometry, and, in recent years, physics. Within algebra he invented the abstract theory of real fields and was led to a solution of Hilbert's problem number 17. He also contributed important papers to the theory of hypercomplex systems which are now called associative algebras. He helped extend and generalize earlier work of Leonard Dickson and Joseph H. M. Wedderburn. Even this list does not exhaust the many important contributions he made during this immensely productive period.

In 1929, Artin married Natasha Jasny. She had been one of his students. After their marriage, she continued to follow developments in mathematics and was later an editor of the journal *Communications on Pure and Applied Mathematics*. The young couple had two children, Michael and Karin, while living in Germany, and a third child, Tom, after they moved to the United States. Michael developed into a very important mathematician in his own right. Unfortunately for the Artin family, the early years of their marriage were clouded by the worldwide Depression and the rise to power of the Nazi Party. According to Brauer, after 1933 "It was only a question of time before Artin, with his feeling for individual freedom, his sense of justice, his abhorrence of physical violence, would leave Germany". In 1937, Artin and his family left Germany and emigrated to the United States.

He spent a year at Notre Dame in South Bend, Indiana, and then moved to Indiana University at Bloomington. He continued to give lectures at Notre Dame, and it is there that he published his famous book on Galois theory which we reproduce in this volume. He and his family spent eight happy years in Bloomington. During this time he entered into a productive collaboration with George Whaples, which resulted in several influential joint papers. Two of these, on the product formula, are reproduced in this volume. Another paper concerns rings with the descending chain condition on ideals. Such rings are now called Artinian rings. A small book on this subject, *Rings with Minimal Condition*, was written by Artin during this period [8] (co-authored with C. J. Nesbitt and R. M. Thrall). Bloomington was congenial both professionally and socially. However, when Artin received an offer to join the faculty of Princeton University in 1946, he took it. The high quality of the Princeton Mathematics Department, the proximity of the Institute for Advanced Study, and the prospect of first-rate graduate students must have been lures too difficult to resist. At Princeton, he became interested once again in the foundations of class field theory. A new approach was afforded by the use of group cohomology, which was introduced into the subject by Gerhardt Hochschild, Tadashi Nakayama, and André Weil. This approach was developed and extended further in Artin's seminar at Princeton. In the seminar were two of Artin's most illustrious graduate students, Serge Lang and John Tate. Tate made important contributions to class field theory, which led to significant generalizations of Artin's work on reciprocity. The Artin–Tate notes, derived from these seminars, are simply entitled *Class Field Theory* [18]. They are now out of print, but remain a classic work familiar to almost every researcher in the field.

In 1956, Artin took a sabbatical leave from Princeton and returned to Germany for the first time since his departure under unhappy circumstances in 1937. He spent one semester at Göttingen and another at Hamburg. While at Göttingen he gave a beautiful set of lectures on algebraic number theory. The notes from this course, written by George Würges, were distributed by the University of Göttingen [14]. They are included in this book.

While in Germany, Artin decided to move back permanently. In 1958 he accepted a Professorship at the University of Hamburg. There he remained for the rest of his life. He died suddenly and unexpectedly on December 20, 1962. He had been given only four years to enjoy his new life at the university of his youth.

There does not exist a full-length biography of Emil Artin. After his death a number of articles on his life and work appeared. Some of them are listed in the bibliography at the end of this introduction [20, 21, 22, 25, 28]. The articles of Richard Brauer [20] and Hans Zassenshaus [28], prominent mathematicians who knew Artin well, are especially worth reading. Zassenhaus had been his graduate student, obtaining his Ph.D. in 1934. The article by J. J. O'Connor and E. F. Robertson [25] is very informative. A delightful and useful resource is the section on Hilbert's ninth problem in *The Honor's Class* by Ben H. Yandell [27], especially pages 230 to 245. The biographical material there is based on, in part, interviews with Natasha (Artin) Brunswick and Artin's children, Michael, Karin, and Tom. There are also a number of interesting and evocative photographs.

The picture that emerges from all these works is that of an exceptionally deep and talented man with manifold interests in art, music, science, and literature, in addition to his profound and abiding love of mathematics. Whatever he did, he did with passion. He was an excellent musician, playing the flute as well as a number of keyboard instruments. He was also a serious amateur astronomer who built his own telescope, grinding the mirror to a near-perfect parabolic shape[1]. All these characteristics might exist in a very inward-directed individual, but Artin loved to communicate with others. He was well known for the excellence of his teaching. Both in lectures and in books he was a master at communicating his vision, his special way of seeing his subject. He exerted tremendous influence in mathematics, both through his creative output and his communicative skills.

It is worth quoting Artin himself (from a 1953 book review of Bourbaki's *Algebra*).

> We all believe that Mathematics is an art. The author in a book, the lecturer in a classroom, tries to convey the structural beauty of mathematics to his readers, to his listeners. In this attempt, he must always fail. Mathematics is logical to be sure; each conclusion is drawn from previously derived statements. Yet the whole of it, the real piece of art, is not linear; worse than that its perception should be instantaneous. We have all experienced on some rare occasions the feeling of elation in realizing that we have enabled our listeners to see at a moment's glance the whole architecture and all of its ramifications.

This quote is given in the article of Brauer [20]. Brauer prefaces Artin's quote with these words of his own.

> There are a number of books and sets of lecture notes by Artin. Each of them presents a novel approach. There are always new ideas and new results. It was a compulsion for Artin to present each argument in its purest form, to replace computation by conceptual arguments, to strip the theory of unnecessary ballast. What was the decisive point for him was to show the beauty of the subject to the reader.

Part 2

Gathered together in this volume are a number of examples of Artin's expository excellence. Three of these are short books: *The Gamma Function*, *Galois Theory*, and *Theory of Algebraic Numbers*. They are each expository in nature, but they all provide examples of Artin's deep feeling for the beauty of his subject, as well as his originality and unique point of view. In addition to these books, we have included a number of papers. Most of these are research papers, but we have felt justified in including them because in these works the writing is so clear, elegant, and insightful. Moreover, the selected works do not make great demands on the background of the reader. For most of them, it suffices to know the material that a good mathematics major learns at a good university.

A word of warning is in order. Artin was a great artist. His writings are not easy to fully appreciate at first glance. The true value of his work becomes apparent only after a certain amount of effort is devoted to its study. What is true of everything presented here is that such effort will be greatly rewarded!

[1]Karin Tate believes the adjective "perfect" would be more appropriate. For the story behind this, see the endnote at the conclusion of the introduction.

The Gamma Function [16] is a translation by Michael Butler (1964) of the German original *Einführung in die Theorie der Gammafunktion*, which appeared in 1931. As Edwin Hewitt asserts in his editor's preface, "...it has been read with joy and fascination by many thousands of mathematicians and students of mathematics." Artin wrote the book to fill a gap he perceived in the literature. The gamma function is important in many parts of mathematics, yet it is often treated as a mere side issue in many beginning books on analysis. In this small book, he develops almost all of the useful properties of the gamma function using nothing beyond elementary calculus. A novel feature is the extensive use made of the notion of log convexity. This approach is due to Harald Bohr and Johannes Mollerup.

Galois Theory [6] is the result of many years of thought devoted to the foundations of the subject. In the thirties, the proof of the main theorem of Galois theory depended on the primitive element theorem. If L/K is a finite, separable extension of fields, the primitive element theorem asserts that L contains an element θ such that $L = K(\theta)$. According to Zassenhaus [28], Artin did not like this state of affairs at all. "He took offense at the central role played by the existence of a primitive element ...". He finally came up with a new proof that was heavily dependent on elementary linear algebra (Part 1 of his book is a condensed treatment of the necessary linear algebra). Primitive elements play no role at all. A novel feature is the use of mappings from a group to a field as elements of a vector space. The new approach did far more than eliminate the use of primitive elements. It led to many new developments, such as the creation of a non-commutative Galois theory.

It must be admitted that although Artin's book *Galois Theory* is beautiful, it is also austere. It is a highly abstract treatment which presaged the use of high abstraction throughout mathematics from the fifties and beyond.

The Theory of Algebraic Numbers [14] is based on lectures given by Artin during his sabbatical semester at Göttingen in 1956–57. The notes were taken by Gerhard Würges and translated and distributed by George Striker. In the late fifties and early sixties one could get a copy by sending $2.50 to George Striker in Göttingen. The book is an introduction to the theory of algebraic numbers using the methods of valuation theory. It provides all the basic results up to and including the finiteness of class number and the Dirichlet unit theorem. The exposition is heavily influenced by the paper *Axiomatic Characterization of Fields by the Product Formula for Valuations* [9], which he wrote with G. Whaples in 1945 (see below). Although not as polished a work as each of the two books discussed above, this is a charming and readable introduction to algebraic numbers as seen through the eyes of a master. A notable feature is the variety of calculations worked out in detail. Known for his love of abstraction, Artin also liked very much to work out detailed examples. Although not too evident in his published works, this tendency is on display here.[2]

The first paper we discuss is the joint paper with G. Whaples mentioned above. After briefly reviewing the theory of archimedean and non-archimedean valuations, the authors point out that, when properly normalized, the set of all valuations of both number fields and function fields satisfy a product formula. This is easily

[2]There are a few errors in the original manuscripts of [6] and [14], which are corrected in the present edition. *Editor's note.*

established. The point of the paper is to show that, conversely, the product formula only holds in these two cases. One rather weak axiom in addition to the product formula is necessary for the proof. With another reasonable axiom, the function field case is shown to hold if and only if the constant field is finite. Number fields and function fields in one variable over a finite constant field are exactly the fields for which class field theory is known to hold. This paper shows that, in some sense, the product formula lies at the base of this imposing structure. Another interesting feature is that the authors deduce the finiteness of class number and the Dirichlet unit theorem (for S-units) without resorting to Minkowski's theorem about lattice points. The paper also points the way to the later systematic use of ideles and adeles in class field theory and beyond. We also include a follow-up paper, wherein the authors are able to weaken further one of the axioms at the base of the theory.

We now come to a set of three papers titled *Kennzeichnung des Körper der reellen algebraischen Zahlen* [1], *Algebraischen Konstruktion reeller Körper* [3], and *Eine Kennzeichnung der reell abgeschlossenen Körper* [5]. The last two were written jointly with Otto Schreier. All three concern the theory of real fields and real closed fields. They are extremely interesting in their own right, but it is worth noting that the material in the first two led Artin to his solution of Hilbert's problem 17 in 1927. We will come back to this after saying something about these papers.

As is obvious from the titles, all three papers are written in German. To make these papers more accessible, we have translated them into English. The translations are fairly literal, with an occasional change in the notation. Also, twice, we have substituted a proof using Zorn's lemma for the original proof which uses the well ordering principle (on the grounds that Zorn's lemma is more likely to be familiar to the modern reader). Otherwise, the translations are straightforward.

In the first paper in this series, Artin attempts to find a purely algebraic characterization of the field P of all real algebraic numbers, i.e., the subset of the real numbers consisting of those which are algebraic over $\mathbb{Q}$. The construction of the real numbers involves limits, so he looks for a characteristic property of the real algebraic numbers that is purely algebraic. Such a property should be preserved under isomorphism. He observes that $\Omega = P(i)$ is the set of all algebraic numbers (the real and imaginary parts of an algebraic number are both algebraic). The paper is then devoted to proving that if $K \subset \Omega$ is a proper subfield of finite codimension, then $\Omega = K(i)$. Moreover, there is an automorphism of Ω that takes K onto P. This is the algebraic characterization sought for; up to isomorphism P is the only subfield of Ω of finite codimension. This short paper is a pleasure to read—one beautiful step after another until the goal is achieved.

In the second paper, the authors search for a purely algebraic definition of the notion of a "real field". They settle on a property of the real numbers that Artin used in the paper about real algebraic numbers discussed above. Namely, they say that a field K is a real field if whenever a sum of squares in K is zero, then each summand must be zero. Equivalently, K is real if -1 cannot be written as a sum of squares in K. This is a purely algebraic property. A field is said to be real closed if it is real but possesses no proper algebraic extension field that is also a real field. They introduce the notion of an ordered field and observe that an ordered field must be a real field. More difficult is the result that a real closed field can be ordered in one and only one way (an element is positive if and only if it is a square). They show that if P is real closed, then $P(i)$ is algebraically closed. Artin's previous paper is then generalized as follows. Suppose Ω is algebraically closed and of characteristic zero. Let K be a proper subfield of finite codimension. Then, K is a real closed field and $\Omega = K(i)$. They go on to show that for polynomials over real closed fields, all the standard theorems of calculus continue to hold, e.g., the intermediate value theorem, Rolle's theorem, the mean value theorem, and the theorems of Sturm. In the next section of the paper, it is proven that if K is a real field and Ω is an algebraically closed extension, then there is a real closed field P between K and Ω such that $\Omega = P(i)$. Finally, this is sharpened to show that an ordered field K can be imbedded in a real closed field in such a way that the order on K is induced from the unique order on P. The last part of the paper concerns the Archimedean axiom and generalizations of it, applications of theory of real closed fields to algebraic number fields, and the construction of an ingenious series of examples that cast new light on the main results.

The aim of the final paper in this series is to show that a field K is real closed if and only if it is of finite codimension in its algebraic closure and is not itself algebraically closed. This had already been shown if K has characteristic zero. The main point then is to show that if K has finite characteristic p, then it cannot be of finite codimension in its algebraic closure Ω unless $K = \Omega$. The new ingredient necessary for the proof is the theory of what has come to be known as Artin-Schreier equations. These are equations of the form $x^p - x - a = 0$, where a is an element of the base field K. Any root of such an equation that is not in K generates a cyclic extension field of degree p over K. Conversely, any cyclic extension of degree p is generated by the root of such an equation. In this paper, cyclic extensions of degree p^2 are dealt with as a tower of two Artin-Schreier extensions. The whole theory is a characteristic p replacement for Kummer theory. It was subsequently considerably generalized by Ernst Witt. The algebraic characterization of real closed fields is an important result. However, it is fair to say that the Artin-Schreier theory of cyclic extensions plays a more central role in mathematics than the reason for its invention.

As mentioned earlier, Artin used the theory of real fields to resolve, positively, Hilbert's problem number 17. This concerns rational functions $f(x_1, x_2, \ldots, x_n)$ in several variables with real coefficients (the usual real numbers). Suppose this function has a non-negative value whenever a real n-tuple is substituted for the variables, assuming the function is defined there. Must it be a sum of squares of rational functions of the same type? Artin showed that the answer is yes in his paper *Uber die Zerlegung definiter Funktionen in Quadrate* [4] published in 1927. Forty years later Albrecht Pfister proved an important refinement: the number of squares necessary is at most 2^n (see [26]). It is interesting to note that Pfister's proof uses Artin's result. He shows, if f is a sum of squares, then it is a sum of at most 2^n squares.

Almost everything we have discussed so far belongs to algebra. Artin's mathematical interests were very far-reaching and he made contributions in other areas as well, e.g., in topology. A classical problem in topology is the question of classifying knots. This subject was actively investigated in the 19th century and remains of great interest today. Artin's contribution was to invent and investigate a somewhat simpler notion. In 1925 he wrote a foundational paper, *Theorie der Zöpfe* [2] (in English, *Theory of Braids*). Braids are somewhat simpler objects than knots. Artin was able to give a satisfactory classification procedure for them. Closed braids turn out to be links, disjoint unions of knots. Thus, the theory of braids turns out to be a tool in the study of knots and links. In fact, braids and the associated braid groups turn up in many different areas of mathematics and even in physics. Artin's invention is today a flourishing subject. Perhaps the easiest way to get a feel for what the subject is about is to read Artin's 1950 article *The Theory of Braids*, which appeared in *The American Scientist* [11]. That article, as well as the more technical 1947 article *Theory of Braids* [10], appear in this volume in reverse chronological order. The latter article is definitely a research article, but it is very well written and accessible to anyone who is comfortable with the definition of the fundamental group of a topological space.

For later developments, together with applications to low-dimensional topology, the reader should consult *Braids, Links, and the Mapping Class Group* by Joan Birman [23]. At the beginning of this book, Birman states, "It is a tribute to Artin's extraordinary insight as a mathematician that the definition he proposed in 1925 ... for the equivalence of geometric braids could ultimately be broadened and generalized in many different directions without destroying the essential features of the theory". The 2002 book of Seiichi Kamada, *Braid and Knot Theory in Dimension Four* [24], provides an up-to-date treatment together with a huge bibliography (996 items). The first fifty pages are devoted to classical braids and links.

We now come to two articles in the area of analysis. The first, *On the Theory of Complex Functions* [7], concerns the foundations of complex analysis. It presents a very general version of Cauchy's theorem and the Cauchy integral formula based on a novel approach to homology theory via winding numbers. In the introduction to his famous graduate-level text *Complex Analysis*, Lars Ahlfors states that the whole structure of his book was deeply influenced by Artin's insight. Later in the text, he states

> It was E. Artin who discovered the characterization of homology by vanishing winding numbers ties in precisely with what is needed for Cauchy's theorem. This idea has led to a remarkable simplification of earlier proofs.

The second article in analysis that we include here is entitled *A proof of the Krein–Milman Theorem* [12]. This is a letter from Artin to his former Ph.D. student, Max Zorn. The letter was published in the *Piccayune Sentinel* of Indiana University in 1950. This slim newsletter was published from time to time by Zorn himself. It contained news about mathematics and mathematicians. The Krein–Milman theorem states, roughly speaking, that a compact convex subset of a real vector space is the closed convex hull of its extreme points. This basic and useful result is proven by Artin in slightly more than two pages. These pages include the relevant definitions and axioms and all the details of the proof. Needless to say, the version of the theorem presented is very general and the proof is exceedingly elegant.

One bit of explanation is in order. On the second page, Artin uses the phrase "The unmentionable Lemma shows ...". This may be puzzling at first. The lemma referred to is Zorn's Lemma, the ubiquitous lemma about partially ordered sets that is discussed in every modern text on graduate algebra.

The last work that is included in this book returns us to the subject of algebra. In it we find an overview of the influence of J. H. M. Wedderburn on the development of modern algebra. This article was published in the *Bulletin of the AMS* in 1950 [13] and is almost as interesting for what it says about Artin's attitudes toward modern algebra as for what it says about Wedderburn. Artin is always the champion of abstract methods and this comes across clearly as he traces the history of the theory of associative algebras. After stating Wedderburn's most famous result, that a simple, finite-dimensional algebra is a full matrix algeba over a division ring, Artin proceeds to present a beautiful generalization due to Claude Chevalley and (independently) to Nathan Jacobson. This generalization is known today as the Jacobson density theorem. After that he discusses another celebrated theorem of Wedderburn: every finite division ring is a field. He relates how this led him to conjecture that a polynomial with coefficients in a finite field with no constant term and having degree smaller than the number of variables must have a non-trivial zero. Chevalley proved this conjecture in 1935, shortly after it was made. Artin went on to conjecture other results of a similar nature for p-adic and global fields. These conjectures led to interesting research by Lang, Ax, Kochen, and others in the 1960s.

This completes our survey of the contents of this book. For readers who want to read more of Artin's expository work, we suggest consulting "The Collected Papers of Emil Artin" [17], the beautiful and valuable book *Geometric Algebra* [15], and the less well known, but worthwhile *Introduction to Algebraic Topology* (written with Hel Braun) [19].

Endnote: It was pointed out earlier that Artin was a serious amateur astronomer who built his own telescope, grinding the mirror to "a near perfect parabolic shape". Karin Tate believes the adjective "perfect" would be more appropriate. To back this up, she related the following charming story. "Of course nothing is perfect. It's just that to make the surface parabolic after grinding out a spherical section certain critical moves have to be made with the grinding stone, and I remember with amusement my father's trying over and over to achieve this. He ground the mirror in our basement, and each time he tried to get a parabolic surface he would do an elaborate test with candles. Mike and I were fascinated by this, and he would tolerate our presence only if we stood still. "Don't move the air!" he would say. Many attempts later, my mother suggested that maybe it was too cold in the basement and that he should bring everything upstairs and try again. My father did this, muttering that he didn't think it would work, but that he'd try it anyway. Of course, it did work, much to my mother's glee. So that is why I made the comment I did - to my father, the mirror was perfect!"

References

[1] E. Artin, *Kennzeichnung des Körper der reellen algebraischen Zahlen*, Hamb. Abh. **3** (1924), 469–492.

[2] E. Artin, *Theorie der Zöpfe*, Hamb. Abh. **4** (1925), 47–72.

[3] E. Artin (with O. Schreier), *Algebraische Konstruktion reeller Körper*, Hamb. Abh. **5** (1926), 85–99.

[4] E. Artin, *Über die Zerlegung definiter Funktionen in Quadrate*, Hamb. Abh. **4** (1925), 47–72.

[5] E. Artin (with O. Schreier), *Eine Kennzeichnung der reell abgeschlossenen Körper*, Hamb. Abh. **5** (1927), 225–231.

[6] E. Artin, *Galois Theory* (Second Edition), Notre Dame Mathematical Lectures No. 2, Notre Dame, IN, 1942.

[7] E. Artin, *On the Theory of Complex Functions*, Notre Dame Mathematical Lectures No. 4, Notre Dame, IN, 1944.

[8] E. Artin (with C. J. Nesbitt and R. M. Thrall), *Rings with Minimum Condition*, U. of Michigan Press, Ann Arbor, MI, 1944.
[9] E. Artin (with G. Whaples), *Axiomatic Characterization of Fields by the Product Formula for Valuations*, Bull. Amer. Math. Soc. **51** (1945), 469–492.
[10] E. Artin, *Theory of Braids*, Ann of Math. **48** (1947), 101–126.
[11] E. Artin, *The Theory of Braids*, American Scientist **38** (1950), 112–119.
[12] E. Artin, *A Proof of the Krein–Milman Theorem* (a letter to Max Zorn), Piccayune Sentinel, Indiana U. (1950).
[13] E. Artin, *The Influence of J. H. M. Wedderburn on the Development of Modern Algebra*, Bull. Amer. Math. Soc. **56** (1950), 65–72.
[14] E. Artin, *Theory of Algebraic Numbers*, (notes by G. Würges, translated and distributed by G. Striker) Göttingen Mathematisches Institut, 1956.
[15] E. Artin, *Geometric Algebra*, Interscience, New York, 1957.
[16] E. Artin, *The Gamma Function* (translated by Michael Butler), Holt, Rinehart, and Winston, New York, 1964.
[17] E. Artin, *The Collected Papers of Emil Artin*, S. Lang and J. Tate (editors), Addison-Wesley, Reading, MA, 1965.
[18] E. Artin and J. Tate, *Class Field Theory*, Benjamin, New York, 1967.
[19] E. Artin (with H. Braun), *Algebraic Topology*, Charles Merrill Publishing Co., Columbus, OH, 1969.
[20] R. Brauer, *Emil Artin*, Bull. Amer. Math. Soc. **73** (1967), 27–43.
[21] H. Cartan, *Emil Artin*, Hamb. Abh. **28** (1965), 1–5.
[22] C. Chevalley, *Emil Artin (1898-1962)*, Bull. Soc. Math. France **92** (1965), 1–5.
[23] J. Birman, *Braids, Links, and Mapping Class Groups*, Ann. Math. Studies vol. 82, Princeton Univ. Press, Princeton, NJ, 1975.
[24] S. Kamada, *Braid and Knot theory in Dimension Four*, Amer. Math. Soc., Providence, RI, 2002.
[25] J. J. O'Connor and E. F. Robertson, `http://www-groups.dcs.st-and.ac.uk/~history/Mathematicians/Artin.html`, 2000.
[26] A. Pfister, *Sums of squares in real function fields*, Actes Congrès Intern. Math., Tome 1, 1970, 297–300.
[27] B. Yandell, *The Honor's Class*, A. K. Peters, Natick, MA, 2002.
[28] H. Zassenhaus, *Emil Artin, his life and work*, Notre Dame J. Formal Logic **5** (1964), 1–9.

Books

by Emil Artin

Translated by

Michael Butler

The Gamma Function

Emil Artin

Professor of Mathematics
Hamburg University

HOLT, RINEHART AND WINSTON

New York • Chicago • San Francisco
Toronto • London

Editor's Preface

A generation has passed since the late Emil Artin's little classic on the gamma function appeared in the *Hamburger Mathematische Einzelschriften*. Since that time, it has been read with joy and fascination by many thousands of mathematicians and students of mathematics. In the United States (and presumably elsewhere as well), it has for many years been hard to find, and dog-eared copies and crude photocopies have been passed from hand to hand. Professor Artin's monograph has given many a student his first look at genuine analysis—the delicacy of its arguments, the precision of its results. Artin had a deep feeling for these aspects of analysis, and he treated them with a master's hand. His undergraduate lectures in the calculus, for example, were filled with elegant constructions and theorems which, alas, Artin never had time to put into printed form. We may be all the more grateful for this beautiful essay, and for its appearance in a new English edition. Various changes made by Artin himself have been incorporated in the present edition. In particular a small error following formula (59) (this edition) was corrected on the basis of a suggestion by Professor Børge Jessen.

Finally, thanks are due to the translator, Mr. Michael Butler, and to the firm of B. G. Teubner for English-language rights.

EDWIN HEWITT

Seattle, Washington
May, 1964

Preface

I have written this monograph with the hope of filling in a certain gap which has often been felt to exist in the mathematical literature. Despite the importance of the gamma function in many different parts of mathematics, calculus books often treat this function in a very sketchy and complicated fashion. I feel that this monograph will help to show that the gamma function can be thought of as one of the elementary functions, and that all of its basic properties can be established using elementary methods of the calculus.

As far as prerequisites are concerned, the reader need only be well acquainted, with calculus, including improper integrals. Some of the more important concepts needed will even be introduced and discussed again in the first chapter. With this background the reader should have no trouble understanding everything but the later parts of the last two chapters, which do assume some knowledge of Fourier series. But then, these parts of the monograph can be passed over on a first reading without any difficulty whatsoever.

The following parts of the theory will *not* be discussed:

(1) Extension to complex variables. For those familiar with the theory of complex variables, it will suffice to point out that for the most part the expressions used are analytic, and hence they retain their validity in the complex case because of the principle of analytic continuation. The only parts of the theory that really need to be changed are those dealing with approximations. This certainly should not be much of an obstacle.

(2) Hölder's theorem showing that the gamma function does not satisfy any algebraic differential equation.

(3) Kummer's series and the integral representation of $\log \Gamma(x)$.

(4) The formula for the logarithmic derivative of $\Gamma(x)$. All the necessary expressions for this can easily be worked out by the reader.

I have chosen the integral as my original definition of the gamma function because this approach saves us the trouble of proving the convergence of Gauss' product. Any other analytic expression having the characteristic properties of the gamma function could just as well have been used. The whole theory will then be deduced using the concept of log convexity. This method comes from Bohr and Mollerup.[1]

Emil Artin

[1] H. Bohr and J. Mollerup, *Laerebog i matematisk Analyse* (Kopenhagen 1922), vol. III, 149-164.

Contents

[1]

Convex Functions

Let $f(x)$ be a real-valued function defined on an open interval $a < x < b$ of the real line. For each pair x_1, x_2 of distinct numbers in the interval we form the difference quotient

$$\varphi(x_1, x_2) = \frac{f(x_1) - f(x_2)}{x_1 - x_2} = \varphi(x_2, x_1), \tag{1.1}$$

and for each triple of distinct numbers x_1, x_2, x_3 the quotient

$$\begin{aligned} \Psi(x_1, x_2, x_3) &= \frac{\varphi(x_1, x_3) - \varphi(x_2, x_3)}{x_1 - x_2} \\ &= \frac{(x_3 - x_2)f(x_1) + (x_1 - x_3)f(x_2) + (x_2 - x_1)f(x_3)}{(x_1 - x_2)(x_2 - x_3)(x_3 - x_1)}. \end{aligned} \tag{1.2}$$

The value of the function $\Psi(x_1, x_2, x_3)$ does not change when the arguments x_1, x_2, x_3 are permuted.

$f(x)$ is called *convex* (on the interval (a, b)) if, for every number x_3 of our interval, $\varphi(x_1, x_3)$ is a monotonically increasing function of x_1. This means, of course, that for any pair of numbers $x_1 > x_2$ distinct from x_3 the inequality $\varphi(x_1, x_3) \geqslant \varphi(x_2, x_3)$ holds; in other words, that $\Psi(x_1, x_2, x_3) \geqslant 0$. Since the value of Ψ is not changed by permuting the arguments, the convexity of $f(x)$ is equivalent to the inequality

$$\Psi(x_1, x_2, x_3) \geqslant 0 \tag{1.3}$$

for all triples of distinct numbers in our interval.

Suppose $g(x)$ is another function that is defined and convex on the same interval. By adding (1.3) to the corresponding inequality for $g(x)$, we can easily see that the sum $f(x) + g(x)$ is also convex. More generally, suppose $f_1(x)$, $f_2(x)$, $f_3(x)$ $\cdots$ is a sequence of functions that are all defined and convex on the same interval. Furthermore, suppose that the limit $\lim_{n\to\infty} f_n(x) = f(x)$ exists and is finite for all x in the interval. By forming the inequality (1.3) for $f_n(x)$ with arbitrary but fixed numbers x_1, x_2, x_3, and then taking the limit as $n \to \infty$, we see that $f(x)$ is likewise convex. This proves the following theorem:

Theorem 1.1

The sum of convex functions is again convex. The limit function of a convergent sequence of convex functions is convex. A convergent infinite series whose terms are all convex has a convex sum.

The last statement of this theorem follows from the fact that each partial sum of the series is a convex function and the sum of the series is merely the limit of these partial sums.

We are now going to investigate some important properties of a function $f(x)$ defined and convex on the open interval (a, b). For a fixed x_0 in the interval let x_1 range over all numbers $> x_0$ and x_2 range over all numbers $< x_0$. We have

$$\varphi(x_1, x_0) \geqslant \varphi(x_2, x_0). \tag{1.4}$$

If x_2 is kept fixed and x_1 decreases approaching x_0, the left side of Eq. (1.4) will decrease but always remain greater than the right side. This implies that the "right-handed" derivative of $f(x)$ exists; that is to say, the limit

$$\lim_{\substack{x_1 > x_0 \\ x_1 \to x_0}} \varphi(x_1, x_0) = \lim_{\substack{x_1 > x_0 \\ x_1 \to x_0}} \frac{f(x_1) - f(x_0)}{x_1 - x_0},$$

for which we shall use the intuitive notation $f'(x_0 + 0)$. Furthermore, the inequality (1.4) also shows that

$$f'(x_0 + 0) \geqslant \varphi(x_2, x_0).$$

If we let x_2 increase, approaching x_0, we see that the "left-handed" derivative $f'(x_0 - 0)$ also exists, and that

$$f'(x_0 + 0) \geqslant f'(x_0 - 0). \tag{1.5}$$

Given two numbers $x_0 < x_1$ in our interval, we can choose x_2, x_3 such that $x_0 < x_2 < x_3 < x_1$. Then

$$\varphi(x_2, x_0) \leqslant \varphi(x_3, x_0) = \varphi(x_0, x_3) \leqslant \varphi(x_1, x_3) = \varphi(x_3, x_1).$$

If we let x_2 approach x_0 and x_3 approach x_1, we obtain

$$f'(x_0 + 0) \leqslant f'(x_1 - 0) \qquad \text{for} \qquad x_0 < x_1. \tag{1.6}$$

This proves that the one-sided derivatives of a convex function always exist and that they satisfy the inequalities (1.5) and (1.6). We shall refer to the properties (1.5) and (1.6) by saying that the one-sided derivatives are monotonically increasing.

In order to show the converse, we must generalize the ordinary mean-value theorem to cover the case of functions for which only the one-sided derivatives exist. The analogue to Rolle's theorem is the following:

Theorem 1.2

Let $f(x)$ be a function, defined and continuous on $a \leqslant x \leqslant b$, whose one-sided derivatives exist in the open interval $a < x < b$. Suppose $f(a) = f(b)$. Then there exists a value ξ with $a < \xi < b$ such that one of the values $f'(\xi + 0)$ and $f'(\xi - 0)$ is $\geqslant 0$ and the other $\leqslant 0$.

Proof

(1) If $f(x)$ takes on its maximum ξ in the interior of our interval, then

$$\frac{f(\xi + h) - f(\xi)}{h}$$

is $\leqslant 0$ for positive h, $\geqslant 0$ for negative h. Taking limits, we get $f'(\xi + 0) \leqslant 0$, $f'(\xi - 0) \geqslant 0$.

(2) If the minimum ξ is taken on in the interior, we obtain similarly $f'(\xi + 0) \geqslant 0, f'(\xi - 0) \leqslant 0$.

(3) If both maximum and minimum are at a or b, then $f(x)$ is constant, $f'(x) = 0$, and ξ can be taken anywhere in the interior. This completes the proof.

The substitute for the mean-value theorem is the following:

Theorem 1.3

Let $f(x)$ be defined and continuous on $a \leqslant x \leqslant b$ and have one-sided derivatives in the interior. Then there exists a value ξ in the interior such that $(f(b) - f(a))/(b - a)$ lies between $f'(\xi - 0)$ and $f'(\xi + 0)$.

Proof

The function

$$F(x) = f(x) - \frac{f(b) - f(a)}{b - a}(x - a)$$

is continuous, has one-sided derivatives

$$F'(x \pm 0) = f'(x \pm 0) - \frac{f(b) - f(a)}{b - a},$$

and $F(a) = f(a)$, $F(b) = f(a)$. According to our extension of Rolle's theorem, there is a ξ in the interior such that one of the values

$$f'(\xi + 0) - \frac{f(b) - f(a)}{b - a} \quad \text{or} \quad f'(\xi - 0) - \frac{f(b) - f(a)}{b - a}$$

is $\geqslant 0$, the other $\leqslant 0$. This completes the proof.

We are now in a position to prove the desired converse. Let $f(x)$ be a function defined on the open interval $a < x < b$. Suppose $f(x)$ has one-sided derivatives that are monotonically increasing. We contend that $f(x)$ is convex.

Let x_1, x_2, x_3 be distinct numbers in our interval. Since the value of Ψ does not change under permutation of the subscripts, we may assume that $x_2 < x_3 < x_1$. According to the mean-value theorem, we can find ξ, η with $x_2 < \eta < x_3 < \xi < x_1$ such that $\varphi(x_1, x_3)$ lies between $f'(\xi - 0)$ and $f'(\xi + 0)$, and $\varphi(x_2, x_3)$ between $f'(\eta - 0)$ and $f'(\eta + 0)$. Therefore (1.5) implies that

$$\varphi(x_1, x_3) \geqslant f'(\xi - 0) \qquad \text{and} \qquad \varphi(x_2, x_3) \leqslant f'(\eta + 0).$$

From Eq. (1.2) we obtain

$$\Psi(x_1, x_2, x_3) \geqslant \frac{f'(\xi - 0) - f'(\eta + 0)}{x_1 - x_2}.$$

Finally we conclude from (1.6) that

$$\Psi(x_1, x_2, x_3) \geqslant 0,$$

and this is the contention.

Theorem 1.4

$f(x)$ is an convex function if, and only if, $f(x)$ has monotonically increasing one-sided derivatives.

Corollary

Let $f(x)$ be a twice differentiable function. Then $f(x)$ is convex if, and only if, $f''(x) \geqslant 0$ for all x of our interval.

Proof

$f'(x)$ is monotonically increasing if, and only if, $f''(x) \geqslant 0$.

We now return to Eq. (1.2) and select for x_3 the midpoint $(x_1 + x_2)/2$ of x_1 and x_2. Assuming for a moment that $x_2 < x_1$, we have

$$x_3 - x_2 = x_1 - x_3 = \tfrac{1}{2}(x_1 - x_2).$$

The numerator of $\Psi(x_1, x_2, x_3)$ becomes

$$(x_1 - x_2)\left(\tfrac{1}{2}f(x_1) + \tfrac{1}{2}f(x_2) - f(x_3)\right),$$

and the denominator is positive. For a convex function we obtain the inequality

$$f\left(\frac{x_1 + x_2}{2}\right) \leqslant \tfrac{1}{2}\left(f(x_1) + f(x_2)\right), \tag{1.7}$$

which is symmetric in x_1 and x_2 and therefore also holds for $x_1 < x_2$. For $x_1 = x_2$ it is trival.

We shall call a function defined on an interval *weakly convex* if it satisfies the inequality (1.7) for all x_1, x_2 of the interval. It is obvious that the sum of two weakly convex functions, both defined on the same interval, is again weakly convex. It is also obvious that the limit function of a sequence of weakly convex functions, all defined on the same interval, is weakly convex.

Let $f(x)$ be weakly convex. The inequality (1.7) can be generalized to

$$f\left(\frac{x_1 + x_2 + \cdots + x_n}{n}\right) \leqslant \frac{1}{n}\left(f(x_1) + f(x_2) + \cdots + f(x_n)\right). \tag{1.8}$$

Proof

(1) We first show that if (1.8) holds for a certain integer n, then it also holds for $2n$. Indeed, suppose $x_1, x_2, \cdots, x_{2n}$ are numbers in our interval. Replacing x_1 and x_2 in Eq. (1.7) by

$$\frac{x_1 + \cdots + x_n}{n} \quad \text{and} \quad \frac{x_{n+1} + \cdots + x_{2n}}{n},$$

respectively, we have

$$f\left(\frac{x_1 + \cdots + x_{2n}}{2n}\right) \leqslant \tfrac{1}{2}\left(f\left(\frac{x_1 + \cdots + x_n}{n}\right) + f\left(\frac{x_{n+1} + \cdots + x_{2n}}{n}\right)\right).$$

Applying the inequality (1.8) to both terms on the right-hand side, we get the desired formula

$$f\left(\frac{x_1 + \cdots + x_{2n}}{2n}\right) \leqslant \frac{1}{2n}\left(f(x_1) + f(x_2) + \cdots + f(x_{2n})\right).$$

(2) Next we show that if (1.8) holds for $n + 1$, then it also holds for n. With n numbers $(x_1, x_2, \cdots, x_n)$ the number

$$x_{n+1} = \frac{1}{n}(x_1 + \cdots + x_n)$$

also belongs to our interval. If (1.8) holds for $n + 1$, then

$$f(x_{n+1}) = f\left(\frac{nx_{n+1} + x_{n+1}}{n+1}\right) = f\left(\frac{x_1 + \cdots + x_n + x_{n+1}}{n+1}\right)$$

$$\leqslant \frac{1}{n+1}\left(f(x_1) + \cdots + f(x_n) + f(x_{n+1})\right).$$

Transposing the term $1/(n + 1)f(x_{n+1})$ to the left side, we obtain (1.8) for the n given numbers.

(3) We now combine steps (1) and (2) to attain the desired result. If (1.8) holds for any integer n, then step (2) implies that it also holds for all smaller integers. Because of step (1) the contention is true for arbitrarily large integers. Therefore it must be true for all n. This completes the proof.

We wish to prove the following theorem:

Theorem 1.5

A function is convex if, and only if, it is continuous and weakly convex.

Proof

(1) A convex function is continuous since it has one-sided derivatives. It is also weakly convex, as has already been shown.

(2) Suppose that $f(x)$ is weakly convex, that there are $x_2 < x_1$ numbers in our interval, and that $0 \leqslant p \leqslant n$ are two arbitrary integers. Apply (1.8) to the case where p of the n numbers have the value x_1 and the remaining $n - p$ numbers have the value x_2. We obtain

$$f\left(\frac{p}{n} x_1 + \left(1 - \frac{p}{n}\right) x_2\right) \leqslant \frac{p}{n} f(x_1) + \left(1 - \frac{p}{n}\right) f(x_2). \tag{1.9}$$

Assume now that $f(x)$ is continuous and let t be any real number such that $0 \leqslant t \leqslant 1$. Select a sequence of rational numbers between 0 and 1 that converges to t. Every term of this sequence is of the form p/n for suitable integers p and n; therefore Eq. (1.9) can be applied. Since $f(x)$ is continuous, we can go to the limit. We obtain

$$f(tx_1 + (1 - t)\, x_2) \leqslant tf(x_1) + (1 - t) f(x_2). \tag{1.10}$$

For any distinct numbers (x_1, x_2, x_3) in our interval we must show that $\psi(x_1, x_2, x_3) \geqslant 0$. Since ψ is symmetric, we may assume that $x_2 < x_3 < x_1$. The denominator of Eq. (1.2) is positive.

We set $t = (x_3 - x_2)/(x_1 - x_2)$; then

$$0 < t < 1, \qquad 1 - t = \frac{x_1 - x_3}{x_1 - x_2}$$

and

$$tx_1 + (1 - t)\, x_2 = \frac{(x_3 - x_2)\, x_1 + (x_1 - x_3)\, x_2}{x_1 - x_2} = x_3 .$$

Hence Eq. (1.10) implies that

$$f(x_3) \leqslant \frac{x_3 - x_2}{x_1 - x_2} f(x_1) + \frac{x_1 - x_3}{x_1 - x_2} f(x_2),$$

which shows that the numerator of ψ is $\geqslant 0$. This completes the proof.

Numerous inequalities useful in analysis can be obtained from Eq. (1.8) by a suitable choice for $f(x)$. As an example, consider $f(x) = -\log x$ for $x > 0$. We have $f''(x) = 1/x^2$ and our function is convex. Therefore Eq. (1.8) implies that

$$-\log\left(\frac{x_1 + \cdots + x_n}{n}\right) \leqslant -\frac{1}{n}(\log x_1 + \log x_2 + \cdots + \log x_n),$$

hence

$$\log\left(\frac{x_1 + \cdots + x_n}{n}\right) \geqslant \log \sqrt[n]{x_1 + \cdots + x_n}\,,$$

and consequently

$$\sqrt[n]{x_1 + \cdots + x_n} \leqslant \frac{x_1 + \cdots + x_n}{n}\,.$$

We now introduce an important concept closely related to that of convexity. A function $f(x)$ defined and positive on a certain interval is called *log convex* (*weakly log convex*) if the function $\log f(x)$ is convex (weakly convex). The condition that $f(x)$ be positive is obviously necessary, for otherwise the function $\log f(x)$ could not be formed. As an immediate consequence of our previous results, we have the following:

Theorem 1.6

A product of log-convex (weakly log-convex) functions is again log convex (weakly log convex). A convergent sequence of log-convex weakly log-convex) functions has a log-convex (weakly log-convex) limit function, provided the limit is positive.

Instead of the condition that the limit function be positive, we could require that the sequence of the logarithms of the individual terms be convergent.

Theorem 1.7

Suppose $f(x)$ is a twice differentiable function. If the inequalities

$$f(x) > 0, \qquad f(x)f''(x) - (f'(x))^2 \geqslant 0$$

hold, then $f(x)$ is log convex.

This theorem follows directly from the fact that the second derivative of $\log f(x)$ has the value

$$\frac{f(x)f''(x) - (f'(x))^2}{(f(x))^2}\,.$$

The properties of log-convex functions mentioned thus far are all mo less immediate consequences of the definition. The following remark theorem, however, is a much deeper result:

Theorem 1.8

Suppose $f(x)$ and $g(x)$ are functions, defined on a common interv If both are weakly log convex, then their sum $f(x) + g(x)$ is also weak log convex. If both are log convex, then $f(x) + g(x)$ is log convex.

Proof

It suffices to prove the first statement. The second then follows immedi with the addition of continuity.

Both $f(x)$ and $g(x)$ are positive. If x_1 , x_2 are numbers in our interval, th

$$\left(f\left(\frac{x_1 + x_2}{2}\right)\right)^2 \leqslant f(x_1) f(x_2) \qquad \text{and} \qquad \left(g\left(\frac{x_1 + x_2}{2}\right)\right)^2 \leqslant g(x_1) g(x_2)$$

We have to show that

$$\left(f\left(\frac{x_1 + x_2}{2}\right) + g\left(\frac{x_1 + x_2}{2}\right)\right)^2 \leqslant (f(x_1) + g(x_1))\,(f(x_2) + g(x_2)).$$

In other words the proof of our theorem amounts to showing that if a_1 , b_1 a_2 , b_2 , c_2 are positive real numbers with $a_1c_1 - b_1^2 \geqslant 0$ and $a_2c_2 - b_2^2$ then

$$(a_1 + a_2)\,(c_1 + c_2) - (b_1 + b_2)^2 \geqslant 0.$$

Consider the quadratic form $a_1x^2 + 2b_1xy + c_1y^2$ where $a_1 > 0$. We

$$a_1(a_1x^2 + 2b_1xy + c_1y^2) = (a_1x + b_1y)^2 + (a_1c_1 - b_1^2)\,y^2.$$

If $a_1c_1 - b_1^2 \geqslant 0$, the quadratic form never takes on a negative value, w ever x, y may be. On the other hand, if $a_1c_1 - b_1^2 < 0$, the quadratic form t on the negative value $a_1c_1 - b_1^2$ for $y = 1$, $x = -(b_1/a_1)$.

Our conditions imply that neither

$$a_1x^2 + 2b_1xy + c_1y^2 \qquad \text{nor} \qquad a_2x^2 + 2b_2xy + c_2y^2$$

takes on negative values. Therefore

$$(a_1 + a_2)\,x^2 + 2(b_1 + b_2)\,xy + (c_1 + c_2)\,y^2$$

will not take on negative values. Consequently

$$(a_1 + a_2)\,(c_1 + c_2) - (b_1 + b_2)^2 \geqslant 0.$$

This completes the proof of the theorem.

The reader who enjoys working with identities can check the validity of the following, for an alternate proof:

$$a_1a_2((a_1 + a_2)(c_1 + c_2) - (b_1 + b_2)^2)$$

$$= a_2(a_1 + a_2)(a_1c_1 - b_1^2) + a_1(a_1 + a_2)(a_2c_2 - b_2^2) + (a_1b_2 - a_2b_1)^2.$$

If

$$a_1 > 0, \qquad a_2 > 0, \qquad a_1c_1 - b_1^2 \geqslant 0, \qquad \text{and} \qquad a_2c_2 - b_2^2 \geqslant 0,$$

the right side is $\geqslant 0$ and the conclusion follows.

Other important facts can be obtained by combining our previous results. Suppose $f(t, x)$ is a function of the two variables x and t, which is defined and continuous for t in the interval $a \leqslant t \leqslant b$ and x in some arbitrary interval. Furthermore, for any fixed value of t, suppose that $f(t, x)$ is a log-convex, twice differentiable function of x. For every integer n we can build the function

$$F_n(x) = h(f(a, x) + f(a + h, x) + f(a + 2h, x) + \cdots + f(a + (n - 1)h, x)),$$

where $h = (b - a)/n$. Being the sum of log-convex functions, $F_n(x)$ is also log convex. As n approaches infinity, the functions $F_n(x)$ converge to the integral

$$\int_a^b f(t, x)\,dt;$$

hence this integral is also log convex.

Suppose that $f(t, x)$ only satisfies our conditions in the interior of the t interval, or that the upper bound of the interval is infinite. If the improper integral

$$\int_a^b f(t, x)\,dt$$

exists, then it is log convex. This follows directly from the fact that an improper integral is the limit of proper integrals over subintervals. Hence, as the limit function of log-convex functions, it is also log convex.

In this book we will only have to test integrals of the form

$$\int_a^b \varphi(t)\, t^{x-1}\,dt$$

for log convexity, where $\varphi(t)$ is a positive continuous function in the interior of the integration interval. If we take the logarithm of the integrand and then differentiate twice with respect to x, we get 0.

Theorem 1.9

If $\varphi(t)$ is a positive continuous function defined on the interior of the integration interval, then

$$\int_a^b \varphi(t)\, t^{x-1}\, dt$$

is a log-convex function of x for every interval on which the proper or improper integral exists.

The following theorem is quite obvious:

Theorem 1.10

If $f(x)$ is log convex on a certain interval, and if c is any real number $\neq 0$, then both the functions $f(x + c)$ and $f(cx)$ are log convex on the corresponding intervals.

[2]

The Euler Integrals and the Gauss Product Formula

The theory of the gamma function was developed in connection with the problem of generalizing the factorial function of the natural numbers, that is, the problem of finding an expression that has the value $n!$ for positive integers n, and that can be extended to arbitrary real numbers at the same time. In looking for such an expression, we come upon the following well-known improper integral:

$$\int_0^\infty e^{-t}\, t^n\, dt = n!$$

This suggests replacing the integer n on the left side by an arbitrary real number (provided the integral still converges) and defining $x!$ for arbitrary x as the value of this integral. Rather than doing precisely that, we will follow the custom of introducing a function that has the value $(n-1)!$ for positive integers n. Namely,

$$\Gamma(x) = \int_0^\infty e^{-t}\, t^{x-1}\, dt. \tag{2.1}$$

We still must determine the values of x for which this integral converges. The integrand is smaller than t^{x-1} when t is positive; therefore

$$\int_\epsilon^1 e^{-t}\, t^{x-1}\, dt < \int_\epsilon^1 t^{x-1}\, dt = \frac{1}{x} - \frac{\epsilon^x}{x}.$$

For $x > 0$,

$$\int_\epsilon^1 e^{-t}\, t^{x-1}\, dt$$

is bounded from above by $1/x$. If we hold x fixed and let ϵ decrease, the value of the integral increase monotonically. This means that

$$\int_0^1 e^{-t}\, t^{x-1}\, dt = \lim_{\epsilon\to 0} \int_\epsilon^1 e^{-t}\, t^{x-1}\, dt$$

exists for all positive x.

When t is positive, every term of the series for e^t is positive, and the inequality $e^t > t^n/n!$ holds for all integers n. Hence $e^{-t} < n!/t^n$, which gives another inequality for the integrand, namely, $e^{-t}t^{x-1} < n!/t^{n+1-x}$. Therefore, by holding x fixed and choosing $n > x + 1$, we can make $n!/(n - x)$ an upper bound for

$$\int_1^\delta e^{-t}\, t^{x-1}\, dt.$$

But the value of this integral increases as δ increases, and thus

$$\int_1^\infty e^{-t}\, t^{x-1}\, dt = \lim_{\delta \to \infty} \int_1^\delta e^{-t}\, t^{x-1}\, dt$$

exists. This implies that our definition, Eq. (2.1), is meaningful for all positive real x.

If we replace x by $x + 1$ in Eq. (2.1) and integrate the approximating integral by parts, we get

$$\begin{aligned} \int_\epsilon^\delta e^{-t}\, t^x\, dt &= - e^{-t}\, t^x \Big|_\epsilon^\delta + x \int_\epsilon^\delta e^{-t}\, t^{x-1}\, dt \\ &= e^{-\epsilon}\epsilon^x - e^{-\delta}\delta^x + x \int_\epsilon^\delta e^{-t}\, t^{x-1}\, dt. \end{aligned}$$

The term $e^{-\delta}\delta^x$ is again smaller than $n!/\delta^{n-x}$. As ϵ approaches 0 and δ approaches $+\infty$, the first two terms on the right side vanish, and we have the formula

$$\Gamma(x + 1) = x\Gamma(x). \tag{2.2}$$

This functional equation is basic for the development of the rest of the theory. It represents a generalization of the identity $n! = n(n - 1)!$ for nonintegral values of n. Suppose the value of the gamma function is known on the interval $0 < x \leqslant 1$. With the help of Eq. (2.2) we can easily calculate its value on the interval $1 < x \leqslant 2$, then again on the next interval of length 1, and so on. By repeated application of Eq. (2.2), we get

$$\Gamma(x + n) = (x + n - 1)\,(x + n - 2) \cdots (x + 1)\, x\Gamma(x) \tag{2.3}$$

for every positive integer n.

Equation (2.1) is only a definition for positive x. Now we want to extend this definition to include negative real numbers. If x lies in the interval $-n < x < -n + 1$, we define the value of the gamma function at x by

$$\Gamma(x) = \frac{1}{x(x + 1) \cdots (x + n - 1)}\, \Gamma(x + n). \tag{2.4}$$

If x is a negative integer or 0, the right side of Eq. (2.4) is meaningless. We will consider $\Gamma(x)$ as undefined for these particular numbers. Otherwise, the left side of Eq. (2.4) is well defined, since the argument $(x + n)$ on the right lies in the interval 0 to 1. This extended definition is obviously so constructed that the functional equation, Eq. (2.2), always holds.

The two paragraphs above clearly show that Eq. (2.2) does not determine the gamma function uniquely. If $f(x)$ is any arbitrary function defined on the interval $0 < x \leqslant 1$, we can set

$$f(x + n) = (x + n - 1)(x + n - 2) \cdots (x + 1)\, x f(x) \tag{2.5}$$

for $0 < x \leqslant 1$, and

$$f(x) = \frac{1}{x(x + 1) \cdots (x + n - 1)} f(x + n)$$

for $-n < x < -n + 1$. Thus $f(x)$ is defined for all real numbers, with the exception of 0 and the negative integers, in such a way that the functional equation $f(x + 1) = x f(x)$ always holds. This certainly makes our definitions, Eqs. (2.1) and (2.4), seem rather arbitrary. If we keep our original problem in mind, it is quite natural to want Eq. (2.2) to hold. It is the appropriate generalization of an elementary property of the factorial function. But an infinite number of arbitrary functions can be found that share this property with the gamma function. What singles out $\Gamma(x)$ from all the other possible functions we could have defined? One glance at the integral in Eq. (2.1) shows that

$$\Gamma(1) = 1,$$

and therefore

$$\Gamma(n) = (n - 1)!. \tag{2.6}$$

Furthermore, $\Gamma(x)$ is continuous and differentiable. (This will be proved later on.) But even so, an infinite number of other functions that also have these properties can be found.

Our integral in Eq. (2.1), however, has another property that catches the eye. It is log convex. This fact follows immediately from our conclusions concerning integrals and log convexity in Chapter 1. Intuitively, it means that the curve $y = \log \Gamma(x)$ is very smooth. Strange as it may seem, this property is enough to single out $\Gamma(x)$ from all the other solutions of the functional equation $f(x + 1) = x f(x)$.

We shall now prove the following:

Theorem 2.1

If a function $f(x)$ satisfies the following three conditions, then it is identical in its domain of definition with the gamma function:

(1) $f(x + 1) = xf(x)$.

(2) The domain of definition of $f(x)$ contains all $x > 0$, and is log convex for these x.

(3) $f(1) = 1$.

Proof

The existence of a function with these properties (the gamma function) has already been proved.

Suppose $f(x)$ is a function that satisfies our three conditions. Then Eq. (2.5) is valid because of condition (1), and $f(n) = (n - 1)!$ for all integers $n > 0$ because of condition (3). It suffices to show that $f(x)$ agrees with $\Gamma(x)$ on the interval $0 < x \leqslant 1$. If this is the case, then $f(x)$ must agree with $\Gamma(x)$ everywhere because of condition (1). Let x be a real number, $0 < x \leqslant 1$, and n an integer $\geqslant 2$. The inequality

$$\frac{\log f(-1 + n) - \log f(n)}{(-1 + n) - n} \leqslant \frac{\log f(x + n) - \log f(n)}{(x + n) - n} \leqslant \frac{\log f(1 + n) - \log f(n)}{(1 + n) - n}$$

expresses the monotonic growth of the difference quotient for particular values, and is therefore valid because of condition (2). Since $f(n) = (n - 1)!$, we have

$$\log (n - 1) \leqslant \frac{\log f(x + n) - \log (n - 1)!}{x} \leqslant \log n$$

or

$$\log (n - 1)^x (n - 1)! \leqslant \log f(x + n) \leqslant \log n^x(n - 1)!.$$

But the logarithm is a monotonic function; hence

$$(n - 1)^x (n - 1)! \leqslant f(x + n) \leqslant n^x(n - 1)!.$$

With the help of Eq. (2.5), we get the following inequality for $f(x)$ itself:

$$\frac{(n - 1)^x (n - 1)!}{x(x + 1) \cdots (x + n - 1)} \leqslant f(x) \leqslant \frac{n^x(n - 1)!}{x(x + 1) \cdots (x + n - 1)}$$

$$= \frac{n^x n!}{x(x + 1) \cdots (x + n)} \frac{x + n}{n}.$$

Since this inequality holds for all $n \geqslant 2$, we can replace n by $n + 1$ on the left side. Thus

$$\frac{n^x n!}{x(x + 1) \cdots (x + n)} \leqslant f(x) \leqslant \frac{n^x n!}{x(x + 1) \cdots (x + n)} \frac{x + n}{n}.$$

An easy calculation gives the inequality

$$f(x)\frac{n}{x+n} \leqslant \frac{n^x n!}{x(x+1)\cdots(x+n)} \leqslant f(x).$$

As n approaches infinity, we get

$$f(x) = \lim_{n\to\infty} \frac{n^x n!}{x(x+1)\cdots(x+n)}.$$

But $\Gamma(x)$ is also a function that satisfies our three conditions. Hence the relation we have just derived is still valid if we put $\Gamma(x)$ instead of $f(x)$ on the left side. This completes the proof of the theorem.

As a corollary we have the formula

$$\Gamma(x) = \lim_{n\to\infty} \frac{n^x n!}{x(x+1)\cdots(x+n)}. \tag{2.7}$$

Actually Eq. (2.7) was only proved for the interval $0 < x \leqslant 1$. To show that it holds in general, we denote the function under the limit sign by $\Gamma_n(x)$. It is easy to see that

$$\Gamma_n(x+1) = x\Gamma_n(x)\frac{n}{x+n+1}, \qquad \Gamma_n(x) = \frac{1}{x}\frac{x+n+1}{n}\Gamma_n(x+1).$$

These two expressions help clarify the following fact: As n approaches infinity, if the limit in Eq. (2.7) exists for a number x, it also exists for $x + 1$. Conversely, if it exists for $x + 1$ and $x \neq 0$, it also exists for x. Hence the limit exists for exactly those values of x for which $\Gamma(x)$ is defined. If we denote the limit in Eq. (2.7) by $f(x)$, we get the equation $f(x + 1) = xf(x)$. Since $f(x)$ already agrees with $\Gamma(x)$ on the interval $0 < x \leqslant 1$, it must also agree everywhere else. Equation (2.7) was derived by Gauss, and it is often used as the fundamental definition of the gamma function.

Another form of Eq. (2.7) which is important in the theory of functions was derived by Weierstrass. A simple manipulation shows that

$$\Gamma_n(x) = e^{x(\log n - 1/1 - 1/2 - \cdots - 1/n)}\frac{1}{x}\frac{e^{x/1}}{1+x/1}\frac{e^{x/2}}{1+x/2}\cdots\frac{e^{x/n}}{1+x/n}.$$

But the limit

$$C = \lim_{n\to\infty}\left(\frac{1}{1}+\frac{1}{2}+\cdots+\frac{1}{n}-\log n\right)$$

exists.* It is often called Euler's constant. Therefore, we have

$$\Gamma(x) = e^{-Cx}\frac{1}{x}\lim_{n\to\infty}\prod_{i=1}^{n}\frac{e^{x/i}}{1+x/i} = e^{-Cx}\frac{1}{x}\prod_{i=1}^{\infty}\frac{e^{x/i}}{1+x/i}, \tag{2.8}$$

where $\prod$ is the product symbol.

We shall now show that the function $\Gamma(x)$ can be differentiated as often as we please; that is, $\Gamma(x)$ has derivatives of arbitrarily high order. Because of the functional equation (2.2), it suffices to prove the assertion for positive x. But $\Gamma(x) > 0$ for $x > 0$, therefore $\log \Gamma(x)$ is defined. From (2.8) we get

$$\log \Gamma(x) = -Cx - \log x + \lim_{n\to\infty}\sum_{i=1}^{n}\left(\frac{x}{i} - \log\left(1+\frac{x}{i}\right)\right)$$

$$= -Cx - \log x + \sum_{i=1}^{\infty}\left(\frac{x}{i} - \log\left(1+\frac{x}{i}\right)\right). \tag{2.9}$$

We will now proceed to prove our assertion for the function $\log \Gamma(x)$. The conclusion for $\Gamma(x)$ will then follow immediately from $\Gamma(x) = e^{\log \Gamma(x)}$. Can we take the derivative of the series in Eq. (2.9) term by term? This can be done if the new series—the one obtained by termwise differentiation—is uniformly convergent. Should this be the case, the left side of Eq. (2.9) is differentiable. The differentiated series obtained from Eq. (2.9) is

$$-C - \frac{1}{x} + \sum_{i=1}^{\infty}\left(\frac{1}{i} - \frac{1}{x+i}\right) = -C - \frac{1}{x} + \sum_{i=1}^{\infty}\frac{x}{i(x+i)}.$$

* To prove this we can set

$$C_n = \frac{1}{1} + \frac{1}{2} + \cdots + \frac{1}{n} - \log n \qquad \text{and} \qquad D_n = C_n - \frac{1}{n},$$

which gives us

$$C_{n+1} - C_n = \frac{1}{n+1} - \log\left(1+\frac{1}{n}\right),$$

and

$$D_{n+1} - D_n = \frac{1}{n} - \log\left(1+\frac{1}{n}\right).$$

The elementary inequality

$$\frac{1}{n+1} < \log\left(1+\frac{1}{n}\right) < \frac{1}{n}$$

shows that the sequence C_n decreases monotonically, whereas the D_n increases monotonically. Furthermore $D_n < C_n$; hence, $D_1 = 0$ is a lower bound for the C_n. In other words, the sequence C_n converges to a limit.

Because x is positive, the general term of this series is smaller than x/i^2. If we restrict x to an arbitrary interval $0 < x \leqslant r$, then the general term is smaller than r/i^2. This number is completely independent of x. The series

$$\sum_{i=1}^{\infty} \frac{r}{i^2} = r\left(\frac{1}{1^2} + \frac{1}{2^2} + \frac{1}{3^2} + \cdots\right)$$

converges; therefore, our differentiated series converges uniformly. But this means $\log \Gamma(x)$ is differentiable, and that, on the interval in question,

$$\frac{d}{dx} \log \Gamma(x) = \frac{\Gamma'(x)}{\Gamma(x)} = -C - \frac{1}{x} + \sum_{i=1}^{\infty} \left(\frac{1}{i} - \frac{1}{x+i}\right). \tag{2.10}$$

But the choice of this interval was arbitrary, which implies that Eq. (2.10) holds for all $x > 0$.

Suppose we take the derivative once again. We get the series

$$\frac{1}{x^2} + \sum_{i=1}^{\infty} \frac{1}{(x+i)^2}.$$

The general term is smaller than $1/i^2$ because $x > 0$. This series obviously converges uniformly for positive x. Repeated differentiation leads to ever better converging series, all of which converge uniformly for positive x. This shows that $\log \Gamma(x)$ can be differentiated as often as we please. We have the formula

$$\frac{d^{k-1}}{dx^{k-1}} \left(\frac{\Gamma'(x)}{\Gamma(x)}\right) = \sum_{i=0}^{\infty} \frac{(-1)^k (k-1)}{(x+i)^k}, \qquad k \geqslant 2. \tag{2.11}$$

There is no trouble in extending the validity of Eq. (2.11) to include negative x. This is easily done with the help of Eq. (2.2). We merely determine the functional equation that the left side of Eq. (2.11) satisfies, and then show that the right side also satisfies it. Both these steps are quite obvious.

The case $k = 2$ is of special interest. The function

$$\frac{d}{dx}\left(\frac{\Gamma'(x)}{\Gamma(x)}\right)$$

is always positive; therefore, the following inequality holds for all x:

$$\Gamma(x)\,\Gamma''(x) - (\Gamma'(x))^2 > 0 \qquad \text{or} \qquad \Gamma(x)\,\Gamma''(x) > (\Gamma'(x))^2 \geqslant 0.$$

This shows that the functions $\Gamma(x)$ and $\Gamma''(x)$ are either both positive or both negative for each particular value of x. Consequently the function $|\,\Gamma(x)\,|$ is convex. When is $\Gamma(x)$ positive and when is it negative? We already know that $\Gamma(x)$ is positive for positive x. It follows from Eq. (2.4) that $\Gamma(x)$ has the sign

$(-1)^n$ on the interval $-n < x < -n + 1$. Furthermore, Eq. (2.4) shows that $\Gamma(x)$ has a very large absolute value whenever x gets close to zero or a negative integer. If the well-known values of $\Gamma(x)$ for the positive integers are also taken into account, we get a good idea of what the curve $y = \Gamma(x)$ looks like. (The reader is encouraged to draw a sketch.)

The integral of Eq. (2.1) is due to Euler, and it is referred to as *Euler's second integral.* He also discovered another integral related to the gamma function, which is called *Euler's first integral:*

$$B(x, y) = \int_0^1 t^{x-1}(1-t)^{y-1}\,dt. \tag{2.12}$$

This time we have a function of the two variables x and y. We want to prove that the integral exists whenever x and y are positive. First, we write our integral as the sum of two integrals, one from 0 to $\frac{1}{2}$, and the other from $\frac{1}{2}$ to 1. The integrand of the first is always smaller than $t^{x-1}(1-t)^{-1}$ and hence smaller than $2t^{x-1}$. In the second the integrand is always smaller than $t^{-1}(1-t)^{y-1}$ and hence smaller than $2(1-t)^{y-1}$. The method we used to prove the existence of Euler's second integral can now be applied to these two integrals. The rest of the details are left to the reader.

If we replace x by $x + 1$ in Eq. (2.12) and write the integral in the form

$$B(x+1, y) = \int_0^1 (1-t)^{x+y-1}\left(\frac{t}{1-t}\right)^x dt,$$

we can integrate by parts. We get

$$\int_\epsilon^{1-\delta} (1-t)^{x+y-1}\left(\frac{t}{1-t}\right)^x dt$$

$$= -\frac{(1-t)^{x+y}}{x+y}\left(\frac{t}{1-t}\right)^x \Bigg|_\epsilon^{1-\delta} + \int_\epsilon^{1-\delta} \frac{x}{x+y}(1-t)^{x+y}\left(\frac{t}{1-t}\right)^{x-1}\frac{1}{(1-t)^2}\,dt$$

$$= \frac{(1-\epsilon)^y\,\epsilon^x - \delta^y(1-\delta)^x}{x+y} + \frac{x}{x+y}\int_\epsilon^{1-\delta} t^{x-1}(1-t)^{y-1}\,dt.$$

If we let ϵ and δ converge to zero, we get the following functional equation:

$$B(x+1, y) = \frac{x}{x+y}\,B(x, y).$$

Now we hold y fixed and consider the integral of Eq. (2.12) as a function of x. In order to obtain a function that satisfies the functional equation (2.2) we set

$$f(x) = B(x, y)\,\Gamma(x+y).$$

This function obviously satisfies condition (1) in Theorem 2.1. Furthermore, $f(x)$ is the product of two log-convex functions and is therefore log convex itself. The log convexity of $B(x, y)$, regarded as a function of x, follows immediately from our theorem on the log convexity of integrals. $\Gamma(x + y)$ is obviously log convex. This means that $f(x)$ also satisfies condition (2) in Theorem 2.1. Condition (3), however, does not hold. We have

$$B(1, y) = \int_0^1 (1 - t)^{y-1}\, dt = \frac{1}{y},$$

and therefore

$$f(1) = \frac{1}{y}\, \Gamma(1 + y) = \Gamma(y).$$

But this is not really a serious difficulty. Given any function $g(x)$ that satisfies conditions (1) and (2), we can always construct a function that also satisfies (3). Condition (2) implies that $g(1) > 0$, which means we can form the quotient $g(x)/g(1)$. This function satisfies all three conditions and therefore is $\Gamma(x)$. In other words, we have

$$g(x) = g(1)\, \Gamma(x).$$

In our particular case we get

$$f(x) = \Gamma(y)\, \Gamma(x).$$

But this means we have evaluated the integral in Eq. (2.12):

$$\frac{\Gamma(x)\, \Gamma(y)}{\Gamma(x + y)} = \int_0^1 t^{x-1}(1 - t)^{y-1}\, dt. \tag{2.13}$$

This formula holds for all positive x and y.

By setting $x = \frac{1}{2}$ and $y = \frac{1}{2}$ in Eq. (2.13), we get an integral that can easily be evaluated. The substitution $t = \sin^2 \varphi$ gives us

$$(\Gamma(\tfrac{1}{2}))^2 = 2 \int_0^{\pi/2} d\varphi = \pi.$$

But $\Gamma\frac{1}{2}$ is positive; therefore, we have the following remarkable and important identity:

$$\Gamma(\tfrac{1}{2}) = \sqrt{\pi}\,. \tag{2.14}$$

Using Eq. (2.14) and the functional equation (2.1), we can easily calculate the value of $\Gamma(n + \frac{1}{2})$ for integral n.

[3]

Large Values of x and the Multiplication Formula

Can we find an elementary function that gives an accurate approximation of $\Gamma(x)$ for large values of x? If the growth of $n!$ is estimated, it is found to increase with n faster than $n^n e^{-n}$, but not quite as fast as $n^{n+1}e^{-n}$.* In other words, the growth of $\Gamma(n)$ is caught between $n^{n-1}e^{-n}$ and $n^n e^{-n}$. This suggests that we consider a function of the form

$$f(x) = x^{x-1/2}\, e^{-x}\, e^{\mu(x)}, \tag{3.1}$$

in order to study the behavior of $\Gamma(x)$ for large x. Our goal is to make $f(x)$ satisfy the basic conditions for the gamma function by choosing $\mu(x)$ in an appropriate way.

If we replace x by $x + 1$ in Eq. (3.1) and divide the resulting expression by Eq. (3.1), we get

$$\frac{f(x+1)}{f(x)} = \left(1 + \frac{1}{x}\right)^{x+1/2} xe^{-1}\, e^{\mu(x+1)-\mu(x)}.$$

This shows that $f(x)$ satisfies condition (1) in Theorem 2.1 if, and only if,

$$\mu(x) - \mu(x+1) = (x + \tfrac{1}{2}) \log\left(1 + \frac{1}{x}\right) - 1, \tag{3.2}$$

holds for $\mu(x)$.

* If we consider the elementary inequalities

$$\left(1 + \frac{1}{k}\right)^k < e < \left(1 + \frac{1}{k}\right)^{k+1}$$

for $k = 1, 2, \cdots, (n-1)$, and multiply them together, we get

$$\frac{n^{n-1}}{(n-1)!} < e^{n-1} < \frac{n^n}{(n-1)!}.$$

This leads to the approximation

$$en^n e^{-n} < n! < en^{n+1}e^{-n}.$$

We denote the right side of Eq. (3.2) by $g(x)$. A function $\mu(x)$ with this property is easy to find. If we set

$$\mu(x) = \sum_{n=0}^{\infty} g(x+n), \tag{3.3}$$

then Eq. (3.2) holds, provided the infinite series in Eq. (3.3) converges. Let us postpone the proof of convergence for a moment and consider condition (2) of theorem 2.1.

The factor $x^{x-1/2}\, e^{-x}$ in Eq. (3.1) is log convex because the second derivative of its logarithm, $1/x + \frac{1}{2}x^2$, is always positive when x is positive. If we can show that the factor $e^{\mu(x)}$ is log convex, in other words that $\mu(x)$ is convex, then $f(x)$ also satisfies condition (2). This means that the function $f(x)$ determined by the particular $\mu(x)$ we defined in Eq. (3.3) will agree with $\Gamma(x)$ to within a constant factor. Our $\mu(x)$ is convex if the general term of the series $g(x+n)$ is convex. To show this, it suffices to prove the convexity of $g(x)$ itself. But we have

$$g''(x) = \frac{1}{2x^2(x+1)^2} > 0.$$

The convergence of the series in Eq. (3.3) still remains to be shown. We will combine this with an approximation of the function $\mu(x)$. Let us begin by considering the expansion

$$\tfrac{1}{2} \log \frac{1+y}{1-y} = \frac{y}{1} + \frac{y^3}{3} + \frac{y^5}{5} + \cdots,$$

which is valid for $|y| < 1$. Now we replace y by $1/(2x+1)$. The resulting expansion is valid for positive x because $1/(2x+1) < 1$ whenever $x > 0$. We multiply this equation by $2x+1$ and bring the first term on the right side over to the left side:

$$(x + \tfrac{1}{2}) \log \left(1 + \frac{1}{x}\right) - 1 = g(x)$$

$$= \frac{1}{3(2x+1)^2} + \frac{1}{5(2x+1)^4} + \frac{1}{7(2x+1)^6} + \cdots.$$

This expression again shows that $g(x)$ is convex, since every term on the right side is convex. Now we can approximate $g(x)$. If the integers 5, 7, 9, $\cdots$ are all replaced by 3, then the value of the right side increases. The result is an infinite geometric series, having $1/(3(2x+1)^2)$ as its first term and $1/(2x+1)^2$ as its ratio. Its sum is

$$\frac{1}{3(2x+1)^2}\,\frac{1}{1-(1/(2x+1)^2)} = \frac{1}{12x(x+1)} = \frac{1}{12x} - \frac{1}{12(x+1)}.$$

But $g(x)$ is positive, hence

$$0 < g(x) < \frac{1}{12x} - \frac{1}{12(x+1)}.$$

Since every term of the series in Eq. (3.3) is positive, it suffices to show the convergence of

$$\sum_{n=0}^{\infty} \left(\frac{1}{12(x+n)} - \frac{1}{12(x+n+1)} \right),$$

which converges trivially to the limit $1/12x$. This not only proves our assertion, it also gives the approximation

$$0 < \mu(x) < \frac{1}{12x}.$$

In other words,

$$\mu(x) = \frac{\theta}{12x}$$

where θ is a number independent of x between 0 and 1.

By a suitable choice of the constant a, we get

$$\Gamma(x) = ax^{x-1/2}\, e^{-x+\mu(x)} = ax^{x-1/2}\, e^{-x+\theta/12x}. \tag{3.4}$$

If we let x be an integer n and multiply the expression by n, we get the approximation

$$n! = an^{n+1/2}\, e^{-n+\theta/12n}. \tag{3.5}$$

We are now going to find the exact value of this constant a and determine some other important constants at the same time.

Let p be a positive integer. We consider the function

$$f(x) = p^x \Gamma\left(\frac{x}{p}\right) \Gamma\left(\frac{x+1}{p}\right) \cdots \Gamma\left(\frac{x+p-1}{p}\right),$$

for $x > 0$. The second derivative of $\log p^x$ is zero, and each of the functions $\Gamma((x+i)/p)$ is obviously log convex. This implies that $f(x)$ is also log convex. If we replace x by $x+1$, p^x takes on the factor p, $\Gamma((x+i)/p)$ goes over into the next factor, and $\Gamma((x+p-1)/p)$ becomes

$$\Gamma\left(\frac{x}{p}+1\right) = \frac{x}{p}\Gamma\left(\frac{x}{p}\right).$$

In other words, $f(x)$ is multiplied by x. Our function again satisfies the conditions (1) and (2) in Theorem 2.1; therefore,

$$p^x \Gamma\left(\frac{x}{p}\right) \Gamma\left(\frac{x+1}{p}\right) \cdots \Gamma\left(\frac{x+p-1}{p}\right) = a_p \Gamma(x), \tag{3.6}$$

where a_p is a constant depending on p. For $x = 1$ in Eq. (3.6), we have

$$a_p = p\Gamma\left(\frac{1}{p}\right)\Gamma\left(\frac{2}{p}\right)\cdots\Gamma\left(\frac{p}{p}\right). \tag{3.7}$$

If we set $x = k/p$ in Eq. (2.7), then a simple manipulation gives

$$\Gamma\left(\frac{k}{p}\right) = \lim_{n\to\infty} \frac{n^{k/p} n!\, p^{n+1}}{k(k+p)(k+2p)\cdots(k+np)}.$$

Now we set $k = 1, 2, \cdots, p$, one after the other, and multiply all these expressions together. Factors of the form $(k + hp)$ appear in the denominator, where k runs from 1 to p, and h runs from 0 to n. For $h = 0$ we get the numbers from 1 to p; for $h = 1$, the numbers from $p + 1$ to $2p$; and so on. The product in the denominator is obviously $(np + p)!$. The final result is

$$a_p = p \lim_{n\to\infty} \frac{n^{(p+1)/2}(n!)^p\, p^{np+1}}{(np+p)!}.$$

The well-known infinite product

$$1 = \lim_{n\to\infty}\left(1+\frac{1}{np}\right)\left(1+\frac{2}{np}\right)\cdots\left(1+\frac{p}{np}\right),$$

which can be written as

$$1 = \lim_{n\to\infty} \frac{(np+p)!}{(np)!\,(np)^p},$$

can now be applied. If we multiply this last expression with the above identity for a_p, we obtain

$$a_p = p \lim_{n\to\infty} \frac{(n!)^p\, p^{np}}{(np)!\, n^{(p-1)/2}}.$$

But Eq. (3.5) implies that

$$(n!)^p = a^p n^{np+p/2} e^{-np} e^{\theta_1 p/12n},$$

$$(np)! = a(np)^{np+1/2} e^{-np} e^{\theta_2/12np}.$$

After making the appropriate substitutions above, we obtain

$$a_p = \sqrt{p}\, a^{p-1} \lim_{n\to\infty} e^{(\theta_1 p/12n)-(\theta_2/12np)},$$

and finally

$$a_p = \sqrt{p}\, a^{p-1}. \tag{3.8}$$

By evaluating a_2 with the help of Eq. (3.7) and then comparing the result with Eq. (3.8), we get

$$a_2 = 2\Gamma(\tfrac{1}{2})\,\Gamma(1) = 2\sqrt{\pi} = a\sqrt{2}.$$

But this determines the exact values of our constants:

$$a = \sqrt{2\pi} \qquad \text{and} \qquad a_p = p^{1/2}(2\pi)^{(p-1)/2}.$$

Now we gather together all the important expressions from this chapter:

$$\Gamma(x) = \sqrt{2\pi}\, x^{x-1/2}\, e^{-x+\mu(x)},$$

$$\mu(x) = \sum_{n=0}^{\infty} (x + n + \tfrac{1}{2}) \log\left(1 + \frac{1}{x+n}\right) - 1 = \frac{\theta}{12x}, \qquad 0 < \theta < 1,$$

$$n! = \sqrt{2\pi}\, n^{n+1/2}\, e^{-n+\theta/12n}. \tag{3.9}$$

$$\Gamma\left(\frac{x}{p}\right)\Gamma\left(\frac{x+1}{p}\right)\cdots\Gamma\left(\frac{x+p-1}{p}\right) = \frac{(2\pi)^{(p-1)/2}}{p^{x-1/2}}\,\Gamma(x). \tag{3.10}$$

In particular, for $p = 2$

$$\Gamma\left(\frac{x}{2}\right)\Gamma\left(\frac{x+1}{2}\right) = \frac{\sqrt{\pi}}{2^{x-1}}\,\Gamma(x). \tag{3.11}$$

The formulas in Eq. (3.9), which describe the behavior of $\Gamma(x)$ for large values of x, are called *Stirling's formulas*. If our approximation of $\mu(x)$ is used, the accuracy of the formula for $\Gamma(x)$ will increase as x increases. This is also true for estimates of $n!$ The relative accuracy for $n \geqslant 10$ is already quite high.

The functional equation (3.10), discovered by Gauss, is called *Gauss' multiplication formula*. By replacing x by px in Eq. (3.10), we obtain an expression for $\Gamma(px)$ as the product of factors, each of the form $\Gamma(x + (k/p))$. This fact gave rise to the name "multiplication formula." The most important special case is $p = 2$. It was discovered by Legendre and is often referred to as *Legendre's relation*.

[4]

The Connection with sin x

The gamma function satisfies another very important functional equation. In order to derive it, we set

$$\varphi(x) = \Gamma(x)\,\Gamma(1-x)\sin \pi x. \tag{4.1}$$

This function is only defined for nonintegral arguments. If we replace x by $x + 1$, then $\Gamma(x)$ becomes $x\Gamma(x)$. The function $\Gamma(1 - x)$ becomes

$$\Gamma(-x) = \frac{\Gamma(1-x)}{-x},$$

and $\sin \pi x$ changes its sign. This means that $\varphi(x)$ is left fixed, and is therefore periodic of period 1:

$$\varphi(x+1) = \varphi(x). \tag{4.2}$$

The Legendre relation can be written in the form

$$\Gamma\left(\frac{x}{2}\right)\Gamma\left(\frac{x+1}{2}\right) = b2^{-x}\,\Gamma(x),$$

where b is a constant. Actually, the exact value of b was determined in Chapter 3. But this extra information need not (and will not) be assumed here. As far as we are concerned now, b is just some particular constant.

In the expression above, we replace x by $1 - x$:

$$\Gamma\left(\frac{1-x}{2}\right)\Gamma\left(1-\frac{x}{2}\right) = b2^{x-1}\,\Gamma(1-x).$$

Now we consider

$$\varphi\left(\frac{x}{2}\right)\varphi\left(\frac{x+1}{2}\right) = \Gamma\left(\frac{x}{2}\right)\Gamma\left(1-\frac{x}{2}\right)\sin\frac{\pi x}{2}\,\Gamma\left(\frac{x+1}{2}\right)\Gamma\left(\frac{1-x}{2}\right)\cos\frac{\pi x}{2}$$

$$= \frac{b^2}{4}\,\Gamma(x)\,\Gamma(1-x)\sin \pi x,$$

and we get the relation

$$\varphi\left(\frac{x}{2}\right)\varphi\left(\frac{x+1}{2}\right) = d\varphi(x), \tag{4.3}$$

where d is a constant depending on b. The exact value of d is not important here.

Since both $\Gamma(x)$ and $\sin x$ can be differentiated as often as we please, $\varphi(x)$ also has this property. Because of the functional equation (2.2), we can write

$$\varphi(x) = \frac{\Gamma(1+x)}{x}\,\Gamma(1-x)\sin \pi x$$
$$= \Gamma(1+x)\,\Gamma(1-x)\left(\pi - \frac{\pi^3x^2}{3!} + \frac{\pi^5x^4}{5!} - \cdots\right),$$

where the power series converges for all values of x. But the right side of this equation is also defined for $x = 0$, and it represents a function having derivatives of all orders at this point. This suggests that we extend our definition, Eq. (4.1), by giving $\varphi(x)$ the value π at $x = 0$. Because our function is periodic, we define π to be the value of $\varphi(x)$ for all integral arguments. $\varphi(x)$ is now continuous everywhere and has derivatives of all orders at every point. The relation of Eq. (4.3) was only proven for nonintegral x. But if we let x approach an arbitrary integer, the validity of Eq. (4.3) for all x follows from continuity. As far as the sign of $\varphi(x)$ is concerned, Eq. (4.1) shows that $\varphi(x)$ is positive on the interval $0 < x < 1$. Because of Eq. (4.2), this is now true for all x.

Our goal is to prove that $\varphi(x)$ is a constant. Let $g(x)$ denote the second derivative of $\log \varphi(x)$. $g(x)$ is also periodic of period 1. Because of Eq. (4.3), it satisfies the functional equation

$$\tfrac{1}{4}\left(g\left(\frac{x}{2}\right) + g\left(\frac{x+1}{2}\right)\right) = g(x). \tag{4.4}$$

Since $g(x)$ is continuous on the interval $0 \leqslant x \leqslant 1$, it is bounded on this interval, say, $|\, g(x) \,| \leqslant M$. But this inequality holds for all x, because $g(x)$ is periodic.

The following argument shows that $g(x)$ vanishes. Equation (4.4) gives us the inequality

$$|\, g(x) \,| \leqslant \tfrac{1}{4}\left|\, g\left(\frac{x}{2}\right) \right| + \tfrac{1}{4}\left|\, g\left(\frac{x+1}{2}\right) \right| \leqslant \frac{M}{4} + \frac{M}{4} = \frac{M}{2}.$$

This means that the upper bound can be pushed down from M to $M/2$. If we repeat the process again, we get $M/4$ as an upper bound, and so on. In other words, the upper bound for $g(x)$ can be made as small as we please. This implies that $g(x) = 0$. But $g(x)$ was the second derivative of $\log \varphi(x)$, hence $\log \varphi(x)$ is a linear function. Furthermore, $\log \varphi(x)$ is periodic, which means that $\log \varphi(x)$ must be a constant. Therefore $\varphi(x)$ is also a constant. We already know one value of $\varphi(x)$, namely, $\varphi(0) = \pi$. This implies that $\varphi(x) = \pi$ for all x.

Recalling the definition of $\varphi(x)$ in Eq. (4.1), we get

$$\Gamma(x)\,\Gamma(1-x) = \frac{\pi}{\sin \pi x}, \tag{4.5}$$

which is often called *Euler's functional equation.* If we set $x = \frac{1}{2}$ in Eq. (4.5), we get a new proof of Eq. (2.14). The exact value of the constant in Legendre's relation was never used; therefore, this proof of Eq. (2.14) is independent of what was done in Chapter 3.

With the help of Eq. (2.2) we can write Eq. (4.5) in the form

$$\sin \pi x = \frac{\pi}{-x\Gamma(x)\,\Gamma(-x)}.$$

Now we substitute the expressions for $\Gamma(x)$ and $\Gamma(-x)$ given by Weierstrass' product formula, Eq. (2.8). This gives the following representation of $\sin \pi x$ as an infinite product:

$$\sin \pi x = \pi x \prod_{i=1}^{\infty} \left(1 - \frac{x^2}{i^2}\right). \tag{4.6}$$

If we had assumed this product development of $\sin \pi x$ to start with (there are other proofs for the expression), we could have derived Eq. (4.5) by direct calculation. The approach we took is preferable, inasmuch as it gives us Eq. (4.6) at the same time. For the importance of Eq. (4.6) in analysis, we refer the reader to books on function theory.

Let us formulate the main result of this section as a theorem:

Theorem 4.1

Every positive periodic function that has a continuous second derivative and satisfies the functional equation (4.3) is a constant.

[5]

Applications to Definite Integrals

There is extensive literature that deals with more-or-less trivial manipulations of the two Euler integrals. We can only mention a few of the important results here.

If we set $e^{-t} = \tau$ in Eq. (2.1), and then write t instead of τ again, we get

$$\Gamma(x) = \int_0^1 \left(\log \frac{1}{t}\right)^{x-1} dt. \tag{5.1}$$

Similarly, the substitution $t^x = \tau$ turns Eq. (2.1) into

$$\Gamma(x) = \int_0^\infty e^{-t^{1/x}} \frac{1}{x}\, dt.$$

If we replace x by $1/x$, then

$$\Gamma\left(1 + \frac{1}{x}\right) = \int_0^\infty e^{-t^x}\, dt. \tag{5.2}$$

The case $x = 2$ in Eq. (5.2) is of special interest:

$$\int_0^\infty e^{-t^2}\, dt = \tfrac{1}{2}\sqrt{\pi}. \tag{5.3}$$

The number of similar manipulations can be extended indefinitely. We will mention one other example of importance in analytic number theory. If $a > 0$, the substitution $t = a\tau$ leads to the integral

$$\frac{\Gamma(x)}{a^x} = \int_0^\infty e^{-at}\, t^{x-1}\, dt. \tag{5.4}$$

Now we turn our attention to the first Euler integral, Eq. (2.13). The substitutions $t = \tau/(\tau + 1)$ and $\sin^2 \varphi$, respectively, give the expressions

$$\int_0^\infty \frac{t^{x-1}}{(1 + t)^{x+y}}\, dt = \frac{\Gamma(x)\, \Gamma(y)}{\Gamma(x + y)}, \tag{5.5}$$

and

$$\int_0^{\pi/2} (\sin \varphi)^{2x-1} (\cos \varphi)^{2y-1}\, d\varphi = \frac{1}{2} \frac{\Gamma(x)\, \Gamma(y)}{\Gamma(x + y)}. \tag{5.6}$$

If we set $y = (1 - x)$ in Eq. (2.13), Eq. (4.5) gives us the following interesting special cases:

$$\int_0^1 t^{x-1}(1-t)^{-x}\,dt = \frac{\pi}{\sin \pi x}, \qquad 0 < x < 1, \tag{5.7}$$

$$\int_0^\infty \frac{t^{x-1}}{1+t}\,dt = \frac{\pi}{\sin \pi x}, \qquad 0 < x < 1, \tag{5.8}$$

$$2\int_0^{\pi/2} (\operatorname{tg}\varphi)^{2x-1}\,d\varphi = \frac{\pi}{\sin \pi x}, \qquad 0 < x < 1. \tag{5.9}$$

If x and y are both rational numbers, Eq. (2.13) is the intergal of an algebraic function. Suppose we set $x = m/n$ and $y = \frac{1}{2}$ in Eq. (2.13), and make the substitution $t = \tau^n$. We get

$$\int_0^1 \frac{t^{m-1}}{\sqrt{1-t^n}}\,dt = \frac{\Gamma(m/n)\sqrt{\pi}}{n\Gamma(m/n+1/2)}. \tag{5.10}$$

For $m = 1$ and $n = 4$ or $n = 3$, we get the following numerical results, with the help of Eqs. (3.11) and (4.5):

$$\int_0^1 \frac{dt}{\sqrt{1-t^4}} = \frac{(\Gamma(1/4))^2}{\sqrt{32\pi}}, \tag{5.11}$$

$$\int_0^1 \frac{dt}{\sqrt{1-t^3}} = \frac{(\Gamma(1/3))^3}{\sqrt{3}\,\sqrt[3]{16}\pi}, \tag{5.12}$$

which shows a connection between these particular numbers and the elliptical integrals.

An integral representation for the error $\mu(x)$ in Stirling's formula can also be found. Since we have

$$\int_0^1 \frac{1/2-t}{t+x}\,dt = (x+\tfrac{1}{2})\log\left(1+\frac{1}{x}\right) - 1,$$

the series in Eq. (3.9) can be written in the form

$$\mu(x) = \sum_{n=0}^{\infty} \int_0^1 \frac{1/2-t}{t+n+x}\,dt.$$

If we define the following noncontinuous function:

$$H(t) = \begin{cases} \frac{1}{2} - t, & \text{for} \quad 0 < t < 1 \\ 0, & \text{for} \quad t = 0 \\ \text{otherwise periodic of period 1} \end{cases}$$

it follows that

$$\mu(x) = \sum_{n=0}^{\infty} \int_0^1 \frac{H(t)}{t + n + x}\, dt = \sum_{n=0}^{\infty} \int_n^{n+1} \frac{H(t)}{t + x}\, dt = \lim_{n\to\infty} \int_0^n \frac{H(t)}{t + x}\, dt.$$

Because the integrand is an oscillating function that approaches zero as t approaches infinity, the integral

$$\mu(x) = \int_0^\infty \frac{H(t)}{t + x}\, dt \tag{5.13}$$

exists.

Equation (5.13) is the first step in deriving the so-called *Stirling series*, a refinement of Stirling's formula. We now introduce the functions

$$H_{2n}(t) = 2\,(-1)^{n-1} \sum_{i=1}^{\infty} \frac{\cos 2i\pi t}{(2i\pi)^{2n}}\,,$$

$$H_{2n-1}(t) = 2\,(-1)^{n} \sum_{i=1}^{\infty} \frac{\sin 2i\pi t}{(2i\pi)^{2n-1}}\,. \tag{5.14}$$

We know that $-H_1(t) = H(t)$, because $-H_1(t)$ is the Fourier series for $H(t)$. For $n \geqslant 2$, the series $H_n(t)$ is absolutely and uniformly convergent for all x; for $n = 1$, the series converges uniformly in every closed interval that contains no integer. Therefore Eq. (5.14) implies that

$$H'_{n+1}(t) = H_n(t). \tag{5.15}$$

When $n \geqslant 2$, this holds for all x; when $n = 1$, for all nonintegral x. Because $H_{n+1}(t)$ is always continuous, Eq. (5.15) implies that

$$\int_0^t H_n(t)\, dt = H_{n+1}(t) - H_{n+1}(0). \tag{5.16}$$

The functions $H_n(t)$ are periodic of period 1, so it suffices to study them on the interval $0 \leqslant t < 1$. Because $H_1(t)$ is a polynomial on this interval, $H_n(t)$ must also be a polynomial there. We maintain that the coefficients of these polynomials are all rational numbers. This can be shown by induction. Our assertion is true when $n = 1$; we assume that it also holds for $H_n(t)$. This implies that

$$\int_0^t H_n(t)\, dt,$$

is a rational polynomial; thus it suffices to prove that $H_{n+1}(0)$ is a rational number. When $n + 1$ is odd, Eq. (5.14) implies $H_{n+1}(0) = 0$. When $n + 1$ is even, we can set $H_{n+1}(t) = \varphi(t) + H_{n+1}(0)$, where $\varphi(t)$ is a rational polynomial. If we integrate this expression, then using Eq. (5.16) we get

$$H_{n+2}(t) - H_{n+2}(0) = H_{n+2}(t) = \int_0^t \varphi(t)\,dt + H_{n+1}(0)\,t,$$

because $n + 2$ is odd. But Eq. (5.14) implies that $H_{n+2}(1)$ vanishes; thus we have the expression

$$0 = \int_0^1 \varphi(t)\,dt + H_{n+1}(0), \tag{5.17}$$

which enables us to calculate $H_{n+1}(0)$ and show that it is rational.

We are now in a position to write Eq. (5.13) in the desired form. Repeated integration by parts gives

$$\mu(x) = \frac{H_2(0)}{x} + \frac{H_3(0)\,1!}{x^2} + \frac{H_4(0)\,2!}{x^3} + \cdots + \frac{H_n(0)\,(n-2)!}{x^{n-1}} - \int_0^\infty \frac{H_n(t)\,(n-1)!}{(t+x)^n}\,dt. \tag{5.18}$$

An easy calculation shows that the sum of the last two terms in Eq. (5.18) is equal to

$$\int_0^\infty \frac{(H_n(0) - H_n(t))\,(n-1)!}{(t+x)^n}\,dt\,.$$

If n is even, $H_n(0) - H_n(t)$ has the same sign (plus or minus) as $H_n(0)$ for all t; thus the integral above also has this sign. Furthermore, Eq. (5.14) implies that the numbers $H_{2n+1}(0)$ all vanish, and that the signs of the numbers $H_{2n}(0)$ alternate between plus and minus. This shows that the partial sums

$$\frac{H_2(0)\,0!}{x} + \frac{H_4(0)\,2!}{x^3} + \cdots + \frac{H_{2n}(0)\,(2n-2)!}{x^{2n-1}}$$

are alternately larger and smaller than $\mu(x)$. In other words, for every n there exists a number θ, $0 < \theta < 1$, so that

$$\mu(x) = \frac{H_2(0)\,0!}{x} + \frac{H_4(0)\,2!}{x^3} + \cdots + \frac{H_{2n-2}(0)\,(2n-4)!}{x^{2n-3}} + \theta\,\frac{H_{2n}(0)\,(2n-2)!}{x^{2n-1}}, \qquad 0 < \theta < 1. \tag{5.19}$$

We can not take the limit in Eq. (5.19) because the series diverges. But so long as n does not get too large, it gives us a very useful approximation. When $n = 8$, for example, we get the approximation

$$\mu(x) = \frac{1}{12x} - \frac{1}{360x^3} + \frac{1}{1260x^5} - \frac{\theta}{1680x^7}\,. \tag{5.10}$$

We now have a method for computing $\Gamma(x)$. In the interval $4 \leqslant x \leqslant 5$, for instance, we can compute $\mu(x)$ to within six decimal places by using Eq. (5.20). The first formula in Eq. (3.9) gives the value of $\log \Gamma(x)$ with the same accuracy. Using the functional equation (2.2), we finally get the value of $\Gamma(x)$ in the interval from 1 to 2. This enables us to compute $\Gamma(x)$ for arbitrary x. When x is very large, Eq. (3.9) can be used directly.

Anyone familiar with the so-called Bernoulli numbers will recognize the connection with our $H_n(0)$. The reader is left to pursue this topic on his own.

[6]

Determining $\Gamma(x)$ *by Functional Equations*

We have aquainted ourselves with three different functional equations for the gamma function: the functional equation (2.2), the multiplication formula, and Euler's formula. To what extent is the gamma function determined by one, or a combination, of these equations?

Suppose for the time being that $f(x)$ is an arbitrary function and that $\varphi(x)$ denotes the quotient $f(x)/\Gamma(x)$. If $f(x)$ satisfies a functional equation of the type in Eqs. (2.2), (3.10) or (4.5), then $\varphi(x)$ will clearly satisfy the corresponding equation among the following:

$$\varphi(x+1) = \varphi(x), \tag{6.1}$$

$$\varphi\left(\frac{x}{p}\right)\varphi\left(\frac{x+1}{p}\right)\cdots\varphi\left(\frac{x+p-1}{p}\right) = \varphi(x), \tag{6.2}$$

$$\varphi(x)\,\varphi(1-x) = 1. \tag{6.3}$$

If $f(x)$ satisfies Legendre's functional equation (3.11), for instance, then

$$\varphi\left(\frac{x}{2}\right)\varphi\left(\frac{x+1}{2}\right) = \varphi(x). \tag{6.4}$$

For the sake of simplicity, we will assume from now on that $f(x)$ satisfies the functional equation (2.2); consequently, $\varphi(x)$ is periodic of period 1, that is, $\varphi(x)$ satisfies Eq. (6.1). We will also assume that $f(x)$, along with $\varphi(x)$, is continuous for all x. As a result of Eq. (6.1), the continuity of $\varphi(x)$ for positive x implies continuity at zero and at negative integers, provided $\varphi(x)$ is defined for these values in the right way. If Eqs. (6.2) or (6.3) also hold, they are valid for all x because of continuity.

If we assume further that $f(x)$ is always positive when x is positive, then the logarithm of $\varphi(x)$ is continuous. If we denote $\log \varphi(x)$ by $g(x)$, the corresponding functional equations (in addition to $g(x+1) = g(x)$) are

$$g\left(\frac{x}{p}\right) + g\left(\frac{x+1}{p}\right) + \cdots + g\left(\frac{x+p-1}{p}\right) = g(x), \tag{6.5}$$

$$g(x) = -g(1-x) = -g(-x), \tag{6.6}$$

$$g\left(\frac{x}{2}\right) + g\left(\frac{x+1}{2}\right) = g(x). \tag{6.7}$$

If $f(x)$ has a continuous second derivative, then so does $\varphi(x)$. We will assume this to be the case, and also that $\varphi(x)$ satisfies Eq. (6.4). But then $\varphi(x)$ must be a constant. This follows from the theorem proven at the end of Chapter 4. Because of Eq. (6.4), the value of this constant must be 1. In other words, we have $f(x) = \Gamma(x)$.

Theorem 6.1

The gamma function is the only solution of both Eq. (2.2) and Eq. (3.11) that is positive for positive x and possesses a continuous second derivative.

This theorem demonstrates the significance of the Legendre functional equation. Our next step is to find out whether the assumption of a continuous second derivative can be weakened. As a matter of fact, we will be able to show that a continuous first derivative is sufficient.

Suppose we start with a preliminary observation. Equation (6.5) represents an infinite number of functional equations, one for each value of p. But these equations are not independent of each other. Assume, for instance, that Eq. (6.5) holds for the integers p_1 and p_2. If we consider it for the integer p_1, but with the argument $(x + k)/p_2$, we get

$$\sum_{i=0}^{p_1-1} g\left(\frac{x+k+ip_2}{p_1p_2}\right) = g\left(\frac{x+k}{p_2}\right).$$

Now we take the sum over k from zero to $p_2 - 1$. On the right side we get $g(x)$, since Eq. (6.5) holds for p_2. But $k + ip_2$ runs over all integers from zero to $p_1p_2 - 1$. This yields the equation

$$\sum_{j=0}^{p_1p_2-1} g\left(\frac{x+j}{p_1p_2}\right) = g(x);$$

therefore Eq. (6.5) also holds for the product p_1p_2.

With this in mind, let us assume that Eq. (6.7) holds. Then Eq. (6.5) is valid for all integers of the form 2^n, and hence for arbitrarily large values of p. More generally, if Eq. (6.5) holds for an integer p, it also holds for the powers of that integer, and hence for certain arbitrarily large integers.

Now we take the derivative of Eq. (6.5)

$$\frac{1}{p}\left(g'\left(\frac{x}{p}\right) + g'\left(\frac{x+1}{p}\right) + \cdots + g'\left(\frac{x+p-1}{p}\right)\right) = g'(x), \tag{6.8}$$

and partition the positive x axis into intervals of length $1/p$, beginning with the origin. The arguments in the left side of Eq. (6.8) fall into p distinct consecutive intervals. Since $g'(x)$ is periodic of period 1, we have a value of the function from each of the intervals between zero and 1. But Eq. (6.8) holds for arbitrarily large values of p. As p approaches infinity, the left side of Eq. (6.8) converges to the integral

$$\int_0^1 g'(x)\, dx = g(1) - g(0) = 0.$$

This means that $g'(x) = 0$ for all x; consequently $g(x)$ is a constant. The value of this constant is zero, as can be seen from any of the equations in (6.5) which happen to hold. This proves the following theorem:

Theorem 6.2

The gamma function is the only continuously differentiable function that is positive for positive values of x, and that satisfies both Eq. (2.2) and Eq. (3.10) for some value of p.

Now we might be tempted to conjecture that continuity alone suffices. This is not true at all. The function

$$g(x) = \sum_{n=1}^{\infty} \frac{1}{2^n} \sin(2^n \pi x), \tag{6.9}$$

is continuous, because the series converges uniformly. It is also periodic, and it is easy to see that it satisfies Eq. (6.7). But it is not identically equal to zero. It is not even a constant:

$$g(x) = 0, \qquad g(\tfrac{1}{4}) = \tfrac{1}{2}.$$

If we assume mere continuity, what other properties must also be assumed to make Theorem 6.2 valid? Equation (6.6) is not sufficient; the function in Eq. (6.9) also satisfies Eq. (6.6). A finite number of Eq. (6.5) is not sufficient either. Similar counter examples can always be constructed. But what happens if we assume that $g(x)$ satisfies Eq. (6.5) for *all* integers p? As a conclusion to our study of the gamma function it will be shown that this property is sufficient.

In order to do this we will make use of some facts about Fourier series. Let $f(x)$ be an integrable function of period 1. Setting (for any α)

$$c_\nu = \int_0^1 f(x)\, e^{-2\pi i\nu x}\, dx = \int_\alpha^{\alpha+1} f(x)\, e^{-2\pi i\nu x}\, dx,$$

we associate with $f(x)$ the series

$$\sum_{\nu=-\infty}^{+\infty} c_\nu e^{2\pi i\nu x},$$

regardless of whether or not this series converges. We denote this association by

$$f(x) \sim \sum_{\nu=-\infty}^{+\infty} c_\nu e^{2\pi i \nu x}.$$

We see immediately that the Fourier series of a sum of two functions is the sum of the two Fourier series.

The Fourier series of $f(x + \alpha)$ has the coefficients

$$d_\nu = \int_0^1 f(x + \alpha)\, e^{-2\pi i \nu x}\, dx = \int_\alpha^{\alpha+1} f(x)\, e^{-2\pi i \nu (x-\alpha)}\, dx$$

$$= e^{2\pi i \nu \alpha} \int_\alpha^{\alpha+1} f(x)\, e^{-2\pi i \nu x}\, dx;$$

hence $d_\nu = e^{2\pi i \nu \alpha} c_\nu$. We see that the Fourier series of $f(x + \alpha)$ is obtained from that of $f(x)$ by making the substitution $x \to x + \alpha$.

Let $k \geqslant 1$ be an integer. The Fourier coefficient d_ν of the function $f(kx)$ is given by

$$d_\nu = \int_0^1 f(kx)\, e^{-2\pi i \nu x}\, dx = \frac{1}{k} \int_0^k f(x)\, e^{-2\pi i \nu x/k}\, dx$$

$$= \frac{1}{k} \sum_{m=1}^{k} \int_{m-1}^{m} f(x)\, e^{-2\pi i \nu x/k}\, dx = \frac{1}{k} \sum_{m=1}^{k} \int_0^1 f(x)\, e^{-2\pi i \nu (x+m-1/k)}\, dx$$

$$= \frac{1}{k} \left(\sum_{m=1}^{k} e^{-2\pi i \nu (m-1)/k} \right) \int_0^1 f(x)\, e^{-2\pi i \nu x/k}\, dx.$$

The sum in the last expression above is a geometric series with the ratio $e^{-2\pi i \nu/k}$. If $\nu = \mu k$ is divisible by k, each term of the sum is 1, and we obtain

$$d_{\mu k} = \int_0^1 f(x)\, e^{-2\pi i \mu x}\, dx = c_\mu .$$

If ν is not divisible by k, $e^{-2\pi i \nu/k} \neq 1$, and the formula for the sum of the geometric series shows that $d_\nu = 0$. Therefore

$$f(kx) \sim \sum_{\mu=-\infty}^{+\infty} c_\mu e^{2\pi i \nu k x}.$$

The Fourier series for $f(kx)$ is obtained from that of $f(x)$ by merely substituting kx for x.

If $f(x)$ is continuous and the Fourier series converges for a particular value x_0, the value of the Fourier series for $x = x_0$ is the function value $f(x_0)$. The reader will find a proof of this fact in most books on Fourier series. As a matter of fact a consequence of Fejer's theorem is that the Fourier series of a continuous function $f(x)$ is always summable (even if not convergent) to $f(x)$.

We are now in a position to proceed with the proof at hand. Suppose our function $g(x)$ satisfies Eq. (6.5) for all integers p. Replacing x by px, we obtain from Eq. (6.5)

$$g(x) + g\left(x + \frac{1}{p}\right) + \cdots + g\left(x + \frac{p-1}{p}\right) = g(px). \tag{6.10}$$

If we let

$$g(x) \sim \sum_{\nu=-\infty}^{+\infty} c_\nu e^{2\pi i\nu x},$$

then

$$g\left(x + \frac{m}{p}\right) \sim \sum_{\nu=-\infty}^{+\infty} c_\nu e^{2\pi i\nu x} e^{2\pi i\nu m/p},$$

and

$$g(px) \sim \sum_{\nu=-\infty}^{+\infty} c_\nu e^{2\pi i\nu px}. \tag{6.11}$$

Substituting this into Eq. (6.10), we see that

$$\sum_{m=0}^{p-1} \sum_{\nu=-\infty}^{+\infty} c_\nu e^{2\pi i\nu x} e^{2\pi i\nu m/p} = \sum_{\nu=-\infty}^{+\infty} c_\nu e^{2\pi i\nu x} \left(\sum_{m=0}^{p-1} e^{2\pi i\nu m/p}\right)$$

is the Fourier series of the left side.

Just as before, we obtain for

$$\sum_{m=0}^{p-1} e^{2\pi i\nu m/p}$$

the value p if ν is divisible by p and zero if ν is not divisible by p. The Fourier series of $g(px)$ is, consequently,

$$\sum_{\mu=-\infty}^{+\infty} pc_{\mu p}\, e^{2\pi i\mu px}.$$

Comparing this result with Eq. (6.11), we find that

$$c_\mu = pc_{\mu p}\,.$$

In particular for the cases $\mu = 1, -1, 0$

$$c_p = \frac{c_1}{p}; \qquad c_{-p} = \frac{c_{-1}}{p}; \qquad c_0 = 0.$$

Since we have assumed that Eq. (6.5) holds for all integers $p \geqslant 1$, we get

$$g(x) \sim \sum_{\nu=1}^{\infty} \left(\frac{c_1}{\nu} e^{2\nu\pi i x} + \frac{c_{-1}}{\nu} e^{-2\nu\pi i x} \right).$$

But $g(x)$ is a real-valued function, and hence

$$g(x) \sim \sum_{\nu=1}^{\infty} \left(\frac{a}{\nu} \sin 2\nu\pi x + \frac{b}{\nu} \cos 2\nu\pi x \right).$$

The terms

$$\sum_{\nu=1}^{\infty} \frac{\sin 2\nu\pi x}{\nu} \quad \text{and} \quad \sum_{\nu=1}^{\infty} \frac{\cos 2\nu\pi x}{\nu}$$

are the Fourier series for the functions $\pi H(x)$ and $-\log(2 \sin \pi x)$, respectively, where $H(x)$ denotes the function introduced in Chapter 5. Therefore, at every point of continuity,

$$g(x) = a\pi H(x) - b \log(2 \sin \pi x).$$

The function $H(x)$ is bounded, the function $-\log(2 \sin \pi x)$ is not. This implies that $b = 0$ for continuous $g(x)$. But $H(x)$ is not continuous; hence $a = 0$. This proves the following theorem:

Theorem 6.3

The gamma function is the only continuous function that is positive for positive x, and that satisfies Eq. (2.2) and Eqs. (3.10) for all values of p.

Index

Galois Theory

E. Artin

Contents

CHAPTER I

LINEAR ALGEBRA

A. Fields

A field is a set of elements in which a pair of operations called multiplication and addition is defined analogous to the operations of multiplication and addition in the real number system (which is itself an example of a field). In each field F there exist unique elements called o and 1 which, under the operations of addition and multiplication, behave with respect to all the other elements of F exactly as their correspondents in the real number system. In two respects, the analogy is not complete; 1) multiplication is not assumed to be commutative in every field, and 2) a field may have only a finite number of elements.

More exactly, a field is a set of elements which, under the above mentioned operation of addition, forms an additive abelian group and for which the elements, exclusive of zero, form a multiplicative group and, finally, in which the two group operations are connected by the distributive law. Furthermore, the product of o and any element is defined to be o.

If multiplication in the field is commutative, then the field is called a commutative field.

B. Vector Spaces

If V is an additive abelian group with elements $A, B, \ldots$, F a field with elements $a, b, \ldots$, and if for each $a \in F$ and $A \in V$ the product aA denotes an element of V, then V is called a *(left) vector space over* F if the following assumptions hold:

1) $a(A + B) = aA + aB$
2) $(a + b)A = aA + bA$
3) $a(bA) = (ab)A$
4) $1A = A$

The reader may readily verify that if V is a vector space over F, then $\text{o}A = 0$ and $a0 = 0$ where o is the zero element of F and 0 that of V. For example, the first relation follows from the equations:

$$aA = (a + \text{o})A = aA + \text{o}A$$

Sometimes products between elements of F and V are written in the form Aa in which case V is called a *right vector space over* F to distinguish it from the previous case where multiplication by field elements is from the left. If, in the discussion, left and right vector spaces do not occur simultaneously, we shall simply use the term "vector space."

C. Homogeneous Linear Equations

If in a field F, a_{ij}, $i = 1, 2, \ldots, m$, $j = 1, 2, \ldots, n$, are $m \cdot n$ elements, it is frequently necessary to know conditions guaranteeing the existence of elements in F such that the following equations are satisfied:

$$(1) \qquad \begin{aligned} a_{11}x_1 + a_{12}x_2 + \cdots + a_{1n}x_n &= 0 \\ \cdots\cdots\cdots\cdots\cdots\cdots \\ a_{m1}x_1 + a_{m2}x_2 + \cdots + a_{mn}x_n &= 0 \end{aligned}$$

The reader will recall that such equations are called *linear homogeneous equations*, and a set of elements, $x_1, x_2, \ldots, x_n$ of F, for which all the above equations are true, is called a solution of the system. If not all of the elements $x_1, x_2, \ldots, x_n$ are 0 the solution is called *non-trivial*; otherwise, it is called *trivial*.

THEOREM 1. *A system of linear homogeneous equations always has a non-trivial solution if the number of unknowns exceeds the number of equations.*

The proof of this follows the method familiar to most high school students, namely, successive elimination of unknowns. If no equations in $n > 0$ variables are prescribed, then our unknowns are unrestricted and we may set them all $= 1$.

We shall proceed by complete induction. Let us suppose that each system of k equations in more than k unknowns has a non-trivial solution when $k < m$. In the system of equations (1) we assume that $n > m$, and denote the expression $a_{i1}x_1 + \cdots + a_{in}x_n$ by L_i, $i = 1, 2, \ldots, m$. We seek elements $x_1, \ldots, x_n$ not all 0 such that $L_1 = L_2 = \cdots = L_m = 0$. If $a_{ij} = 0$ for each i and j, then any choice of $x_1, \ldots, x_n$ will serve as a solution. If not all a_{ij} are 0, then we may assume that $a_{11} \neq 0$, for the order in which the equations are written or in which the unknowns are numbered has no influence on the existence or non-existence of a simultaneous solution. We can find a non-trivial solution to our given system of equations, if and only if we can find a non-trivial solution to the following system

$$\begin{aligned} &L_1 = 0 \\ &L_2 - a_{21}a_{11}^{-1}L_1 = 0 \\ &\cdots\cdots\cdots\cdots \\ &L_m - a_{m1}a_{11}^{-1}L_1 = 0 \end{aligned}$$

For, if $x_1, \ldots, x_n$ is a solution of these latter equations then, since $L_1 = 0$, the second term in each of the remaining equations is 0 and, hence, $L_2 = L_3 = \cdots = L_m = 0$. Conversely, if (1) is satisfied, then the new system is clearly satisfied. The reader will notice that the new system was set up in such a way as to "eliminate" x_1 from the last equations. Furthermore, if a non-trivial solution of the last $n-1$ equations, when viewed as equations in $x_2, \ldots, x_n$, exists then taking $x_1 = -a_{11}^{-1}(a_{12}x_2 + a_{13}x_3 + \cdots + a_{1n}x_n)$ would give us a solution to the whole system. However, the last $m - 1$ equations have a solution by our inductive assumption, from which the theorem follows.

REMARK. If the linear homogeneous equations had been written in the form $\sum x_i a_{ij} = 0$, $i = 1, 2, \ldots, n$, the above theorem would still hold and with the same proof although with the order in which terms are written changed in a few instances.

D. Dependence and Independence of Vectors

In a vector space V over a field F, the vectors $A_1, \dots, A_n$ are called *dependent* if there exist elements $x_1, \dots, x_n$ not all o of F such that $x_1A_1+x_2A_2+\cdots+x_nA_n = 0$. If the vectors $A_1, \dots, A_m$ are not dependent, they are called *independent.*

The *dimension* of a vector space V over a field F is the maximum number of independent elements in V. Thus, the dimension of V is n if there are n independent elements in V, but no set of more than n independent elements.

A system $A_1, \dots, A_m$ of elements in V is called a *generating system* of V if each element A of V can be expressed linearly in terms of $A_1, \dots, A_m$, i.e., $A = \sum_{i=1}^{m} a_iA_i$ for a suitable choice of a_i, $i = 1, \dots, m$, in F.

THEOREM 2. *In any generating system the maximum number of independent vectors is equal to the dimension of the vector space.*

Let $A_1, \dots, A_m$ be a generating system of a vector space V of dimension n. Let r be the maximum number of independent elements in the generating system. By a suitable reordering of the generators we may assume $A_1, \dots, A_r$ independent. By the definition of dimension. It follows that $r \leq n$. For each j, $A_1, \dots, A_r, A_{r+j}$ are dependent, and in the relation

$$a_1A_1 + a_2A_2 + \cdots + a_rA_r + a_{r+j}A_{r+j} = 0$$

expressing this, $a_{r+j} \neq$ o, for the contrary would assert the dependence of $A_1, \dots,$ A_r. Thus,

$$A_{r+j} = -a_{r+j}^{-1}[a_1A_1 + a_2A_2 + \cdots + a_rA_r].$$

It follows that $A_1, \dots, A_r$ is also a generating system since in the linear relation for any element V the terms involving A_{r+j}, $j \neq 0$, can all be replaced by linear expressions in $A_1, \dots, A_r$.

Now, let $B_1, \dots, B_t$ be any system of vectors in V where $t > r$, then there exist a_{ij} such that $B_j = \sum_{i=1}^{r} a_{ij}A_i$, $j = 1, 2, \dots, t$, since the A_i's form a generating system. If we can show that $B_1, \dots, B_t$ are dependent, this will give us $r \geq n$, and the theorem will follow from this together with the previous inequality $r \leq n$. Thus, we must exhibit the existence of a non-trivial solution out of F of the equation

$$x_1B_1 + x_2B_2 + \cdots + x_tB_t = 0.$$

To this end, it will be sufficient to choose the x_i's so as to satisfy the linear equations $\sum_{j=1}^{t} x_ja_{ij} =$ o, $i = 1, 2, \dots, r$, since these expressions will be the coefficients of A_i when in $\sum_{j=1}^{t} x_jB_j$ the B_j's are replaced by $\sum_{i=1}^{r} a_{ij}A_i$ and terms are collected. A solution to the equations $\sum_{j=1}^{t} x_ja_{ij} =$ o, $i = 1, 2, \dots, r$, always exists by Theorem 1.

REMARK. Any n independent vectors $A_1, \dots, A_n$ in an n dimensional vector space form a generating system. For any vector A, the vectors $A, A_1, \dots, A_n$ are dependent and the coefficient of A, in the dependence relation, cannot be zero. Solving for A in terms of $A_1, \dots, A_n$, exhibits $A_1, \dots, A_n$ as a generating system.

A subset of a vector space is called a *subspace* if it is a subgroup of the vector space and if, in addition, the multiplication of any element in the subset by any element of the field is also in the subset. If $A_1, \dots, A_s$ are elements of a vector space V, then the set of all elements of the form $a_1A_1 + \cdots + a_sA_s$ clearly forms a subspace of V. It is also evident, from the definition of dimension, that the dimension of any subspace never exceeds the dimension of the whole vector space.

An s-tuple of elements $(a_1, \ldots, a_s)$ in a field F will be called a *row vector*. The totality of such s-tuples form a vector space if we define

α) $(a_1, a_2, \ldots, a_s) = (b_1, b_2, \ldots, b_s)$ if and only if $a_i = b_i$, $i = 1, \ldots, s$,
β) $(a_1, a_2, \ldots, a_s) + (b_1, b_2, \ldots, b_s) = (a_1 + b_1, a_2 + b_2, \ldots, a_s + b_s)$,
γ) $b(a_1, a_2, \ldots, a_s) = (ba_1, ba_2, \ldots, ba_s)$, for b an element of F.

When the s-tuples are written vertically, $\begin{pmatrix} a_1 \\ \vdots \\ a_s \end{pmatrix}$ they will be called *column vectors*.

THEOREM 3. *The row (column) vector space F^n of all n-tuples from a field F is a vector space of dimension n over F.*

The n elements

$$\begin{aligned} \varepsilon_1 &= (1, 0, 0, \ldots, 0) \\ \varepsilon_2 &= (0, 1, 0, \ldots, 0) \\ &\ldots\ldots\ldots\ldots\ldots \\ \varepsilon_n &= (0, 0, \ldots, 0, 1) \end{aligned}$$

are independent and generate F^n. Both remarks follow from the relation $(a_1, a_2, \ldots, a_s) = \sum a_i \varepsilon_i$.

We call a rectangular array

$$\begin{pmatrix} a_{11} & a_{12} & \cdots & a_{1n} \\ a_{22} & a_{22} & \cdots & a_{2n} \\ \cdots & \cdots & \cdots & \cdots \\ a_{m1} & a_{m2} & \cdots & a_{mn} \end{pmatrix}$$

of elements of a field F a *matrix*. By the *right row rank* of a matrix, we mean the maximum number of independent row vectors among the rows $(a_{i1}, \ldots, a_{in})$ of the matrix when multiplication by field elements is from the right. Similarly, we define left row rank, right column rank and left column rank.

THEOREM 4. *In any matrix the right column rank equals the left row rank and the left column rank equals the right row rank. If the field is commutative, these four numbers are equal to each other and are called the rank of the matrix.*

Call the column vectors of the matrix $C_1, \ldots, C_n$ and the row vectors $R_1, \ldots, R_m$. The column vector 0 is $\begin{pmatrix} 0 \\ \vdots \\ 0 \end{pmatrix}$ and any dependence $C_1x_1 + C_2x_2 + \cdots + C_nx_n = 0$ is equivalent to a solution of the equations

$$\begin{aligned} a_{11}x_1 + a_{12}x_2 + \cdots + a_{1n}x_n &= 0 \\ \ldots\ldots\ldots\ldots\ldots\ldots\ldots\ldots & \\ a_{m1}x_1 + a_{m2}x_2 + \cdots + a_{mn}x_n &= 0 \end{aligned} \tag{1}$$

Any change in the order in which the rows of the matrix are written gives rise to the same system of equations and, hence, does not change the column rank of the matrix, but also does not change the row rank since the changed matrix would have the same set of row vectors. Call c the right column rank and r the left row rank of the matrix. By the above remarks we may assume that the first r rows are independent row vectors. The row vector space generated by all the rows of the matrix has, by Theorem 1, the dimension r and is even generated by the first r rows. Thus, each row after the rth is linearly expressible in terms of the first r rows. Consequently, any solution of the first r equations in (1) will be a solution of the entire system since any of the last $n-r$ equations is obtainable as a linear combination of the first r. Conversely, any solution of (1) will also be a solution of the first r equations. This means that the matrix

$$\begin{pmatrix} a_{11} & a_{12} & \dots & a_{1n} \\ \dots & \dots & \dots & \dots \\ a_{r1} & a_{r2} & \dots & a_{rn} \end{pmatrix}$$

consisting of the first r rows of the original matrix has the same right column rank as the original. It has also the same left row rank since the r rows were chosen independent. But the column rank of the amputated matrix cannot exceed r by Theorem 3. Hence, $c \leq r$. Similarly, calling c' the left column rank and r' the right row rank, $c' \leq r'$. If we form the transpose of the original matrix, that is, replace rows by columns and columns by rows, then the left row rank of the transposed matrix equals the left column rank of the original. If then to the transposed matrix we apply the above considerations we arrive at $r \leq c$ and $r' \leq c'$.

E. Non-homogeneous Linear Equations

The system of non-homogeneous linear equations

$$\begin{aligned} a_{11}x_1 + a_{12}x_2 + \dots + a_{1n}x_n &= b_1 \\ a_{21}x_2 + \dots\dots\dots + a_{2n}x_n &= b_2 \\ \dots\dots\dots\dots\dots\dots\dots\dots \\ a_{m1}x_1 + a_{m2}x_2 + \dots + a_{mn}x_n &= b_m \end{aligned} \tag{2}$$

has a solution if and only if the column vector $\begin{pmatrix} b_1 \\ \vdots \\ b_m \end{pmatrix}$ lies in the space generated by the vectors $\begin{pmatrix} a_{11} \\ \vdots \\ a_{m1} \end{pmatrix}, \dots, \begin{pmatrix} a_{1n} \\ \vdots \\ a_{mn} \end{pmatrix}$. This means that there is a solution if and only if the right column rank of the matrix $\begin{pmatrix} a_{11} & \dots & a_{1n} \\ \dots & \dots & \dots \\ a_{mn} & \dots & a_{mn} \end{pmatrix}$ is the same as the right column rank of the augmented matrix $\begin{pmatrix} a_{11} & \dots & a_{1n} & b_1 \\ \dots & \dots & \dots & \dots \\ a_{mn} & \dots & a_{mn} & b_m \end{pmatrix}$ since the vector space generated by the original must be the same as the vector space generated by the augmented matrix and in either case the dimension is the same as the rank of the matrix by Theorem 2.

By Theorem 4, this means that the row ranks are equal. Conversely, if the row rank of the augmented matrix is the same as the row rank of the original matrix, the column ranks will be the same and the equations will have a solution.

If the equations (2) have a solution, then any relation among the rows of the original matrix subsists among the rows of the augmented matrix. For equations (2) this merely means that like combinations of equals are equal. Conversely, if each relation which subsists between the rows of the original matrix also subsists between the rows of the augmented matrix, then the row rank of the augmented matrix is the same as the row rank of the original matrix. *In terms of the equations this means that there will exist a solution if and only if the equations are consistent, i.e., if and only if any dependence between the left hand sides of the equations also holds between the right sides.*

THEOREM 5. *If in equations* (2) $m = n$, *there exists a unique solution if and only if the corresponding homogeneous equations*

$$a_{11}x_1 + a_{12}x_2 + \cdots + a_{1n}x_n = 0$$

$$\cdots\cdots\cdots\cdots\cdots\cdots\cdots\cdots$$

$$a_{n1}x_1 + a_{n2}x_2 + \cdots + a_{nn}x_n = 0$$

have only the trivial solution.

If they have only the trivial solution, then the column vectors are independent. It follows that the original n equations in n unknowns will have a unique solution if they have any solution, since the difference, term by term, of two distinct solutions would be a non-trivial solution of the homogeneous equations. A solution would exist since the n independent column vectors form a generating system for the n-dimensional space of column vectors.

Conversely, let us suppose our equations have one and only one solution. In this case, the homogeneous equations added term by term to a solution of the original equations would yield a new solution to the original equations. Hence, the homogeneous equations have only the trivial solution.

F. Determinants[1]

The theory of determinants that we shall develop in this chapter is not needed in Galois theory. The reader may, therefore, omit this section if he so desires.

We assume our field to be commutative and consider the square matrix

$$(1) \qquad \begin{pmatrix} a_{11} & a_{12} & \cdots & a_{1n} \\ a_{22} & a_{22} & \cdots & a_{2n} \\ \cdots & \cdots & \cdots & \cdots \\ a_{n1} & a_{n2} & \cdots & a_{nn} \end{pmatrix}$$

of n rows and n columns. We shall define a certain function of this matrix whose value is an element of our field. The function will be called the determinant and will be denoted by

$$(2) \qquad \begin{vmatrix} a_{11} & a_{12} & \cdots & a_{1n} \\ a_{22} & a_{22} & \cdots & a_{2n} \\ \cdots & \cdots & \cdots & \cdots \\ a_{n1} & a_{n2} & \cdots & a_{nn} \end{vmatrix}$$

[1] Of the preceding theory only Theorem 1 for homogeneous equations and the notion of linear dependence are assumed known.

or by $D(A_1, A_2, \ldots, A_n)$ if we wish to consider it as a function of the column vectors $A_1, A_2, \ldots, A_n$ of (1). If we keep all the columns but A_k constant and consider the determinant as a function of A_k, then we write $D_k(A_k)$ and sometimes even only D.

DEFINITION. A function of the column vectors is a determinant if it satisfies the following three axioms:

1. Viewed as a function of any column A_k it is linear and homogeneous, i.e.,

$$D_k(A_k + A'_k) = D_k(A_k) + D_k(A'_k) \tag{3}$$

$$D_k(cA_k) = c \cdot D_k(A_k) \tag{4}$$

2. Its value is $= 0$ if two adjacent columns A_k and A_{k_1} are equal.
3. Its value is $= 1$ if all A_k are the unit vectors U_k, where

$$U_1 = \begin{pmatrix} 1 \\ 0 \\ 0 \\ \vdots \\ 0 \end{pmatrix}, \quad U_2 = \begin{pmatrix} 0 \\ 1 \\ 0 \\ \vdots \\ 0 \end{pmatrix}, \ldots \; U_n = \begin{pmatrix} 0 \\ 0 \\ 0 \\ \vdots \\ 1 \end{pmatrix} \tag{5}$$

The question as to whether determinants exist will be left open for the present. But we derive consequences from the axioms:

a) If we put $c = 0$ in (4) we get: a determinant is 0 if one of the columns is 0.

b) $D_k(A_k) = D_k(A_k + cA_{k\pm1})$ or a determinant remains unchanged if we add a multiple of one column to an adjacent column. Indeed

$$D_k(A_k + cA_{k\pm1}) = D_k(A_k) + cD_k(A_{k\pm1}) = D_k(A_k)$$

because of axiom 2.

c) Consider the two columns A_k and A_{k+1}. We may replace them by A_k and $A_{k+1} + A_k$; subtracting the second from the first we may replace them by $-A_{k+1}$ and $A_{k+1} + A_k$; adding the first to the second we now have $-A_{k+1}$ and A_k; finally, we factor out -1. We conclude: a determinant changes sign if we interchange two adjacent columns.

d) A determinant vanishes if any two of its columns are equal. Indeed, we may bring the two columns side by side after an interchange of adjacent columns and then use axiom 2. In the same way as in b) and c) we may now prove the more general rules:

e) Adding a multiple of one column to another does not change the value of the determinant.

f) Interchanging any two columns changes the sign of D.

g) Let $(\nu_1, \nu_2, \ldots, \nu_n)$ be a permutation of the subscripts $(1, 2, \ldots, n)$. If we rearrange the columns in $D(A_{\nu_1}, A_{\nu_2}, \ldots, A_{\nu_n})$ until they are back in the natural order, we see that

$$D(A_{\nu_1}, A_{\nu_2}, \ldots, A_{\nu_n}) = \pm D(A_1, A_2, \ldots, A_n).$$

Here $\pm$ is a definite sign that does not depend on the special values of the A_k. If we substitute U_k for A_k we see that $D(U_{\nu_1}, U_{\nu_2}, \ldots, U_{\nu_n}) = \pm 1$ and that the sign depends only on the permutation of the unit vectors.

Now we replace each vector A_k by the following linear combination A'_k of $A_1, A_2, \ldots, A_n$:

$$A'_k = b_{1k}A_1 + b_{2k}A_2 + \cdots + b_{nk}A_n. \tag{6}$$

In computing $D(A_1', A_2', \ldots, A_n')$ we first apply axiom 1 on A_1' breaking up the determinant into a sum, then in each term we do the same with A_2' and so on. We get

$$(7)\quad D(A_1', A_2', \ldots, A_n') = \sum_{\nu_1, \nu_2, \ldots, \nu_n} D(b_{\nu_1 1} A_{\nu_1}, b_{\nu_2 2} A_{\nu_2}, \ldots, b_{\nu_n n} A_{\nu_n}) = \sum_{\nu_1, \nu_2, \ldots, \nu_n} b_{\nu_1 1} \cdot b_{\nu_2 2} \cdot \ldots \cdot b_{\nu_n n} D(A_{\nu_1}, A_{\nu_2}, \ldots, A_{\nu_n})$$

where each ν_i runs independently from 1 to n. Should two of the indices ν_i be equal, then $D(A_{\nu_1}, A_{\nu_2}, \ldots, A_{\nu_n}) = 0$; we need therefore keep only those terms in which $(\nu_1, \nu_2, \ldots, \nu_n)$ is a permutation of $(1, 2, \ldots, n)$. This gives

$$(8)\quad D(A_1', A_2', \ldots, A_n') = D(A_1, A_2, \ldots, A_n) \cdot \sum_{\nu_1, \nu_2, \ldots, \nu_n} \pm b_{\nu_1 1} \cdot b_{\nu_2 2} \cdot \ldots \cdot b_{\nu_n n}$$

where $(\nu_1, \nu_2, \ldots, \nu_n)$ runs through all the permutations of $(1, 2, \ldots, n)$ and where $\pm$ stands for the sign associated with that permutation. It is important to remark that we would have arrived at the same formula (8) if our function D satisfied only the first two of our axioms.

Many conclusions may be derived from (8).

We first assume axiom 3 and specialize the A_k to the unit vectors U_k of (5). This makes $A_k' = B_k$ where B_k is the column vector of the matrix of the b_{ik}. (8) yields now:

$$(9)\quad D(B_1, B_2, \ldots, B_n) = \sum_{\nu_1, \nu_2, \ldots, \nu_n} \pm b_{\nu_1 1} \cdot b_{\nu_2 2} \cdot \ldots \cdot b_{\nu_n n}$$

giving us an explicit formula for determinants and showing that they are uniquely determined by our axioms provided they exist at all.

With expression (9) we return to formula (8) and get

$$(10)\quad D(A_1', A_2', \ldots, A_n') = D(A_1, A_2, \ldots, A_n)\, D(B_1, B_2, \ldots, B_n).$$

This is the so-called multiplication theorem for determinants. At the left of (10) we have the determinant of an n-rowed matrix whose elements c_{ik} are given by

$$(11)\quad c_{ik} = \sum_{\nu=1}^{n} a_{i\nu} b_{\nu k}.$$

c_{ik} is obtained by multiplying the elements of the i-th row of $D(A_1, A_2, \ldots, A_n)$ by those of the k-th column of $D(B_1, B_2, \ldots, B_n)$ and adding.

Let us now replace D in (8) by a function $F(A_1, \ldots, A_n)$ that satisfies only the first two axioms. Comparing with (9) we find

$$F(A_1', A_2', \ldots, A_n') = F(A_1, \ldots, A_n)\, D(B_1, B_2, \ldots, B_n).$$

Specializing A_k to the unit vectors U_k leads to

$$(12)\quad F(B_1, B_2, \ldots, B_n) = c \cdot D(B_1, B_2, \ldots, B_n) \quad \text{with } c = F(U_1, U_2, \ldots, U_n).$$

Next we specialize (10) in the following way: If i is a certain subscript from 1 to $n-1$ we put $A_k = U_k$ for $k \neq i,\ i+1$, $A_i = U_i + U_{i+1}$, $A_{i+1} = 0$. Then $D(A_1, A_2, \ldots, A_n) = 0$ since one column is 0. Thus, $D(A_1', A_2', \ldots, A_n') = 0$, but this determinant differs from that of the elements b_{jk} only in the respect that the $i+1$-st row has been made equal to the i-th. We therefore see:

A determinant vanishes if two adjacent *rows* are equal.

Each term in (9) is a product where precisely one factor comes from a given row, say, the i-th. This shows that the determinant is linear and homogeneous if considered as function of this row. If, finally, we select for each row the corresponding unit vector, the determinant is $= 1$ since the matrix is the same as that in which the columns are unit vectors. This shows that a determinant satisfies our three axioms if we consider it as function of the row vectors. In view of the uniqueness it follows:

A determinant remains unchanged if we transpose the row vectors into column vectors, that is, if we rotate the matrix about its main diagonal.

A determinant vanishes if any two rows are equal. It changes sign if we interchange any two rows. It remains unchanged if we add a multiple of one row to another.

We shall now prove the existence of determinants. For a 1-rowed matrix a_{11}, the element a_{11} itself is the determinant. Let us assume the existence of $(n-1)$-rowed determinants. If we consider the n-rowed matrix (1) we may associate with it certain $(n-1)$-rowed determinants in the following way: Let a_{ik} be a particular element in (1). We cancel the i-th row and the k-th column in (1) and take the determinant of the remaining $(n-1)$-rowed matrix. This determinant multiplied by $(-1)^{i+k}$ will be called the cofactor of a_{ik} and be denoted by A_{ik}. The distribution of the sign $(-1)^{i+k}$ follows the chessboard pattern, namely,

$$\begin{pmatrix} + & - & + & - & \cdots \\ - & + & - & + & \cdots \\ + & - & + & - & \cdots \\ - & + & - & + & \cdots \\ \cdots & \cdots & \cdots & \cdots & \cdots \end{pmatrix}$$

Let i be any number from 1 to n. We consider the following function D of the matrix (1);

$$D = a_{i1}A_{i1} + a_{i2}A_{i2} + \cdots + a_{in}A_{in}. \tag{13}$$

It is the sum of the products of the i-th row and their cofactors.

Consider this D in its dependence on a given column, say, A_k. For $\nu \neq k$, $A_{i\nu}$ depends linearly on A_k and $a_{i\nu}$ does not depend on it; for $\nu = k$, A_{ik} does not depend on A_k but a_{ik} is one element of this column. Thus, axiom 1 is satisfied. Assume next that two adjacent columns A_k and A_{k+1} are equal. For $\nu \neq k, k+1$ we have then two equal columns in $A_{i\nu}$ so that $A_{i\nu} = 0$. The determinants used in the computation of $A_{i,k}$ and $A_{i,k+1}$ are the same but the signs are opposite, hence, $A_{i,k} = -A_{i,k+1}$, whereas $a_{i,k} = a_{i,k+1}$. Thus $D = 0$ and axiom 2 holds. For the special case $A_\nu = U_\nu$ $(\nu = 1, 2, \ldots, n)$ we have $a_{i\nu} = 0$ for $\nu \neq i$ while $a_{ii} = 1$, $A_{ii} = 1$. Hence, $D = 1$ and this is axiom 3. This proves both the existence of an n-rowed determinant as well as the truth of formula (13), the so-called development of a determinant according to its i-th row. (13) may be generalized as follows: In our determinant replace the i-th row by the j-th row and develop according to this new row. For $i \neq j$ that determinant is 0 and for $i = j$ it is D:

$$a_{j1}A_{i1} + a_{j2}A_{i2} + \cdots + a_{jn}A_{in} = \begin{cases} D & \text{for } j = i \\ 0 & \text{for } j \neq i \end{cases} \tag{14}$$

If we interchange the rows and the columns we get the following formula:

$$a_{1h}A_{1k} + a_{2h}A_{2k} + \cdots + a_{nh}A_{nk} = \begin{cases} D & \text{for } h = k \\ 0 & \text{for } h \neq k \end{cases} \tag{15}$$

Now let A represent an n-rowed and B an m-rowed square matrix. By $|A|$, $|B|$ we mean their determinants. Let C be a matrix of n rows and m columns and form the square matrix of $n + m$ rows

$$\begin{pmatrix} A & C \\ 0 & B \end{pmatrix} \tag{16}$$

where 0 stands for a zero matrix with m rows and n columns. If we consider the determinant of the matrix (16) as function of the columns of A only, it satisfies obviously the first two of our axioms. Because of (12) its value is $c \cdot |A|$ where c is the determinant of (16) after substituting unit vectors for the columns of A. This c still depends on B and considered as function of the rows of B satisfies the first two axioms. Therefore the determinant of (16) is $d \cdot |A| \cdot |B|$ where d is the special case of the determinant of (16) with unit vectors for the columns of A as well as of B. Subtracting multiples of the columns of A from C we can replace C by 0. This shows $d = 1$ and hence the formula

$$\begin{vmatrix} A & C \\ 0 & B \end{vmatrix} = |A| \cdot |B|. \tag{17}$$

In a similar fashion we could have shown

$$\begin{vmatrix} A & 0 \\ C & B \end{vmatrix} = |A| \cdot |B|. \tag{18}$$

The formulas (17), (18) are special cases of a general theorem by Lagrange that can be derived from them. We refer the reader to any textbook on determinants since in most applications (17) and (18) are sufficient.

We now investigate what it means for a matrix if its determinant is zero. We can easily establish the following facts:

a) If $A_1, A_2, \ldots, A_n$ are linearly dependent, then $D(A_1, A_2, \ldots, A_n) = 0$. Indeed one of the vectors, say A_k, is then a linear combination of the other columns; subtracting this linear combination from the column A_k reduces it to 0 and so $D = 0$.

b) If any vector B can be expressed as linear combination of $A_1, A_2, \ldots, A_n$ then $D(A_1, A_2, \ldots, A_n) \neq 0$. Returning to (6) and (10) we may select the values for b_{ik} in such a fashion that every $A'_i = U_i$. For this choice the left side in (10) is 1 and hence $D(A_1, A_2, \ldots, A_n)$ on the right side $\neq 0$.

c) Let $A_1, A_2, \ldots, A_n$ be linearly independent and B any other vector. If we go back to the components in the equation $A_1x_1 + A_2x_2 + \cdots + A_nx_n + By = 0$ we obtain n linear homogeneous equations in the $n+1$ unknowns $x_1, x_2, \ldots, x_n, y$. Consequently, there is a non-trivial solution, y must be $\neq 0$ or else the $A_1, A_2, \ldots, A_n$ would be linearly dependent. But then we can compute B out of this equation as a linear combination of $A_1, A_2, \ldots, A_n$.

Combining these results we obtain:

A determinant vanishes if and only if the column vectors (or the row vectors) are linearly dependent.

Another way of expressing this result is:

The set of n linear homogeneous equations

$$a_{i1}x_1 + a_{i2}x_2 + \cdots + a_{in}x_n = 0 \quad (i = 1, 2, \ldots, n)$$

in n unknowns has a non-trivial solution if and only if the determinant of the coefficients is zero.

Another result that can be deduced is:

If $A_1, A_2, \ldots, A_n$ are given, then their linear combinations can represent any other vector B if and only if $D(A_1, A_2, \ldots, A_n) \neq 0$.

Or:

The set of linear equations

$$a_{i1}x_1 + a_{i2}x_2 + \cdots + a_{in}x_n = b_i \quad (i = 1, 2, \ldots, n) \tag{19}$$

has a solution for arbitrary values of the b_i if and only if the determinant of a_{ik} is $\neq 0$. In that case the solution is unique.

We finally express the solution of (19) by means of determinants if the determinant D of the a_{ik} is $\neq 0$.

We multiply for a given k the i-th equation with A_{ik} and add the equations. (15) gives

$$D \cdot x_k = A_{1k}b_1 + A_{2k}b_2 + \cdots + A_{nk}b_n \quad (k = 1, 2, \ldots, n) \tag{20}$$

and this gives x_k. The right side in (12) may also be written as the determinant obtained from D by replacing the k-th column by $b_1, b_2, \ldots, b_n$. The rule thus obtained is known as Cramer's rule.

CHAPTER II

FIELD THEORY

A. Extension Fields

If E is a field and F a subset of E which, under the operations of addition and multiplication in E, itself forms a field, that is, if F is a subfield of E, then we shall call E an *extension* of F. The relation of being an extension of F will be briefly designated by $F \subset E$. If $\alpha, \beta, \gamma, \ldots$ are elements of E, then by $F(\alpha, \beta, \gamma, \ldots)$ we shall mean the set of elements in E which can be expressed as quotients of polynomials in $\alpha, \beta, \gamma, \ldots$ with coefficients in F. It is clear that $F(\alpha, \beta, \gamma, \ldots)$ is a field and is the smallest extension of F which contains the elements $\alpha, \beta, \gamma, \ldots$. We shall call $F(\alpha, \beta, \gamma, \ldots)$ the field obtained after the *adjunction* of the elements $\alpha, \beta, \gamma, \ldots$ to F, or the field *generated* out of F by the elements $\alpha, \beta, \gamma, \ldots$. In the sequel all fields will be assumed commutative.

If $F \subset E$, then ignoring the operation of multiplication defined between the elements of E, we may consider E as a vector space over F. By the *degree* of E over F, written (E/F), we shall mean the dimension of the vector space E over F. If (E/F) is finite, E will be called a *finite extension*.

THEOREM 6. *If F, B, E are three fields such that $F \subset B \subset E$, then*

$$(E/F) = (B/F)(E/B).$$

Let $A_1, A_2, \ldots, A_r$ be elements of E which are linearly independent with respect to B and let $C_1, C_2, \ldots, C_s$ be elements of B which are independent with respect to F. Then the products C_iA_j where $i = 1, 2, \ldots, s$ and $j = 1, 2, \ldots, r$ are elements of E which are independent with respect to F. For[1] if $\sum_{i,j} a_{ij}C_iA_j = 0$, then $\sum_j \left(\sum_i a_{ij}C_i\right) A_j$ is a linear combination of the A_j with coefficients in B and because the A_j were independent with respect to B we have $\sum_i a_{ij}C_i = 0$ for each j. The independence of the C_i with respect to F then requires that each $a_{ij} = 0$. Since there are $r \cdot s$ elements C_iA_j we have shown that for each $r \leq (E/B)$ and $s \leq (B/F)$ the degree $(E/F) \geq r \cdot s$. Therefore, $(E/F) \geq (B/F)(E/B)$. If one of the latter numbers is infinite, the theorem follows. If both (E/B) and (B/F) are finite, say r and s respectively, we may suppose that the A_j and the C_i are generating systems of E and B respectively, and we show that the set of products C_iA_j is a generating system of E over F. Each $A \in E$ can be expressed linearly in terms of the A_j with coefficients in B. Thus, $A = \sum B_jA_j$. Moreover, each B_j being an element of B can be expressed linearly with coefficients in F in terms of the C_i, i.e., $B_j = \sum a_{ij}C_i$, $j = 1, 2, \ldots, r$. Thus, $A = \sum a_{ij}C_iA_j$ and the C_iA_j form an independent generating system of E over F.

COROLLARY. *If $F \subset F_1 \subset F_2 \subset \cdots \subset F_n$, then $(F_n/F) = (F_1/F) \cdot (F_2/F_1) \cdot \ldots \cdot (F_n/F_{n-1})$.*

[1] Henceforth, 0 will denote the zero element of a field.

B. Polynomials

An expression of the form $a_0x^n + a_1x^{n-1} + \cdots + a_n$ is called a *polynomial* in F of degree n if the coefficients $a_0, \ldots, a_n$ are elements of the field F and $a_0 \neq 0$. Multiplication and addition of polynomials are performed in the usual way[2].

A polynomial in F is called *reducible* in F if it is equal to the product of two polynomials in F each of degree at least one. Polynomials which are not reducible in F are called *irreducible* in F.

If $f(x) = g(x) \cdot h(x)$ is a relation which holds between the polynomials $f(x)$, $g(x)$, $h(x)$ in a field F, then we shall say that $g(x)$ *divides* $f(x)$ in F, or that $g(x)$ is a *factor* of $f(x)$. It is readily seen that the degree of $f(x)$ is equal to the sum of the degrees of $g(x)$ and $h(x)$, so that if neither $g(x)$ nor $h(x)$ is a constant then each has a degree less than $f(x)$. It follows from this that by a finite number of factorizations a polynomial can always be expressed as a product of irreducible polynomials in a field F.

For any two polynomials $f(x)$ and $g(x)$ the division algorithm holds, i.e., $f(x) = q(x) \cdot g(x) + r(x)$ where $q(x)$ and $r(x)$ are unique polynomials in F and the degree of $r(x)$ is less than that of $g(x)$. This may be shown by the same argument as the reader met in elementary algebra in the case of the field of real or complex numbers. We also see that $r(x)$ is the uniquely determined polynomial of a degree less than that of $g(x)$ such that $f(x) - r(x)$ is divisible by $g(x)$. We shall call $r(x)$ the *remainder* of $f(x)$.

Also, in the usual way, it may be shown that if α is a root of the polynomial $f(x)$ in F then $x - \alpha$ is a factor of $f(x)$, and as a consequence of this that a polynomial in a field cannot have more roots in the field than its degree.

LEMMA. *If $f(x)$ is an irreducible polynomial of degree n in F, then there do not exist two polynomials each of degree less than n in F whose product is divisible by $f(x)$.*

Let us suppose to the contrary that $g(x)$ and $h(x)$ are polynomials of degree less than n whose product is divisible by $f(x)$. Among all polynomials occurring in such pairs we may suppose $g(x)$ has the smallest degree. Then since $f(x)$ is a factor of $g(x) \cdot h(x)$ there is a polynomial $k(x)$ such that

$$k(x) \cdot f(x) = g(x) \cdot h(x).$$

By the division algorithm,

$$f(x) = q(x) \cdot g(x) + r(x)$$

where the degree of $r(x)$ is less than that of $g(x)$ and $r(x) \neq 0$ since $f(x)$ was assumed irreducible. Multiplying

$$f(x) = q(x) \cdot g(x) + r(x)$$

by $h(x)$ and transposing, we have

$$r(x) \cdot h(x) = f(x) \cdot h(x) - q(x) \cdot g(x) \cdot h(x) = f(x) \cdot h(x) - q(x) \cdot k(x) \cdot f(x)$$

from which it follows that $r(x) \cdot h(x)$ is divisible by $f(x)$. Since $r(x)$ has a smaller degree than $g(x)$, this last is in contradiction to the choice of $g(x)$, from which the lemma follows.

[2]If we speak of the set of all polynomials of degree lower than n, we shall agree to include the polynomial 0 in this set, though it has no degree in the proper sense.

As we saw, many of the theorems of elementary algebra hold in any field F. However, the so-called Fundamental Theorem of Algebra, at least in its customary form, does not hold. It will be replaced by a theorem due to Kronecker which guarantees for a given polynomial in F the existence of an extension field in which the polynomial has a root. We shall also show that, in a given field, a polynomial can not only be factored into irreducible factors, but that this factorization is unique up to a constant factor. The uniqueness depends on the theorem of Kronecker.

C. Algebraic Elements

Let F be a field and E an extension field of F. If α is an element of E we may ask whether there are polynomials with coefficients in F which have α as root. α is called *algebraic* with respect to F if there are such polynomials. Now let α be algebraic and select among all polynomials in F which have α as root one, $f(x)$, of lowest degree.

We may assume that the highest coefficient of $f(x)$ is 1. We contend that this $f(x)$ is uniquely determined, that it is irreducible and that each polynomial in F with the root α is divisible by $f(x)$. If, indeed, $g(x)$ is a polynomial in F with $g(\alpha) = 0$, we may divide $g(x) = f(x)q(x) + r(x)$ where $r(x)$ has a degree smaller than that of $f(x)$. Substituting $x = \alpha$ we get $r(\alpha) = 0$. Now $r(\alpha)$ has to be identically 0 since otherwise $r(x)$ would have the root α and be of lower degree than $f(x)$. So $g(x)$ is divisible by $f(x)$. This also shows the uniqueness of $f(x)$. If $f(x)$ were not irreducible, one of the factors would have to vanish for $x = \alpha$ contradicting again the choice of $f(x)$.

We consider now the subset E_0 of the following elements θ of E:

$$\theta = g(\alpha) = c_0 + c_1\alpha + c_2\alpha^2 + \cdots + c_{n-1}\alpha^{n-1}$$

where $g(x)$ is a polynomial in F of degree less than n (n being the degree of $f(x)$). This set E_0 is closed under addition and multiplication. The latter may be verified as follows:

If $g(x)$ and $h(x)$ are two polynomials of degree less than n we put $g(x)h(x) = q(x)f(x) + r(x)$ and hence $g(\alpha)h(\alpha) = r(\alpha)$. Finally we see that the constants $c_0, c_1, \ldots, c_{n-1}$ are uniquely determined by the element θ. Indeed two expressions for the same θ would lead after subtracting to an equation for α of lower degree than n.

We remark that the internal structure of the set E_0 does not depend on the nature of α but only on the irreducible $f(x)$. The knowledge of this polynomial enables us to perform the operations of addition and multiplication in our set E_0. We shall see very soon that E is a field; in fact, E is nothing but the field $F(\alpha)$. As soon as this is shown we have at once the degree, $(F(a)/F)$, determined as n, since the space $F(\alpha)$ is generated by the linearly independent $1, \alpha, \alpha^2, \ldots, \alpha^{n-1}$.

We shall now try to imitate the set E_0 without having an extension field E and an element α at our disposal. We shall assume only an irreducible polynomial

$$f(x) = x^n + a_{n-1}x^{n-1} + \cdots + a_0$$

as given.

We select a symbol ξ and let E_1 be the set of all formal polynomials

$$g(\xi) = c_0 + c_1\xi + \cdots + c_{n-1}\xi^{n-1}$$

of a degree lower than n. This set forms a group under addition. We now introduce besides the ordinary multiplication a new kind of multiplication of two elements $g(\xi)$ and $h(\xi)$ of E_1 denoted by $g(\xi) \times h(\xi)$. It is defined as the remainder $r(\xi)$ of the ordinary product $g(\xi)h(\xi)$ under division by $f(\xi)$. We first remark that any product of m terms $g_1(\xi), g_2(\xi), \dots, g_m(\xi)$ is again the remainder of the ordinary product $g_1(\xi)g_2(\xi)\cdots g_m(\xi)$. This is true by definition for $m = 2$ and follows for every n by induction if we just prove the easy lemma: The remainder of the product of two remainders (of two polynomials) is the remainder of the product of these two polynomials. This fact shows that our new product is associative and commutative and also that the new product $g_1(\xi) \times g_2(\xi) \times \cdots \times g_m(\xi)$ will coincide with the old product $g_1(\xi)g_2(\xi)\cdots g_m(\xi)$ if the latter does not exceed n in degree. The distributive law for our multiplication is readily verified.

The set E_1 contains our field F and our multiplication in E_1 has for F the meaning of the old multiplication. One of the polynomials of E_1 is ξ. Multiplying it i-times with itself, clearly will just lead to ξ^i as long as $i < n$. For $i = n$ this is not any more the case since it leads to the remainder of the polynomial ξ^n. This remainder is

$$\xi^n - f(\xi) = -a_{n-1}\xi^{n-1} - a_{n-2}\xi^{n-2} - \cdots - a_0.$$

We now give up our old multiplication altogether and keep only the new one; we also change notation, using the point (or juxtaposition) as symbol for the new multiplication.

Computing in this sense

$$c_0 + c_1\xi + \cdots + c_{n-1}\xi^{n-1}$$

will readily lead to this element, since all the degrees involved are below n. But

$$\xi^n = -a_{n-1}\xi^{n-1} - a_{n-2}\xi^{n-2} - \cdots - a_0.$$

Transposing we see that $f(\xi) = 0$.

We thus have constructed a set E_1 and an addition and multiplication in E_1 that already satisfies most of the field axioms. E_1 contains F as subfield and ξ satisfies the equation $f(\xi) = 0$. We next have to show: If $g(\xi) \neq 0$ and $h(\xi)$ are given elements of E_1, there is an element

$$X(\xi) = x_0 + x_1\xi + \cdots + x_{n-1}\xi^{n-1}$$

in E_1 such that

$$g(\xi) \cdot X(\xi) = h(\xi).$$

To prove it we consider the coefficients x_i of $X(\xi)$ as unknowns and compute nevertheless the product on the left side, always reducing higher powers of ξ to lower ones. The result is an expression $L_0 + L_1\xi + \cdots + L_{n-1}\xi^{n-1}$ where each L_i is a linear combination of the x_i with coefficients in F. This expression is to be equal to $h(\xi)$; this leads to the n equations with n unknowns:

$$L_0 = b_0, \quad L_1 = b_1, \quad \dots \quad L_{n-1} = b_{n-1}$$

where the b_i are the coefficients of $h(\xi)$. This system will be soluble if the corresponding homogeneous equations

$$L_0 = 0, \quad L_1 = 0, \quad \dots, \quad L_{n-1} = 0$$

have only the trivial solution.

The homogeneous problem would occur if we should ask for the set of elements $X(\xi)$ satisfying $g(\xi) \cdot X(\xi) = 0$. Going back for a moment to the old multiplication this would mean that the ordinary product $g(\xi)X(\xi)$ has the remainder 0, and is therefore divisible by $f(\xi)$. According to the lemma, page 14, this is only possible for $X(\xi) = 0$.

Therefore E_1 is a field.

Assume now that we have also our old extension E with a root α of $f(x)$, leading to the set E_0. We see that E_0 has in a certain sense the same structure as E_1 if we map the element $g(\xi)$ of E_1 onto the element $g(\alpha)$ of E_0. This mapping will have the property that the image of a sum of elements is the sum of the images, and the image of a product is the product of the images.

Let us therefore define: A mapping σ of one field onto another which is one to one in both directions such that $\sigma(\alpha + \beta) = \sigma(\alpha) + \sigma(\beta)$ and $\sigma(\alpha \cdot \beta) = \sigma(\alpha) \cdot \sigma(\beta)$ is called an *isomorphism*. If the fields in question are not distinct — i.e., are both the same field — the isomorphism is called an *automorphism*. Two fields for which there exists an isomorphism mapping one on another are called *isomorphic*. If not every element of the image field is the image under σ of an element in the first field, then σ is called an isomorphism of the first field *into* the second. Under each isomorphism it is clear that $\sigma(0) = 0$ and $\sigma(1) = 1$.

We see that E_0 is also a field and that it is isomorphic to E_1.

We now mention a few theorems that follow from our discussion:

THEOREM 7 (Kronecker). *If $f(x)$ is a polynomial in a field F, there exists an extension E of F in which $f(x)$ has a root.*

Proof: Construct an extension field in which an irreducible factor of $f(x)$ has a root.

THEOREM 8. *Let σ be an isomorphism mapping a field F on a field F'. Let $f(x)$ be an irreducible polynomial in F and $f'(x)$ the corresponding polynomial in F'. If $E = F(\beta)$ and $E' = F'(\beta')$ are extensions of F and F', respectively, where $f(\beta) = 0$ in E and $f'(\beta') = 0$ in E', then σ can be extended to an isomorphism between E and E'.*

Proof: E and E' are both isomorphic to E_0.

D. Splitting Fields

If F, B and E are three fields such that $F \subset B \subset E$, then we shall refer to B as an *intermediate field.*

If E is an extension of a field F in which a polynomial $p(x)$ in F can be factored into linear factors, and if $p(x)$ can not be so factored in any intermediate field, then we call E a *splitting* field for $p(x)$. Thus, if E is a splitting field of $p(x)$, *the roots of $p(x)$ generate E.*

A splitting field is of finite degree since it is constructed by a finite number of adjunctions of algebraic elements, each defining an extension field of finite degree. Because of the corollary on page 13, the total degree is finite.

THEOREM 9. *If $p(x)$ is a polynomial in a field F, there exists a splitting field E of $p(x)$.*

We factor $p(x)$ in F into irreducible factors $f_1(x) \cdot \ldots \cdot f_r(x) = p(x)$. If each of these is of the first degree then F itself is the required splitting field. Suppose then that $f_1(x)$ is of degree higher than the first. By Theorem 7 there is an extension F_1 of F in which $f_1(x)$ has a root. Factor each of the factors $f_1(x), \ldots, f_r(x)$ into irreducible factors in F_1 and proceed as before. We finally arrive at a field in which $p(x)$ can be split into linear factors. The field generated out of F by the roots of $p(x)$ is the required splitting field.

The following theorem asserts that up to isomorphisms, the splitting field of a polynomial is unique.

THEOREM 10. *Let σ be an isomorphism mapping the field F on the field F'. Let $p(x)$ be a polynomial in F and $p'(x)$ the polynomial in F' with coefficients corresponding to those of $p(x)$ under σ. Finally, let E be a splitting field of $p(x)$ and E' a splitting field of $p'(x)$. Under these conditions the isomorphism σ can be extended to an isomorphism between E and E'.*

If $f(x)$ is an irreducible factor of $p(x)$ in F, then E contains a root of $f(x)$. For let $p(x) = (x - \alpha_1)(x - \alpha_2) \cdots (x - \alpha_s)$ be the splitting of $p(x)$ in E. Then $(x - \alpha_1)(x - \alpha_2) \cdots (x - \alpha_s) = f(x) \cdot g(x)$. We consider $f(x)$ as a polynomial in E and construct the extension field $B = E(\alpha)$ in which $f(\alpha) = 0$. Then $(\alpha - \alpha_1) \cdot (\alpha - \alpha_2) \cdot \ldots \cdot (\alpha - \alpha_s) = f(\alpha) \cdot g(\alpha) = 0$ and $\alpha - \alpha_1$ being elements of the field B can have a product equal to 0 only if for one of the factors, say the first, we have $\alpha - \alpha_1 = 0$. Thus, $\alpha = \alpha_1$, and α_1 is a root of $f(x)$.

Now in case all roots of $p(x)$ are in F, then $E = F$ and $p(x)$ can be split in F. This factored form has an image in F' which is a splitting of $p'(x)$, since the isomorphism σ preserves all operations of addition and multiplication in the process of multiplying out the factors of $p(x)$ and collecting to get the original form. Since $p'(x)$ can be split in F', we must have $F' = E'$. In this case, σ itself is the required extension and the theorem is proved if all roots of $p(x)$ are in F.

We proceed by complete induction. Let us suppose the theorem proved for all cases in which the number of roots of $p(x)$ outside of F is less than $n > 1$, and suppose that $p(x)$ is a polynomial having n roots outside of F. We factor $p(x)$ into irreducible factors in F; $p(x) = f_1(x) \cdot f_2(x) \cdot \ldots \cdot f_m(x)$. Not all of these factors can be of degree 1, since otherwise $p(x)$ would split in F, contrary to assumption. Hence, we may suppose the degree of $f_1(x)$ to be $r > 1$. Let $f_1'(x) \cdot f_2'(x) \cdot \ldots \cdot f_m'(x) = p'(x)$ be the factorization of $p'(x)$ into the polynomials corresponding to $f_1(x), \ldots, f_m(x)$ under σ. $f_1'(x)$ is irreducible in F', for a factorization of $f_1'(x)$ in F' would induce[3] under σ^{-1} a factorization of $f_1(x)$, which was however taken to be irreducible.

Let α be a root of $f_1(x)$. By Theorem 8, the isomorphism σ can be extended to an isomorphism σ_1, between the fields $F(\alpha)$ and $F'(\alpha')$.

Since $F \subset F(\alpha)$, $p(x)$ is a polynomial in $F(\alpha)$ and E is a splitting field for $p(x)$ in $F(\alpha)$. Similarly for $p'(x)$. There are now less than n roots of $p(x)$ outside the new ground field $F(\alpha)$. Hence by our inductive assumption σ_1 can be extended from an isomorphism between $F(\alpha)$ and $F'(\alpha')$ to an isomorphism σ_2 between E and E'. Since σ_1 is an extension of σ, and σ_2 an extension of σ_1, we conclude σ_2 is an extension of σ and the theorem follows.

COROLLARY. *If $p(x)$ is a polynomial in a field F, then any two splitting fields for $p(x)$ are isomorphic.*

[3] σ^{-1} is the set theoretic inverse of σ. One easily checks that σ^{-1} is an isomorphism from F' to F. *Editor's note.*

This follows from Theorem 10 if we take $F = F'$ and σ to be the identity mapping, i.e., $\sigma(x) = x$.

As a consequence of this corollary we see that we are justified in using the expression "the splitting field of $p(x)$" since any two differ only by an isomorphism. Thus, if $p(x)$ has repeated roots in one splitting field, so also in any other splitting field it will have repeated roots. The statement "$p(x)$ has repeated roots" will be significant without reference to a particular splitting field.

E. Unique Decomposition of Polynomials into Irreducible Factors

THEOREM 11. *If $p(x)$ is a polynomial in a field F, and if $p(x) = p_1(x) \cdot p_2(x) \cdot \ldots \cdot p_r(x) = q_1(x) \cdot q_2(x) \cdot \ldots \cdot q_s(x)$ are two factorizations of $p(x)$ into irreducible polynomials each of degree at least one, then $r = s$ and after a suitable change in the order in which the q's are written, $p_i(x) = c_i q_i(x)$, $i = 1, 2, \ldots, r$, and $c_i \in F$.*

Let $F(\alpha)$ be an extension of F in which $p_1(\alpha) = 0$. We may suppose the leading coefficients of the $p_i(x)$ and the $q_i(x)$ to be 1, for, by factoring out all leading coefficients and combining, the constant multiplier on each side of the equation must be the leading coefficient of $p(x)$ and hence can be divided out of both sides of the equation. Since $0 = p_1(\alpha) \cdot p_2(\alpha) \cdot \ldots \cdot p_r(\alpha) = p(\alpha) = q_1(\alpha) \cdot \ldots \cdot q_s(\alpha)$ and since a product of elements of $F(\alpha)$ can be 0 only if one of these is 0, it follows that one of the $q_i(\alpha)$, say $q_1(\alpha)$, is 0. This gives (see page 15) $p_1(x) = q_1(x)$. Thus $p_1(x) \cdot p_2(x) \cdot \ldots \cdot p_r(x) = p_1 \cdot q_2(x) \cdot \ldots \cdot q_s(x)$ or $p_1(x) \cdot [p_2(x) \cdot \ldots \cdot p_r(x) - q_2(x) \cdot \ldots \cdot q_s(x)] = 0$. Since the product of two polynomials is 0 only if one of the two is the 0 polynomial. It follows that the polynomial within the brackets is 0 so that $p_2(x) \cdot \ldots \cdot p_r(x) = q_2(x) \cdot \ldots \cdot q_s(x)$. If we repeat the above argument r times we obtain $p_i(x) = q_i(x)$, $= 1, 2, \ldots, r$. Since the remaining q's must have a product 1, it follows that $r = s$.

F. Group Characters

If G is a multiplicative group, F a field and σ a homomorphism mapping G into F, then σ is called a *character* of G in F. By homomorphism is meant a mapping σ such that for α, β any two elements of G, $\sigma(\alpha) \cdot \sigma(\beta) = \sigma(\alpha\beta)$ and $\sigma(\alpha) \neq 0$ for any α. (If $\sigma(\alpha) = 0$ for one element α, then $\sigma(x) = 0$ for each $x \in G$, since $\sigma(\alpha y) = \sigma(\alpha) \cdot \sigma(y) = 0$ and αy takes all values in G when y assumes all values in G).

The characters $\sigma_1, \sigma_2, \ldots, \sigma_n$ are called *dependent* if there exist elements $a_1, a_2, \ldots, a_n$ not all zero in F such that $a_1\sigma_1(x) + a_2\sigma_2(x) + \cdots + a_n\sigma_n(x) = 0$ for each $x \in G$. Such a dependence relation is called *non-trivial.* If the characters are not dependent they are called *independent.*

THEOREM 12. *If G is a group and $\sigma_1, \sigma_2, \ldots, \sigma_n$ are n mutually distinct characters of G in a field F, then $\sigma_1, \sigma_2, \ldots, \sigma_n$ are independent.*

One character cannot be dependent, since $a_1\sigma_1(x) = 0$ implies $a_1 = 0$ due to the assumption that $\sigma_1(\alpha) \neq 0$. Suppose $n > 1$. We make the inductive assumption that no set of less than n distinct characters is dependent. Suppose now that $a_1\sigma_1(x) + a_2\sigma_2(x) + \cdots + a_n\sigma_n(x) = 0$ is a non-trivial dependence between the σ's. None of the elements a_i is zero, else we should have a dependence between less than n characters contrary to our inductive assumption. Since σ_1 and σ_n are distinct, there exists an element α in G such that $\sigma_1(\alpha) \neq \sigma_n(\alpha)$. Multiply the relation between the σ's by a_n^{-1}. We obtain a relation

$$(*) \qquad b_1\sigma_1(x) + \cdots + b_{n-1}\sigma_{n-1}(x) + \sigma_n(x) = 0, \quad b_i = a_n^{-1}a_i \neq 0.$$

Replace in this relation x by αx. We have

$$b_1\sigma_1(\alpha)\sigma_1(x) + \cdots + b_{n-1}\sigma_{n-1}(\alpha)\sigma_{n-1}(x) + \sigma_n(\alpha)\sigma_n(x) = 0,$$

or

$$\sigma_n(\alpha)^{-1}b_1\sigma_1(\alpha)\sigma_1(x) + \cdots + \sigma_n(x) = 0.$$

Subtracting the latter from $(*)$ we have

$$(**) \qquad [b_1 - \sigma_n(\alpha)^{-1}b_1\sigma_1(\alpha)]\sigma_1(x) + \cdots + c_{n-1}\sigma_{n-1}(x) = 0.$$

The coefficient of $\sigma_1(x)$ in this relation is not 0, otherwise we should have $b_1 = \sigma_n(\alpha)^{-1}b_1\sigma_1(\alpha)$, so that

$$\sigma_n(\alpha)b_1 = b_1\sigma_1(\alpha) = \sigma_1(\alpha)b_1$$

and since $b_1 \neq 0$, we get $\sigma_n(\alpha) = \sigma_1(\alpha)$ contrary to the choice of α. Thus, $(**)$ is a non-trivial dependence between $\sigma_1, \sigma_2, \ldots, \sigma_{n-1}$ which is contrary to our inductive assumption.

COROLLARY. *If E and E' are two fields, and $\sigma_1, \sigma_2, \ldots, \sigma_n$ are n mutually distinct isomorphisms mapping E into E', then $\sigma_1, \ldots, \sigma_n$ are independent. (Where "independent" again means there exists no non-trivial dependence $a_1\sigma_1(x) + \cdots + a_n\sigma_n(x) = 0$ which holds for every $x \in E$).*

This follows from Theorem 12, since E without the 0 is a group and the σ's defined. In this group are mutually distinct characters.

If $\sigma_1, \sigma_2, \ldots, \sigma_n$ are isomorphisms of a field E into a field E', then each element a of E such that $\sigma_1(a) = \sigma_2(a) = \cdots = \sigma_n(a)$ is called a *fixed point* of E under $\sigma_1, \sigma_2, \ldots, \sigma_n$. This name is chosen because in the case where the σ's are automorphisms and σ_1 is the identity, i.e., $\sigma_1(x) = x$, we have $\sigma_i(x) = x$ for a fixed point.

LEMMA. *The set of fixed points of E is a subfield of E. We shall call this subfield the fixed field.*

For if a and b are fixed points, then $\sigma_i(a+b) = \sigma_i(a) + \sigma_i(b) = \sigma_j(a) + \sigma_j(b) = \sigma_j(a+b)$ and $\sigma_i(a \cdot b) = \sigma_i(a) \cdot \sigma_i(b) = \sigma_j(a) \cdot \sigma_j(b) = \sigma_j(a \cdot b)$. Finally from $\sigma_i(a) = \sigma_j(a)$ we have $(\sigma_i(a))^{-1} = (\sigma_j(a))^{-1} = \sigma_i(a^{-1}) = \sigma_j(a^{-1})$.

Thus, the sum and product of two fixed points is a fixed point, and the inverse of a fixed point is a fixed point. Clearly, the negative of a fixed point is a fixed point.

THEOREM 13. *If $\sigma_1, \ldots, \sigma_n$ are n mutually distinct isomorphisms of a field E into a field E', and if F is the fixed field of E, then $(E/F) \geq n$.*

Suppose to the contrary that $(E/F) = r < n$. We shall show that we are led to a contradiction. Let $\omega_1, \omega_2, \ldots, \omega_r$ be a generating system of E over F. In the homogeneous linear equations

$$\sigma_1(\omega_1)x_1 + \sigma_2(\omega_1)x_2 + \cdots + \sigma_n(\omega_1)x_n = 0$$
$$\sigma_1(\omega_2)x_1 + \sigma_2(\omega_2)x_2 + \cdots + \sigma_n(\omega_2)x_n = 0$$
$$\cdots\cdots\cdots\cdots\cdots\cdots\cdots\cdots\cdots\cdots\cdots\cdots$$
$$\sigma_1(\omega_r)x_1 + \sigma_2(\omega_r)x_2 + \cdots + \sigma_n(\omega_r)x_n = 0$$

there are more unknowns than equations so that there exists a non-trivial solution which, we may suppose, $x_1, x_2, \ldots, x_n$ denotes. For any element α in E we can find $a_1, a_2, \ldots, a_r$ in F such that $\alpha = a_1\omega_1 + \cdots + \alpha_r\omega_r$. We multiply the first equation by $\sigma_1(a_1)$, the second by $\sigma_1(a_2)$ and so on. Using that $a_i \in F$, hence that $\sigma_1(a_i) = \sigma_j(a_i)$ and also that $\sigma_j(a_i) \cdot \sigma_j(\omega_i) = \sigma_j(a_i\omega_i)$, we obtain

$$\sigma_1(a_1\omega_1)x_1 + \cdots + \sigma_n(a_1\omega_1)x_n = 0$$
$$\cdots\cdots\cdots\cdots\cdots\cdots\cdots\cdots$$
$$\sigma_1(a_r\omega_r)x_1 + \cdots + \sigma_n(a_r\omega_r)x_n = 0.$$

Adding these last equations and using

$$\sigma_i(a_1\omega_1) + \sigma_i(a_2\omega_2) + \cdots + \sigma_i(a_r\omega_r) = \sigma_i(a_1\omega_1 + \cdots + a_r\omega_r) = \sigma_i(\alpha)$$

we obtain

$$\sigma_1(\alpha)x_1 + \sigma_2(\alpha)x_2 + \cdots + \sigma_n(\alpha)x_n = 0.$$

This, however, is a non-trivial dependence relation between $\sigma_1, \sigma_2, \ldots, \sigma_n$ which cannot exist according to the corollary of Theorem 12.

COROLLARY. *If $\sigma_1, \sigma_2, \ldots, \sigma_n$ are automorphisms of the field E, and F is the fixed field, then $(E/F) \geq n$.*

If F is a subfield of the field E, and σ an automorphism of E, we shall say that σ *leaves F fixed* if for each element a of F, $\sigma(a) = a$. If σ and τ are two automorphisms of E, then the mapping $\sigma(\tau(x))$ written briefly $\sigma\tau$ is an automorphism, as the reader may readily verify. [E.g., $\sigma\tau(x \cdot y) = \sigma(\tau(x \cdot y)) = \sigma(\tau(x) \cdot \tau(y)) = \sigma(\tau(x)) \cdot \sigma(\tau(y))$]. We shall call $\sigma\tau$ the *product* of σ and τ. If σ is an automorphism ($\sigma(x) = y$), then we shall call σ^{-1} the mapping of y into x, i.e., $\sigma^{-1}(y) = x$ the inverse of σ. The reader may readily verify that σ^{-1} is an automorphism. The automorphism $I(x) = x$ shall be called the *unit automorphism.*

LEMMA. *If E is an extension field of F, the set G of automorphisms which leave F fixed is a group.*

The product of two automorphisms which leave F fixed clearly leaves F fixed. Also, the inverse of any automorphism in G is in G.

The reader will observe that G, the set of automorphisms which leave F fixed, does not necessarily have F as its fixed field. It may be that certain elements in E which do not belong to F are left fixed by every automorphism which leaves F fixed. Thus, the fixed field of G may be larger than F.

G. Applications and Examples to Theorem 13

Theorem 13 is very powerful as the following examples show:

1) Let k be a field and consider the field $E = k(x)$ of all rational functions of the variable x. If we map each of the functions $f(x)$ of E onto $f(\frac{1}{x})$ we obviously obtain an automorphism of E. Let us consider the following six automorphisms where $f(x)$ is mapped onto $f(x)$ (identity), $f(1-x)$, $f(\frac{1}{x})$, $f(1-\frac{1}{x})$, $f(\frac{1}{1-x})$ and $f(\frac{x}{x-1})$ and call F the fixed point field. F consists of all rational functions satisfying

$$(1) \qquad f(x) = f(1-x) = f\left(\frac{1}{x}\right) = f\left(1 - \frac{1}{x}\right) = f\left(\frac{1}{1-x}\right) = f\left(\frac{x}{x-1}\right).$$

It suffices to check the first two equalities, the others being consequences. The function

$$I = I(x) = \frac{(x^2 - x + 1)^3}{x^2(x-1)^2} \tag{2}$$

belongs to F as is readily seen. Hence, the field $S = k(I)$ of all rational functions of I will belong to F.

We contend: $F = S$ and $(E/F) = 6$.

Indeed, from Theorem 13 we obtain $(E/F) \geq 6$. Since $S \subset F$ it suffices to prove $(E/S) \leq 6$. Now $E = S(x)$. It is thus sufficient to find some 6-th degree equation with coefficients in S satisfied by x. The following one is obviously satisfied:

$$(x^2 - x + 1)^3 - I \cdot x^2(x-1)^2 = 0.$$

The reader will find the study of these fields a profitable exercise. At a later occasion he will be able to derive all intermediate fields.

2) Let k be a field and $E = k(x_1, x_2, \ldots, x_n)$ the field of all rational functions of n variables $x_1, x_2, \ldots, x_n$. If $(\nu_1, \nu_2, \ldots, \nu_n)$ is a permutation of $(1, 2, \ldots, n)$ we replace in each function $f(x_1, x_2, \ldots, x_n)$ of E the variable x_1 by x_{ν_1}, x_2 by x_{ν_2}, $\ldots$, x_n by x_{ν_n}. The mapping of E onto itself obtained in this way is obviously an automorphism and we may construct $n!$ automorphisms in this fashion (including the identity). Let F be the fixed point field, that is, the set of all so-called "symmetric functions." Theorem 13 shows that $(E/F) \geq n!$. Let us introduce the polynomial:

$$f(t) = (t - x_1)(t - x_2) \cdots (t - x_n) = t^n + a_1 t^{n-1} + \cdots + a_n \tag{3}$$

where $a_1 = -(x_1 + x_2 + \cdots + x_n)$; $a_2 = +(x_1x_2 + x_1x_3 + \cdots + x_{n-1}x_n)$ and more generally a_i is $(-1)^i$ times the sum of all products of i different variables of the set $x_1, x_2, \ldots, x_n$. The functions $a_1, a_2, \ldots, a_n$ are called the elementary symmetric functions and the field $S = k(a_1, a_2, \ldots, a_n)$ of all rational functions of $a_1, a_2, \ldots, a_n$ is obviously a part of F. Should we succeed in proving $(E/S) \leq n!$ we would have shown $S = F$ and $(E/F) = n!$.

We construct to this effect the following tower of fields:

$$S = S_n \subset S_{n-1} \subset S_{n-2} \subset \cdots \subset S_2 \subset S_1 = E$$

by the definition

$$S_n = S; \quad S_i = S(x_{i+1}, x_{i+2}, \ldots, x_n) = S_{i+1}(x_{i+1}). \tag{4}$$

It would be sufficient to prove $(S_{i-1}/S_i) \leq i$ or that the generator x_i for S_{i-1} out of S_i satisfies an equation of degree i with coefficients in S_i.

Such an equation is easily constructed. Put

$$F_i(t) = \frac{f(t)}{(t - x_{i+1})(t - x_{i+2}) \cdots (t - x_n)} = \frac{F_{i+1}(t)}{(t - x_{i+1})} \tag{5}$$

and $F_n(t) = f(t)$. Performing the division we see that $F_i(t)$ is a polynomial in t of degree i whose highest coefficient is 1 and whose coefficients are polynomials in the variables $a_1, a_2, \ldots, a_n$ and $x_{i+1}, x_{i+2}, \ldots, x_n$. Only integers enter as coefficients in these expressions. Now x_i is obviously a root of $F_i(t) = 0$.

Now let $g(x_1, x_2, \ldots, x_n)$ be a *polynomial* in $x_1, x_2, \ldots, x_n$. Since $F_1(x_1) = 0$ is of first degree in x_1, we can express x_1 as polynomial of the a_i and of $x_2, x_3, \ldots, x_n$. We introduce this expression in $g(x_1, x_2, \ldots, x_n)$. Since $F_2(x_2) = 0$ we can express

x_2^2 or higher powers as polynomial in $x_3, \ldots, x_n$ and the a_i. Since $F_3(x_3) = 0$ we can express x_3^3 and higher powers as polynomials of $x_4, x_5, \ldots, x_n$ and the a_i. Introducing these expressions in $g(x_1, x_2, \ldots, x_n)$ we see that we can express it as a polynomial in the x_ν and the a_ν such that the degree in x_i is below i. So $g(x_1, x_2, \ldots, x_n)$ is a linear combination of the following $n!$ terms:

$$x_1^{\nu_1} x_2^{\nu_2} \cdots x_n^{\nu_n} \quad \text{where each } \nu_i \leq i - 1. \tag{6}$$

The coefficients of these terms are polynomials in the a_i. Since the expressions (6) are linearly independent in S (this is our previous result), the expression is unique.

This is a *generalization* of the theorem of symmetric functions in its usual form. The latter says that a symmetric polynomial can be written as a polynomial in $a_1, a_2, \ldots, a_n$. Indeed, if $g(x_1, \ldots, x_n)$ is symmetric we have already an expression as linear combination of the terms (6) where only the term 1 corresponding to $\nu_1 = \nu_2 = \cdots = \nu_n = 0$ has a coefficient $\neq 0$ in S, namely, $g(x_1, \ldots, x_n)$. So $g(x_1, x_2, \ldots, x_n)$ is a polynomial in $a_1, a_2, \ldots, a_n$.

But our theorem gives an expression of any polynomial, symmetric or not.

H. Normal Extensions[4]

An extension field E of a field F is called a *normal* extension if the group G of automorphisms of E which leave F fixed has F for its fixed field, and (E/F) is finite.

Although the result in Theorem 13 cannot be sharpened in general, there is one case in which the equality sign will always occur, namely, in the case in which $\sigma_1, \sigma_2, \ldots, \sigma_n$ is a set of automorphisms which form a group. We prove

THEOREM 14. *If $\sigma_1, \sigma_2, \ldots, \sigma_n$ is a group of automorphisms of a field E and if F is the fixed field of $\sigma_1, \sigma_2, \ldots, \sigma_n$, then $(E/F) = n$.*

If $\sigma_1, \sigma_2, \ldots, \sigma_n$ is a group, then the identity occurs, say, $\sigma_1 = I$. The fixed field consists of those elements x which are not moved by any of the σ's, i.e., $\sigma_i(x) = x$, $i = 1, 2, \ldots, n$. Suppose that $(E/F) > n$. Then there exist $n + 1$ elements $\alpha_1, \alpha_2, \ldots, \alpha_{n+1}$ of E which are linearly independent with respect to F. By Theorem 1, there exists a non-trivial solution in E to the system of equations

$$(') \quad \begin{aligned} x_1\sigma_1(\alpha_1) + x_2\sigma_1(\alpha_2) + \cdots + x_{n+1}\sigma_1(\alpha_{n+1}) &= 0 \\ x_1\sigma_2(\alpha_1) + x_2\sigma_2(\alpha_2) + \cdots + x_{n+1}\sigma_2(\alpha_{n+1}) &= 0 \\ \cdots\cdots\cdots\cdots\cdots\cdots\cdots\cdots\cdots\cdots\cdots\cdots \\ x_1\sigma_n(\alpha_1) + x_2\sigma_n(\alpha_2) + \cdots + x_{n+1}\sigma_n(\alpha_{n+1}) &= 0 \end{aligned}$$

We note that the solution cannot lie in F, otherwise, since σ_1 is the identity, the first equation would be a dependence between $\alpha_1, \ldots, \alpha_{n+1}$.

Among all non-trivial solutions $x_1, x_2, \ldots, x_{n+1}$ we choose one which has the least number of elements different from 0. We may suppose this solution to be $a_1, a_2, \ldots, a_r, 0, \ldots, 0$, where the first r terms are different from 0. Moreover, $r \neq 1$ because $a_1\sigma_1(\alpha_1) = 0$ implies $a_1 = 0$ since $\sigma_1(\alpha_1) = \alpha_1 \neq 0$. Also, we may suppose $a_r = 1$, since if we multiply the given solution by a_r^{-1} we obtain a new solution in which the r-th term is 1. Thus, we have

$$(*) \quad a_1\sigma_i(\alpha_1) + a_2\sigma_i(\alpha_2) + \cdots + a_{r-1}\sigma_i(\alpha_{r-1}) + \sigma_i(\alpha_r) = 0$$

[4]What Artin call a normal extension is nowadays called a Galois extension. *Editor's note.*

for $i = 1, 2, \ldots, n$. Since $a_1, \ldots, a_{r-1}$ cannot all belong to F, one of these, say a_1, is in E but not in F. There is an automorphism σ_k for which $\sigma_k(a_1) \neq a_1$. If we use the fact that $\sigma_1, \sigma_2, \ldots, \sigma_n$ form a group, we see $\sigma_k \cdot \sigma_1$, $\sigma_k \cdot \sigma_2$, $\ldots$, $\sigma_k \cdot \sigma_n$ is a permutation of $\sigma_1, \sigma_2, \ldots, \sigma_n$. Applying σ_k to the expressions in $(*)$ we obtain

$$\sigma_k(a_1) \cdot \sigma_k\sigma_j(\alpha_1) + \cdots + \sigma_k(a_{r-1}) \cdot \sigma_k\sigma_j(\alpha_{r-1}) + \sigma_k\sigma_j(\alpha_r) = 0$$

for $j = 1, 2, \ldots, r$, so that from $\sigma_k\sigma_j = \sigma_i$

$$(**) \qquad \sigma_k(a_1) \cdot \sigma_i(\alpha_1) + \cdots + \sigma_k(a_{r-1}) \cdot \sigma_i(\alpha_{r-1}) + \sigma_i(\alpha_r) = 0$$

and if we subtract $(**)$ from $(*)$ we have

$$[a_1 - \sigma_k(a_1)] \cdot \sigma_i(\alpha_1) + \cdots + [a_{r-1} - \sigma_k(a_{r-1})] \cdot \sigma_i(\alpha_{r-1}) = 0$$

which is a non-trivial solution to the system $(')$ having fewer than r elements different from 0, contrary to the choice of r.

COROLLARY 1. *If F is the fixed field for the finite group G, then each automorphism σ that leaves F fixed must belong to G.*

(E/F) = order of $G = n$. Assume there is a σ not in G. Then F would remain fixed under the $n + 1$ elements consisting of σ and the elements of G, thus contradicting the corollary to Theorem 13.

COROLLARY 2. *There are no two distinct finite groups G_1 and G_2 with the same fixed field.*

This follows immediately from Corollary 1.

If $f(x)$ is a polynomial in F, then $f(x)$ is called *separable* if its irreducible factors do not have repeated roots. If E is an extension of the field F, the *element α of E is called separable* if it is root of a separable polynomial $f(x)$ in F, and E is called a *separable extension* if each element of E is separable.

THEOREM 15. *E is a normal extension of F if and only if E is the splitting field of a separable polynomial $p(x)$ in F.*

Sufficiency. Under the assumption that E splits $p(x)$ we prove that E is a normal extension of F.

If all roots of $p(x)$ are in F, then our proposition is trivial, since then $E = F$ and only the unit automorphism leaves F fixed.

Let us suppose $p(x)$ has $n > 1$ roots in E but not in F. We make the inductive assumption that for all pairs of fields with fewer than n roots of $p(x)$ outside of F our proposition holds.

Let $p(x) = p_1(x) \cdot p_2(x) \cdot \ldots \cdot p_r(x)$ be a factorization of $p(x)$ into irreducible factors. We may suppose one of these to have a degree greater than one, for otherwise $p(x)$ would split in F. Suppose $\deg p_1(x) = s > 1$. Let α_1 be a root of $p_1(x)$. Then $(F(\alpha_1)/F) = \deg p_1(x) = s$. If we consider $F(\alpha_1)$ as the new ground field, fewer roots of $p(x)$ than n are outside. From the fact that $p(x)$ lies in $F(\alpha_1)$ and E is a splitting field of $p(x)$ over $F(\alpha_1)$, it follows by our inductive assumption that E is a normal extension of $F(\alpha_1)$. Thus, each element in E which is not in $F(\alpha_1)$ is moved by at least one automorphism which leaves $F(\alpha_1)$ fixed.

$p(x)$ being separable, the roots $\alpha_1, \alpha_2, \ldots, \alpha_s$ of $p_1(x)$ are s distinct elements of E. By Theorem 8 there exist isomorphisms $\sigma_1, \sigma_2, \ldots, \sigma_s$ mapping $F(\alpha_1)$ on $F(\alpha_1), F(\alpha_2), \ldots, F(\alpha_s)$, respectively, which are each the identity on F and map α_1 on $\alpha_1, \alpha_2, \ldots, \alpha_s$ respectively. We now apply Theorem 10. E is a splitting field of $p(x)$ in $F(\alpha_1)$ and is also a splitting field of $p(x)$ in $F(\alpha_i)$. Hence the isomorphism σ_i, which makes $p(x)$ in $F(\alpha_1)$ correspond to the same $p(x)$ in $F(\alpha_i)$, can be extended to an isomorphic mapping of E onto E, that is, to an automorphism of E that we denote again by σ_i. Hence, $\sigma_1, \sigma_2, \ldots, \sigma_s$ are automorphisms of E that leave F fixed and map α_1 onto $\alpha_1, \alpha_2, \ldots, \alpha_n$.

Now let θ be an element that remains fixed under all automorphisms of E that leave F fixed. We know already that it is in $F(\alpha_1)$ and hence has the form

$$\theta = c_0 + c_1\alpha_1 + c_2\alpha_1^2 + \cdots + c_{s-1}\alpha_1^{s-1}$$

where the c_i are in F. If we apply σ_i to this equation we get, since $\sigma_i(\theta) = \theta$:

$$\theta = c_0 + c_1\alpha_i + c_2\alpha_i^2 + \cdots + c_{s-1}\alpha_i^{s-1}.$$

The polynomial $c_{s-1}x^{s-1} + c_{s-2}x^{s-2} + \cdots + c_1x + (c_0 - \theta)$ has therefore the s distinct roots $\alpha_1, \alpha_2, \ldots, \alpha_s$. These are more than its degree. So all coefficients of it must vanish, among them $c_0 - \theta$. This shows θ in F.

Necessity. If E is a normal extension of F, then E is splitting field of a separable polynomial $p(x)$. We first prove the

LEMMA. *If E is a normal extension of F, then E is a separable extension of F. Moreover any element of E is root of an equation over F which splits completely in E.*

Let $\sigma_1, \sigma_2, \ldots, \sigma_s$ be the group G of automorphisms of E whose fixed field is F. Let α be an element of E, and let $\alpha, \alpha_2, \alpha_3, \ldots, \alpha_r$ be the set of distinct elements in the sequence $\sigma_1(\alpha), \sigma_2(\alpha), \ldots, \sigma_s(\alpha)$. Since G is a group,

$$\sigma_j(\alpha_i) = \sigma_j(\sigma_k(\alpha)) = \sigma_j\sigma_k(\alpha) = \sigma_m(\alpha) = \alpha_n.$$

Therefore, the elements $\alpha, \alpha_2, \ldots, \alpha_r$ are permuted by the automorphisms of G. The coefficients of the polynomial $f(x) = (x-\alpha)(x-\alpha_2)\cdots(x-\alpha_r)$ are left fixed by each automorphism of G, since in its factored form the factors of $f(x)$ are only permuted. Since the only elements of E which are left fixed by all the automorphisms of G belong to F, $f(x)$ is a polynomial in F. If $g(x)$ is a polynomial in F which also has α as root, then applying the automorphisms of G to the expression $g(\alpha) = 0$ we obtain $g(\alpha_i) = 0$, so that the degree of $g(x) \geq s$. Hence $f(x)$ is irreducible, and the lemma is established.

To complete the proof of the theorem, let $\omega_1, \omega_2, \ldots, \omega_t$ be a generating system for the vector space E over F. Let $f_i(x)$ be the separable polynomial having ω_i as a root. Then E is the splitting field of $p(x) = f_1(x) \cdot f_2(x) \cdot \ldots \cdot f_t(x)$.

If $f(x)$ is a polynomial in a field F, and E the splitting field of $f(x)$, then we shall call the group of automorphisms of E over F the *group of the equation* $f(x) = 0$. We come now to a theorem known in algebra as the *Fundamental Theorem of Galois Theory* which gives the relation between the structure of a splitting field and its group of automorphisms.

THEOREM 16 (Fundamental Theorem). *If $p(x)$ is a separable polynomial in a field F, and G the group of the equation*

$$p(x) = 0$$

where E is the splitting field of $p(x)$, then: (1) *Each intermediate field, B, is the fixed field for a subgroup G_B of G, and distinct subgroups have distinct fixed fields. We say B and G_B "belong" to each other.* (2) *The intermediate field B is a normal* *extension of F if and only if the subgroup G_B is a normal subgroup of G. In this case the group of automorphisms of B which leaves F fixed is isomorphic to the factor group (G/G_B). (3) For each intermediate field B, we have $(B/F) =$* index of G_B *and $(E/B) =$* order of G_B.

The first part of the theorem comes from the observation that E is splitting field for $p(x)$ when $p(x)$ is taken to be in any intermediate field. Hence, E is a normal extension of each intermediate field B, so that B is the fixed field of the subgroup of G consisting of the automorphisms which leave B fixed. That distinct subgroups have distinct fixed fields is stated in Corollary 2 to Theorem 14.

Let B be any intermediate field. Since B is the fixed field for the subgroup G_B of G, by Theorem 14 we have $(E/B) =$ order of G_B. Let us call $\mathrm{o}(G)$ the order of a group G and $\mathrm{i}(G)$ its index. Then $\mathrm{o}(G) = \mathrm{o}(G_B) \cdot \mathrm{i}(G_B)$. But $(E/F) = \mathrm{o}(G)$, and $(E/F) = (E/B) \cdot (B/F)$ from which $(B/F) = i(G_B)$, which proves the third part of the theorem.

The number $\mathrm{i}(G_B)$ is equal to the number of left cosets of G_B. The elements of G, being automorphisms of E, are isomorphisms of B; that is, they map B isomorphically into some other subfield of E and are the identity on F. The elements of G in any one coset of G_B map B in the same way. For let $\sigma \cdot \sigma_1$ and $\sigma \cdot \sigma_2$ be two elements of the coset σG_B. Since σ_1 and σ_2 leave B fixed, for each α in B we have $\sigma\sigma_1(\alpha) = \sigma(\alpha) = \sigma\sigma_2(\alpha)$. Elements of different cosets give different isomorphisms, for if σ and τ give the same isomorphism, $\sigma(\alpha) = \tau(\alpha)$ for each α in B, then $\sigma^{-1}\tau(\alpha) = \alpha$ for each α in B. Hence, $\sigma^{-1}\tau = \sigma_1$, where σ_1 is an element of G_B. But then $\tau = \sigma\sigma_1$ and $\tau G_B = \sigma\sigma_1 G = \sigma G_B$ so that σ and τ belong to the same coset.

Each isomorphism of B which is the identity on F is given by an automorphism belonging to G. For let σ be an isomorphism mapping B on B' and the identity on F. Then under σ, $p(x)$ corresponds to $p(x)$, and E is the splitting field of $p(x)$ in B and of $p(x)$ in B'. By Theorem 10, σ can be extended to an automorphism σ' of E, and since σ' leaves F fixed it belongs to G. Therefore, the number of distinct isomorphisms of B is equal to the number of cosets of G_B and is therefore equal to (B/F).

The field σB onto which σ maps B has obviously $\sigma G_B \sigma^{-1}$ as corresponding group, since the elements of σB are left invariant by precisely this group.

If B is a normal extension of F, the number of distinct automorphisms of B which leave F fixed is (B/F) by Theorem 14. Conversely, if the number of automorphisms is (B/F) then B is a normal extension, because if F' is the fixed field of all these automorphisms, then $F \subset F' \subset B$, and by Theorem 14, (B/F') is equal to the number of automorphisms in the group, hence $(B/F') = (B/F)$. From $(B/F) = (B/F')(F'/F)$ we have $(F'/F) = 1$ or $F = F'$. Thus, B is a normal extension of F if and only if the number of automorphisms of B is (B/F).

B is a normal extension of F if and only if each isomorphism of B into E is an automorphism of B. This follows from the fact that each of the above conditions are equivalent to the assertion that there are the same number of isomorphisms and automorphisms. Since, for each σ, $B = \sigma B$ is equivalent to $\sigma G_B \sigma^{-1} \subset G_B$, we can finally say that B is a normal extension of F if and only if G_B is a normal subgroup of G.

As we have shown, each isomorphism of B is described by the effect of the elements of some left coset of G_B. If B is a normal extension these isomorphisms are all automorphisms, but in this case the cosets are elements of the factor group (G/G_B). Thus, each automorphism of B corresponds uniquely to an element of (G/G_B) and conversely. Since multiplication in (G/G_B) is obtained by iterating the mappings, the correspondence is an isomorphism between (G/G_B) and the group of automorphisms of B which leave F fixed. This completes the proof of Theorem 16.

I. Finite Fields

It is frequently necessary to know the nature of a finite subset of a field which under multiplication in the field is a group. The answer to this question is particularly simple.

THEOREM 17. *If S is a finite subset $(\neq 0)$ of a field F which is a group under multiplication in F, then S is a cyclic group.*

The proof is based on the following lemmas for abelian groups:

LEMMA 1. *If in an abelian group A and B are two elements of orders a and b, and if c is the least common multiple of a and b, then there is an element C of order c in the group.*

PROOF. (a) If a and b are relatively prime, $C = AB$ has the required order ab. The order of $C^a = B^a$ is b and therefore c is divisible by b. Similarly it is divisible by a. Since $C^{ab} = 1$ it follows $c = ab$.

(b) If d is a divisor of a, we can find in the group an element of order d. Indeed $A^{a/d}$ is this element.

(c) Now let us consider the general case. Let $p_1, p_2, \ldots, p_r$ be the prime numbers dividing either a or b and let

$$a = p_1^{n_1} p_2^{n_2} \cdots p_r^{n_r}$$
$$b = p_1^{m_1} p_2^{m_2} \cdots p_r^{m_r}.$$

Call t_i the larger of the two numbers n_i and m_i. Then

$$c = p_1^{t_1} p_2^{t_2} \cdots p_r^{t_r}.$$

According to (b) we can find in the group an element of order $p_i^{n_i}$ and one of order $p_i^{m_i}$. Thus there is one of order $p_i^{t_i}$. Part (a) shows that the product of these elements will have the desired order c. □

LEMMA 2. *If there is an element C in an abelian group whose order c is maximal (as is always the case if the group is finite) then c is divisible by the order a of every element A in the group, hence $x^c = 1$ is satisfied by each element in the group.*

PROOF. If a does not divide c, the greatest common multiple of a and c would be larger than c and we could find an element of that order, thus contradicting the choice of c. □

We now prove Theorem 17. Let n be the order of S and r the largest order occurring in S. Then $x^r - 1 = 0$ is satisfied for all elements of S. Since this polynomial of degree r in the field cannot have more than r roots, it follows that $r \geq n$. On the other hand $r \leq n$ because the order of each element divides n. S is therefore a cyclic group consisting of $1, \varepsilon, \varepsilon^2, \ldots, \varepsilon^{n-1}$ where $\varepsilon^n = 1$.

Theorem 17 could also have been based on the decomposition theorem for abelian groups having a finite number of generators. Since this theorem will be needed later, we interpolate a proof of it here.

Let G be an abelian group, with group operation written as $+$. The element $g_1, \dots, g_k$ will be said to generate G if each element g of G can be written as sum of multiples of $g_1, \dots, g_k$, $g = n_1 g_1 + \cdots + n_k g_k$. If no set of fewer than k elements generate G, then $g_1, \dots, g_k$ will be called a minimal generating system. Any group having a finite generating system admits a minimal generating system. In particular, a finite group always admits a minimal generating system.

From the identity $n_1(g_1 + mg_2) + (n_2 - n_1 m)g_2 = n_1 g_1 + n_2 g_2$ it follows that if $g_1, g_2, \dots, g_k$ generate G, also $g_1 + mg_2$, $g_2, \dots, g_k$ generate G.

An equation $m_1 g_1 + m_2 g_2 + \cdots + m_k g_k = 0$ will be called a relation between the generators, and $m_1, \dots, m_k$ will be called the coefficients in the relation.

We shall say that the abelian group G is the direct product of its subgroups $G_1, G_2, \dots, G_k$ if each $g \in G$ is uniquely representable as a sum $g = x_1 + x_2 + \cdots + x_k$, where $x_i \in G_i$, $i = 1, \dots, k$.

DECOMPOSITION THEOREM. *Each abelian group having a finite number of generators is the direct product of cyclic subgroups $G_1, \dots, G_n$ where the order of G_i divides the order of G_{i+1}, $i = 1, \dots, n-1$ and n is the number of elements in a minimal generating system. ($G_r, G_{r+1}, \dots, G_n$ may each be infinite, in which case, to be precise, $\mathrm{o}(G_i) \mid \mathrm{o}(G_{i+1})$ for $i = 1, 2, \dots, r-2$).*

We assume the theorem true for all groups having minimal generating systems of $k-1$ elements. If $n = 1$ the group is cyclic and the theorem trivial. Now suppose G is an abelian group having a minimal generating system of k elements. If no minimal generating system satisfies a non-trivial relation, then let $g_1, g_2, \dots, g_k$ be a minimal generating system and $G_1, G_2, \dots, G_k$ be the cyclic groups generated by them. For each $g \in G$, $g = n_1 g_1 + \cdots + n_k g_k$ where the expression is unique; otherwise we should obtain a relation. Thus the theorem would be true. Assume now that some non-trivial relations hold for some minimal generating systems. Among all relations between minimal generating systems, let

$$m_1 g_1 + \cdots + m_k g_k = 0 \tag{1}$$

be a relation in which the smallest positive coefficient occurs. After an eventual reordering of the generators we can suppose this coefficient to be m_1. In any other relation between $g_1, \dots, g_k$,

$$n_1 g_1 + \cdots + n_k g_k = 0 \tag{2}$$

we must have m_1/n_1. Otherwise $n_1 = qm_1 + r$, $0 < r < m$, and q times relation (1) subtracted from relation (2) would yield a relation with a coefficient $r < m_1$. Also in relation (1) we must have m_1/m_i, $i = 2, \dots, k$. For suppose m_1 does not divide one coefficient, say m_2. Then $m_2 = qm_1 + r$, $0 < r < m_1$. In the generating system $g_1 + qg_2, g_2, \dots, g_k$ we should have a relation $m_1(g_1 + qg_2) + rg_2 + m_3 g_3 + \cdots + m_k g_k = 0$ where the coefficient r contradicts the choice of m_1. Hence $m_2 = q_2 m_1$, $m_3 = q_3 m_1$, $\dots$, $m_k = q_k m_1$. The system $\bar{g}_1 = g_1 + q_2 g_2 + \cdots + q_k g_k$, $g_2, \dots, g_k$ is minimal generating, and $m_1 \bar{g}_1 = 0$. In any relation $0 = n_1 \bar{g}_1 + n_2 g_2 + \cdots + n_k g_k$ since m_1 is a coefficient in a relation between $\bar{g}_1, g_2, \dots, g_k$ our previous argument yields $m_1 \mid n_1$, and hence $n_1 \bar{g}_1 = 0$.

Let G' be the subgroup of G generated by $g_2, \ldots, g_k$ and G_1 the cyclic group of order m_1 generated by $\bar{g}_1$. Then G is the direct product of G_1 and G'. Each element g of G can be written

$$g = n_1\bar{g}_1 + n_2 g_2 + \cdots + n_k g_k = n_1\bar{g}_1 + g'.$$

The representation is unique, since $n_1\bar{g}_1 + g' = n_1'\bar{g}_1 + g''$ implies the relation $(n_1 - n_1')\bar{g}_1 + (g' - g'') = 0$, hence $(n_1 - n_1')\bar{g}_1 = 0$, so that $n_1\bar{g}_1 = n_1'\bar{g}_1$ and also $g' = g''$.

By our inductive hypothesis, G' is the direct product of $k-1$ cyclic groups generated by elements $\bar{g}_2, \bar{g}_3, \ldots, \bar{g}_k$ whose respective orders $t_2, \ldots, t_k$ satisfy $t_i \mid t_{i+1}$, $i = 2, \ldots, k-1$. The preceding argument applied to the generators $\bar{g}_1, \bar{g}_2, \ldots, \bar{g}_k$ yields $m_1 \mid t_2$, from which the theorem follows.

By a *finite field* is meant one having only a finite number of elements.

COROLLARY. [5] *The non-zero elements of a finite field form a cyclic group.*

If a is an element of a field F, let us denote the n-fold of a, i.e., the element of F obtained by adding a to itself n times, by na. It is obvious that $n \cdot (m \cdot a) = (nm) \cdot a$ and $(n \cdot a)(m \cdot b) = nm \cdot ab$. If for one element $a \neq 0$, there is an integer n such that $n \cdot a = 0$ then $n \cdot b = 0$ for each b in F, since $n \cdot b = (n \cdot a) \cdot (a^{-1}b) = 0 \cdot a^{-1}b = 0$. If there is a positive integer p such that $p \cdot a = 0$ for each a in F, and if p is the smallest integer with this property, then F is said to have the *characteristic* p. If no such positive integer exists then we say F has characteristic 0. *The characteristic of a field is always a prime number*, for if $p = r \cdot s$ then $pa = rs \cdot a = r \cdot (s \cdot a)$. However, $s \cdot a = b \neq 0$ if $a \neq 0$ and $r \cdot b \neq 0$ since both r and s are less than p, so that $pa \neq 0$ contrary to the definition of the characteristic. If $na = 0$ for $a \neq 0$, then p divides n, for $n = qp + r$ where $0 \leq r < p$ and $na = (qp + r)a = qpa + ra$. Hence $na = 0$ implies $ra = 0$ and from the definition of the characteristic since $r < p$, we must have $r = 0$.

If F is a finite field having q elements and E an extension of F such that $(E/F) = n$, then E has q^n elements. For if $\omega_1, \omega_2, \ldots, \omega_n$ is a basis of E over F, each element of E can be uniquely represented as a linear combination $x_1\omega_1 + x_2\omega_2 + \cdots + x_n\omega_n$ where the x_i belong to F. Since each x_i can assume q values in F, there are q^n distinct possible choices of $x_1, \ldots, x_n$ and hence q^n distinct elements of E. E is finite, hence, *there is an element* α *of* E *so that* $E = F(\alpha)$. (The non-zero elements of E form a cyclic group generated by α).

If we denote by $P \equiv [0, 1, 2, \ldots, p-1]$ the set of multiples of the unit element in a field F of characteristic p, then P is a subfield of F having p distinct elements. In fact, P is isomorphic to the field of integers reduced mod p. If F is a finite field, then the degree of F over P is finite, say $(F/P) = n$, and F contains p^n elements. In other words, *the order of any finite field is a power of its characteristic.*

If F and F' are two finite fields having the same order q, then by the preceding, they have the same characteristic since q is a power of the characteristic. The multiples of the unit in F and F' form two fields P and P' which are isomorphic.

The non-zero elements of F and F' form a group of order $q-1$ and, therefore, satisfy the equation $x^{q-1} - 1 = 0$. The fields F and F' are splitting fields of the equation $x^{q-1} = 1$ considered as lying in P and P' respectively. By Theorem 10, the isomorphism between P and P' can be extended to an isomorphism between F and F'. We have thus proved

[5] This is a corollary of 17, not the Decomposition Theorem. *Editor's note.*

THEOREM 18. *Two finite fields having the same number of elements are isomorphic.*

Differentiation. If $f(x) = a_0 + a_1x + \cdots + a_nx^n$ is a polynomial in a field F, then we define $f' = a_1 + 2a_2x + \cdots + na_nx^{n-1}$. The reader may readily verify that for each pair of polynomials f and g we have

$$(f+g)' = f' + g'$$
$$(f \cdot g)' = fg' + gf'$$
$$(f^n)' = nf^{n-1} \cdot f'$$

THEOREM 19. *The polynomial f has repeated roots if and only if in the splitting field E the polynomials f and f' have a common root. This condition is equivalent to the assertion that f and f' have a common factor of degree greater than 0 in F.*

If α is a root of multiplicity k of $f(x)$ then $f = (x-\alpha)^k Q(x)$ where $Q(\alpha) \neq 0$. This gives $f = (x-\alpha)^k Q'(x) + k(x-\alpha)^{k-1}Q(x) = (x-\alpha)^{k-1}[(x-\alpha)Q'(x) + kQ(x)]$.

If $k > 1$, then α is a root of f' of multiplicity at least $k-1$. If $k = 1$, then $f'(x) = Q(x) + (x-\alpha)Q'(x)$ and $f'(\alpha) = Q(\alpha) \neq 0$. Thus, f and f' have a root α in common if and only if α is a root of f of multiplicity greater than 1.

If f and f' have a root α in common then the irreducible polynomial in F having α as root divides both f and f'. Conversely, any root of a factor common to both f and f' is a root of f and f'.

COROLLARY. *If F is a field of characteristic 0 then each irreducible polynomial in F is separable.*

Suppose to the contrary that the irreducible polynomial $f(x)$ has a root α of multiplicity greater than 1. Then, $f'(x)$ is a polynomial which is not identically zero (its leading coefficient is a multiple of the leading coefficient of $f(x)$ and is not zero since the characteristic is 0) and of degree 1 less than the degree of $f(x)$. But α is also a root of $f'(x)$ which contradicts the irreducibility of $f(x)$.

J. Roots of Unity

If F is a field having any characteristic p, and E the splitting field of the polynomial $x^n - 1$ where p does not divide n, then we shall refer to E as *the field generated out of F by the adjunction of a primitive n^{th} root of unity.*

The polynomial $x^n - 1$ does not have repeated roots in E, since its derivative, nx^{n-1}, has only the root 0 and has, therefore, no roots in common with $x^n - 1$. Thus, E is a normal extension of F. If $\varepsilon_1, \varepsilon_2, \ldots, \varepsilon_n$ are the roots of $x^n - 1$ in E, they form a group under multiplication and by Theorem 17 this group will be cyclic. If $1, \varepsilon, \varepsilon^2, \ldots, \varepsilon^{n-1}$ are the elements of the group, we shall call ε a primitive n^{th} root of unity. The smallest power of ε which is 1 is the nth.

THEOREM 20. *If E is the field generated from F by a primitive n^{th} root of unity, then the group G of E over F is abelian for any n and cyclic if n is a prime number.*

We have $E = F(\varepsilon)$, since the roots of $x^n - 1$ are powers of ε. Thus, if σ and τ are distinct elements of G, $\sigma(\varepsilon) \neq \tau(\varepsilon)$. But $\sigma(\varepsilon)$ is a root of $x^n - 1$ and, hence, a power of ε. Thus, $\sigma(\varepsilon) = \varepsilon^{n_\sigma}$ where n_σ is an integer $1 \leq n_\sigma < n$. Moreover, $\tau\sigma(\varepsilon) = \tau(\varepsilon^{n_\sigma}) = (\tau(\varepsilon))^{n_\sigma} = \varepsilon^{n_\tau \cdot n_\sigma} = \sigma\tau(\varepsilon)$. Thus, $n_{\sigma\tau} = n_\sigma n_\tau \mod n$. Thus, the mapping of σ on n_σ is a homomorphism of G into a multiplicative subgroup of the integers mod n. Since $\tau \neq \sigma$ implies $\tau(\varepsilon) \not\equiv \sigma(\varepsilon)$, it follows that $\tau \neq \sigma$ implies $n_\sigma \not\equiv n_\tau \mod n$. Hence, the homomorphism is an isomorphism. If n is a prime number, the multiplicative group of numbers modulo n forms a cyclic group.

K. Noether Equations

If E is a field, and $G = (\sigma, \tau, \dots)$ a group of automorphisms of E, any set of elements $x_\sigma, x_\tau, \dots$ in E will be said to provide a *solution to Noether's equations* if $x_\sigma \cdot \sigma(x_\tau) = x_{\sigma\tau}$ for each σ and τ in G. If one element $x_\sigma = 0$ then $x_\tau = 0$ for each $\tau \in G$. As τ traces G, $\sigma\tau$ assumes all values in G, and in the above equation $x_{\sigma\tau} = 0$ when $x_\sigma = 0$. Thus, in any solution of the Noether equations no element $x_\sigma = 0$ unless the solution is completely trivial. We shall assume in the sequel that the trivial solution has been excluded.

THEOREM 21. *The system $x_\sigma, x_\tau, \dots$ is a solution to Noether's equations if and only if there exists an element α in E, such that $x_\sigma = \alpha/\sigma(\alpha)$ for each σ.*

For any α, it is clear that $x_\sigma = \alpha/\sigma(\alpha)$ is a solution to the equations, since

$$\alpha/\sigma(\alpha) \cdot \sigma(\alpha/\tau(\alpha)) = \alpha/\sigma(\alpha) \cdot \sigma(\alpha)/\sigma\tau(\alpha) = \alpha/\sigma\tau(\alpha).$$

Conversely, let $x_\sigma, x_\tau, \dots$ be a non-trivial solution. Since the automorphisms $\sigma, \tau, \dots$ are distinct they are linearly independent, and the equation $x_\sigma \cdot \sigma(z) + x_\tau\tau(z) + \dots = 0$ does not hold identically. Hence, there is an element a in E such that $x_\sigma\sigma(a) + x_\tau\tau(a) + \dots = \alpha \neq 0$. Applying σ to α gives

$$\sigma(\alpha) = \sum_{\tau \in G} \sigma(x_\tau) \cdot \sigma\tau(a).$$

Multiplying by x_σ gives

$$x_\sigma \cdot \sigma(\alpha) = \sum_{\tau \in G} x_\sigma\sigma(x_\tau) \cdot \sigma\tau(a).$$

Replacing $x_\sigma\sigma(x_\tau)$ by $x_{\sigma\tau}$ and noting that $\sigma\tau$ assumes all values in G when τ does, we have

$$x_\sigma \cdot \sigma(\alpha) = \sum_{\tau \in G} x_\tau\tau(a) = \alpha$$

so that

$$x_\sigma = \alpha/\sigma(\alpha).$$

A solution to the Noether equations defines a mapping C of G into E, namely, $C(\sigma) = x_\sigma$. If F is the fixed field of G, and the elements x_σ lie in F, then C is a *character* of G. For $C(\sigma\tau) = x_{\sigma\tau} = x_\sigma \cdot \sigma(x_\tau) = x_\sigma x_\tau = C(\sigma) \cdot C(\tau)$ since $\sigma(x_\tau) = x_\tau$ if $x_\tau \in F$. Conversely, each character C of G in F provides a solution to the Noether equations. Call $C(\sigma) = x_\sigma$. Then, since $x_\tau \in F$, we have $\sigma(x_\tau) = x_\tau$. Thus, $x_\sigma \cdot \sigma(x_\tau) = x_\sigma \cdot x_\tau = C(\sigma) \cdot C(\tau) = C(\sigma\tau) = x_{\sigma\tau}$. We therefore have, by combining this with Theorem 20,

THEOREM 22. *If G is the group of the normal field E over F, then for each character C of G into F there exists an element α in E such that $C(\sigma) = \alpha/\sigma(\alpha)$ and, conversely, if $\alpha/\sigma(\alpha)$ is in F for each σ, then $C(\sigma) = \alpha/\sigma(\alpha)$ is a character of G. If r is the least common multiple of the orders of elements of G, then $\alpha^r \in F$.*

We have already shown all but the last sentence of Theorem 22. To prove this we need only show $\sigma(\alpha^r) = \alpha^r$ for each $\sigma \in G$. But $\alpha^r/\sigma(\alpha^r) = (\alpha/\sigma(\alpha))^r = (C(\sigma))^r = C(\sigma^r) = C(I) = 1$.

L. Kummer's Fields

If F contains a primitive n^{th} root of unity, any splitting field E of a polynomial $(x^n - a_1)(x^n - a_2)\cdots(x^n - a_r)$ where $a_i \in F$ for $i = 1, 2, \ldots, r$ will be called a *Kummer extension* of F, or more briefly, a *Kummer field.*

If a field F contains a primitive n^{th} root of unity, the number n is not divisible by the characteristic of F. Suppose, to the contrary, F has characteristic p and $n = qp$. Then $y^p - 1 = (y-1)^p$ since in the expansion of $(y-1)^p$ each coefficient other than the first and last is divisible by p and therefore is a multiple of the p-fold of the unit of F and thus is equal to 0. Therefore $x^n - 1 = (x^q)^p - 1 = (x^q - 1)^p$ and $x^n - 1$ cannot have more than q distinct roots. But we assumed that F has a primitive n^{th} root of unity and $1, \varepsilon, \varepsilon^2, \ldots, \varepsilon^{n-1}$ would be n distinct roots of $x^n - 1$. It follows that n is not divisible by the characteristic of F. For a Kummer field E, none of the. factors $x^n - a_i$, $a_i \neq 0$ has repeated roots since the derivative, nx^{n-1}, has only the root 0 and has therefore no roots in common with $x^n - a_i$. Therefore, the irreducible factors of $x^n - a_i$ are separable, so that *E is a normal extension of F.*

Let α_i be a root of $x^n - a_i$ in E. If $\varepsilon_1, \varepsilon_2, \ldots, \varepsilon_n$ are the n distinct n^{th} roots of unity in F, then $\alpha_i\varepsilon_1, \alpha_i\varepsilon_2, \ldots, \alpha_i\varepsilon_n$ will be n distinct roots of $x^n - a_i$, and hence will be the roots of $x^n - a_i$, so that $E = F(\alpha_1, \alpha_2, \ldots, \alpha_r)$. Let σ and τ be two automorphisms in the group G of E over F. For each α_i, both σ and τ map α_i on some other root of $x^n - a_i$. Thus $\tau(\alpha_i) = \varepsilon_{i\tau}\alpha_i$ and $\sigma(\alpha_i) = \varepsilon_{i\sigma}\alpha_i$ where $\varepsilon_{i\sigma}$ and $\varepsilon_{i\tau}$ are n^{th} roots of unity in the basic field F. It follows that $\tau(\sigma(\alpha_i)) = \tau(\varepsilon_{i\sigma}\alpha_i) = \varepsilon_{i\sigma}\tau(\alpha_i) = \varepsilon_{i\sigma}\varepsilon_{i\tau}\alpha_i = \sigma(\tau(\alpha_i))$. Since σ and τ are commutative over the generators of E, they commute over each element of E. Hence, G is commutative. If $\sigma \in G$, then $\sigma(\alpha_i) = \varepsilon_{i\sigma}\alpha_i$, $\sigma^2(\alpha_i) = \varepsilon_{i\sigma}^2\alpha_i$, etc. Thus, $\sigma^{n_i}(\alpha_i) = \alpha_i$ for n_i such that $\varepsilon_{i\sigma}^{n_i} = 1$. Since the order of an nth root of unity is a divisor of n, we have n_i a divisor of n and the least common multiple m of $n_1, n_2, \ldots, n_r$ is a divisor of n. Since $\sigma^m(\alpha_i) = \alpha_i$ for $i = 1, 2, \ldots, r$ it follows that m is the order of σ. Hence, the order of each element of G is a divisor of n and, therefore, the least common multiple r of the orders of the elements of G is a divisor of n. If ε is a primitive n^{th} root of unity, then $\varepsilon^{n/r}$ is a primitive r^{th} root of unity. These remarks can be summarized in the following

THEOREM 23. *If E is a Kummer field, i.e., a splitting field of $p(x) = (x^n - a_1)(x^n - a_2)\cdots(x^n - a_t)$ where a_i lie in F, and F contains a primitive n^{th} root of unity, then:* (a) *E is a normal extension of F,* (b) *the group G of E over F is abelian,* (c) *the least common multiple of the orders of the elements of G is a divisor of n.*

COROLLARY. *If E is the splitting field of $x^p - a$, and F contains a primitive p^{th} root of unity where p is a prime number, then either $E = F$ and $x^p - a$ is split in F, or $x^p - a$ is irreducible and the group of E over F is cyclic of order p.*

The order of each element of G is, by Theorem 23, a divisor of p and, hence, if the element is not the unit its order must be p. If α is a root of $x^p - a$, then $\alpha, \varepsilon\alpha, \ldots, \varepsilon^{p-1}\alpha$ are all the roots of $x^p - a$ so that $F(\alpha) = E$ and $(E/F) \leq p$.

Hence, the order of G does not exceed p so that if G has one element different from the unit, it and its powers must constitute all of G. Since G has p distinct elements and their behavior is determined by their effect on α, then α must have p distinct images. Hence, the irreducible equation in F for α must be of degree p and is therefore $x^p - a = 0$.

The properties (a), (b) and (c) in Theorem 23 actually characterize Kummer fields.

Let us suppose that E is a normal extension of a field F, whose group G over F is abelian. Let us further assume that F contains a primitive r^{th} root of unity where r is the least common multiple of the orders of elements of G.

The group of characters X of G into the group of rth roots of unity is isomorphic to G. Moreover, to each $\sigma \in G$, if $\sigma \neq 1$, there exists a character $C \in X$ such that $C(\sigma) \neq 1$. Write G as the direct product of the cyclic groups $G_1, G_2, \ldots, G_t$ of orders $m_1 \mid m_2 \mid \cdots \mid m_t$. Each $\sigma \in G$ may be written $\sigma = \sigma_1^{\nu_1}\sigma_2^{\nu_2}\cdots\sigma_t^{\nu_t}$. Call C_i the character sending g_i into ε_i, a primitive $m_i{}^{\text{th}}$ root of unity and σ_j into 1 for $j \neq i$. Let C be any character. If $C(\sigma_i) = \varepsilon_i^{\mu_i}$, then we have $C = C_1^{\mu_1} \cdot C_2^{\mu_2}\cdots C_t^{\mu_t}$. Conversely, $C_1^{\mu_1}\cdots C_t^{\mu_t}$ defines a character. Since the order of C_i is m_i, the character group X of G is isomorphic to G. If $\sigma \neq 1$, then in $\sigma = \sigma_1^{\nu_1}\sigma_2^{\nu_2}\cdots\sigma_t^{\nu_t}$ at least one ν_i, say ν_1, is not divisible by m_1. Thus $C_1(\sigma) = \varepsilon_1^{\nu_1} \neq 1$.

Let A denote the set of those non-zero elements α of E for which $\alpha^r \in F$ and let F_1 denote the non-zero elements of F. It is obvious that A is a multiplicative group and that F_1 is a subgroup of A. Let A^r denote the set of rth powers of elements in A and F_1^r the set of rth powers of elements of F_1. The following theorem provides in most applications a convenient method for computing the group G.

THEOREM 24. *The factor groups* (A/F_1) *and* (A^r/F_1^r) *are isomorphic to each other and to the groups* G *and* X.

We map A on A^r, by making $\alpha \in A$ correspond to $\alpha^r \in A^r$. If $a^r \in F_1^r$, where $a \in F_1$ then $b \in A$ is mapped on a^r if and only if $b^r = a^r$, that is, if b is a solution to the equation $x^r - a^r = 0$. But $a, \varepsilon a, \varepsilon^2 a, \ldots, \varepsilon^{r-1}a$ are distinct solutions to this equation and since ε and a belong to F_1, it follows that b must be one of these elements and must belong to F_1. Thus, the inverse set in A of the subgroup F_1^r of A^r is F_1, so that the factor groups (A/F_1) and (A^r/F_1^r) are isomorphic.

If α is an element of A, then $(\alpha/\sigma(\alpha))^r = a^r/\sigma(\alpha^r) = 1$. Hence, $\alpha/\sigma(\alpha)$ is an rth root of unity and lies in F_1. By Theorem 22, $\alpha/\sigma(\alpha)$ defines a character $C(\sigma)$ of G in F. We map α on the corresponding character C. Each character C is by Theorem 22, image of some α. Moreover, $\alpha \cdot \alpha'$ is mapped on the character $C^*(\sigma) = \alpha\cdot\alpha'/\sigma(\alpha\cdot\alpha') = \alpha\cdot\alpha'/\sigma(\alpha)\cdot\sigma(\alpha') = C(\sigma)\cdot C'(\sigma) = C\cdot C'(\sigma)$, so that the mapping is a homomorphism. The kernel of this homomorphism is the set of those elements α for which $\alpha/\sigma(\alpha) = 1$ for each σ, hence is F_1. It follows, therefore, that (A/F_1) is isomorphic to X and hence also to G. In particular, (A/F_1) is a finite group.

We now prove the equivalence between Kummer fields and fields satisfying (a), (b) and (c) of Theorem 23.

THEOREM 25. *If* E *is an extension field over* F, *then* E *is a Kummer field if and only if* E *is normal, its group* G *is abelian and* F *contains a primitive* r^{th} *root* ε *of unity where* r *is the least common multiple of the orders of the elements of* G.

The necessity is already contained in Theorem 23. We prove the sufficiency. Out of the group A, let $\alpha_1 F_1, \alpha_2 F_1, \ldots, \alpha_t F_1$ be the cosets of F_1. Since $\alpha_i \in A$, we have $\alpha_i^r = a_i \in F$. Thus, α_i is a root of the equation $x^r - a_i = 0$ and since $\varepsilon\alpha_i, \varepsilon^2\alpha_i, \ldots, \varepsilon^{r-1}\alpha_i$, are also roots, $x^r - a_i$ must split in E. We prove that E is the splitting field of $(x^r - a_1)(x^r - a_2)\cdots(x^r - a_t)$ which will complete the proof of the theorem. To this end it suffices to show that $F(\alpha_1, \alpha_2, \ldots, \alpha_t) = E$.

Suppose that $F(\alpha_1, \alpha_2, \ldots, \alpha_t) \neq E$. Then $F(\alpha_1, \ldots, \alpha_t)$ is an intermediate field between F and E, and since E is normal over $F(\alpha_1, \ldots, \alpha_t)$ there exists an automorphism $\sigma \in G$, $\sigma \neq 1$, which leaves $F(\alpha_1, \ldots, \alpha_t)$ fixed. There exists a character C of G for which $C(\sigma) \neq 1$. Finally, there exists an element α in E such that $C(\sigma) = \alpha/\sigma(\alpha) \neq 1$. But $\alpha^r \in F_1$ by Theorem 22, hence $\alpha \in A$. Moreover, $A \subset F(\alpha_1, \ldots, \alpha_t)$ since all the cosets $\alpha_i F_1$ are contained in $F(\alpha_1, \ldots, \alpha_t)$. Since $F(\alpha_1, \ldots, \alpha_t)$ is by assumption left fixed by σ, $\sigma(\alpha) = \alpha$ which contradicts $\alpha/\sigma(\alpha) \neq 1$. It follows, therefore, that $F(\alpha_1, \ldots, \alpha_t) = E$.

COROLLARY. *If E is a normal extension of F, of prime order p, and if F contains a primitive p^{th} root of unity, then E is splitting field of an irreducible polynomial $x^p - a$ in F.*

E is generated by elements $\alpha_1, \ldots, \alpha_n$ where $\alpha_i^p \in F$. Let α_1 be not in F. Then $x^p - a$ is irreducible, for otherwise $F(\alpha_1)$ would be an intermediate field between F and E of degree less than p, and by the product theorem for the degrees, p would not be a prime number, contrary to assumption. $E = F(\alpha_1)$ is the splitting field of $x^p - a$.

M. Simple Extensions

We consider the question of determining under what conditions an extension field is generated by a *single element*, called a *primitive*. We prove the following

THEOREM 26. *A finite extension E of F is primitive over F if and only if there are only a finite number of intermediate fields.*

(a) Let $E = F(\alpha)$ and call $f(x) = 0$ the irreducible equation for α in F. Let B be an intermediate field and $g(x)$ the irreducible equation for α in B. The coefficients of $g(x)$ adjoined to F will generate a field B' between F and B. $g(x)$ is irreducible in B, hence also in B'. Since $E = B'(\alpha)$ we see $(E/B) = (E/B')$. This proves $B' = B$. So B is uniquely determined by the polynomial $g(x)$. But $g(x)$ is a divisor of $f(x)$, and there are only a finite number of possible divisors of $f(x)$ in E. Hence there are only a finite number of possible B's.

(b) Assume there are only a finite number of fields between E and F. Should F consist only of a finite number of elements, then E is generated by one element according to the Corollary on page 29. We may therefore assume F to contain an infinity of elements. We prove: To any two elements α, β there is a γ in E such that $F(\alpha, \beta) = F(\gamma)$. Let $\gamma = \alpha + a\beta$ with a in F but for the moment undetermined. Consider all the fields $F(\gamma)$ obtained in this way. Since we have an infinity of a's at our disposal, we can find two, say a_1 and a_2, such that the corresponding γ's, $\gamma_1 = \alpha + a_1\beta$ and $\gamma_2 = \alpha + a_2\beta$, yield the same field $F(\gamma_1) = F(\gamma_2)$. Since both γ_1 and γ_2 are in $F(\gamma_1)$, their difference (and therefore β) is in this field. Consequently also $\gamma_1 - a_1\beta = \alpha$. So $F(\alpha, \beta) \subset F(\gamma_1)$. Since $F(\gamma_1) \subset F(\alpha, \beta)$ our contention is proved. Select now η in E in such a way that $(F(\eta)/F)$ is as large as possible.

Every element ε of E must be in $F(\eta)$ or else we could find an element δ such that $F(\delta)$ contains both η and ε. This proves $E = F(\eta)$.

THEOREM 27. *If $E = F(\alpha_1, \alpha_2, \ldots, \alpha_n)$ is a finite extension of the field F, and $\alpha_1, \alpha_2, \ldots, \alpha_n$ are separable elements in E, then there exists a primitive θ in E such that $E = F(\theta)$.*

PROOF. Let $f_i(x)$ be the irreducible equation of α_i in F and let B be an extension of E that splits $f_1(x)f_2(x)\ldots f_n(x)$. Then B is normal over F and contains, therefore, only a finite number of intermediate fields (as many as there are subgroups of G). So the subfield E contains only a finite number of intermediate fields. Theorem 26 now completes the proof. □

N. Existence of a Normal Basis

The following theorem is true for any field though we prove it only in the case that F contains an infinity of elements.

THEOREM 28. *If E is a normal extension of F and $\sigma_1, \sigma_2, \ldots, \sigma_n$ are the elements of its group G, there is an element θ in E such that the n elements $\sigma_1(\theta), \sigma_2(\theta), \ldots, \sigma_n(\theta)$ are linearly independent with respect to F.*

According to Theorem 27 there is an α such that $E = F(\alpha)$. Let $f(x)$ be the equation for α, put $\sigma_i(a) = \alpha_i$, $g(x) = \frac{f(x)}{(x-\alpha)f'(\alpha)}$ and $g_i(x) = \sigma_i(g(x)) = \frac{f(x)}{(x-\alpha_i)f'(\alpha_i)}$. $g_i(x)$ is a polynomial in E having α_k as root for $k \neq i$ and hence

$$g_i(x)g_k(x) = 0 \pmod{f(x)} \quad \text{for } i \neq k. \tag{1}$$

In the equation

$$g_1(x) + g_2(x) + \cdots + g_n(x) - 1 = 0 \tag{2}$$

the left side is of degree at most $n-1$. If (2) is true for n different values of x, the left side must be identically 0. Such n values are $\alpha_1, \alpha_2, \ldots, \alpha_n$, since $g_i(\alpha_i) = 1$ and $g_k(\alpha_i) = 0$ for $k \neq i$.

Multiplying (2) by $g_i(x)$ and using (1) shows:

$$(g_i(x))^2 \equiv g_i(x) \pmod{f(x)}. \tag{3}$$

We next compute the determinant

$$D(x) = |\sigma_i\sigma_k(g(x))| \quad i, k = 1, 2, \ldots, n \tag{4}$$

and prove $D(x) \neq 0$. If we square it by multiplying column by column and compute its value (mod $f(x)$) we get from (1), (2), (3) a determinant that has 1 in the diagonal and 0 elsewhere. So

$$(D(x))^2 = 1 \pmod{f(x)}.$$

$D(x)$ can have only a finite number of roots in F. Avoiding them we can find a value a for x such that $D(a) \neq 0$. Now set $\theta = g(a)$. Then the determinant

$$|\sigma_i\sigma_k(\theta)| \neq 0. \tag{5}$$

Consider any linear relation $x_1\sigma_1(\theta) + x_2\sigma_2(\theta) + \cdots + x_n\sigma_n(\theta) = 0$ where the x_i are in F. Applying the automorphism σ_i to it would lead to n homogeneous equations for the n unknowns x_i. (5) shows that $x_i = 0$ and our theorem is proved.

O. Theorem on Natural Irrationalities

Let F be a field, $p(x)$ a polynomial in F whose irreducible factors are separable, and let E be a splitting field for $p(x)$. Let B be an arbitrary extension of F, and let us denote by EB the splitting field of $p(x)$ when $p(x)$ is taken to lie in B. If $\alpha_1, \ldots, \alpha_s$ are the roots of $p(x)$ in EB, then $F(\alpha_1, \ldots, \alpha_s)$ is a subfleld of EB which is readily seen to form a splitting field for $p(x)$ in F. By Theorem 10, E and $F(\alpha_1, \ldots, \alpha_s)$ are isomorphic. There is therefore no loss of generality if in the sequel we take $E = F(\alpha_1, \ldots, \alpha_s)$ and assume therefore that E is a subfield of EB. Also, $EB = B(\alpha_1, \ldots, \alpha_s)$.

Let us denote by $E \cap B$ the intersection of E and B. It is readily seen that $E \cap B$ is a field and is intermediate to F and E.

THEOREM 29. *If G is the group of automorphisms of E over F, and H the group of EB over B, then H is isomorphic to the subgroup of G having $E \cap B$ as its fixed field.*

Each automorphism of EB over B simply permutes $\alpha_1, \ldots, \alpha_s$ in some fashion and leaves B, and hence also F, fixed. Since the elements of EB are quotients of polynomial expressions in $\alpha_1, \ldots, \alpha_s$ with coefficients in B, the automorphism is completely determined by the permutation it effects on $\alpha_1, \ldots, \alpha_s$. Thus, each automorphism of EB over B defines an automorphism of $E = F(\alpha_1, \ldots, \alpha_s)$ which leaves F fixed. Distinct automorphisms, since $\alpha_1, \ldots, \alpha_s$ belong to E, have different effects on E. Thus, the group H of EB over B can be considered as a subgroup of the group G of E over F. Each element of H leaves $E \cap B$ fixed since it leaves even all of B fixed. However, any element of E which is not in $E \cap B$ is not in B, and hence would be moved by at least one automorphism of H. It follows that $E \cap B$ is the fixed field of H.

COROLLARY. *If, under the conditions of Theorem 29, the group G is of prime order, then either $H = G$ or H consists of the unit element alone.*

CHAPTER III

APPLICATIONS
by A. N. Milgram

A. Solvable groups

Before proceeding with the applications we must discuss certain questions in the theory of groups. We shall assume several simple propositions: (a) if N is a normal subgroup of the group G, then the mapping $f(x) = xN$ is a homomorphism of G on the factor group G/N. f is called the natural homomorphism. (b) The image and the inverse image of a normal subgroup under a homomorphism is a normal subgroup. (c) If f is a homomorphism of the group G on G', then setting $N' = f(N)$, and defining the mapping g as $g(xN) = f(x)N'$, we readily see that g is a homomorphism of the factor group G/N on the factor group G'/N'. Indeed, if N is the inverse image of N' then g is an isomorphism.

We now prove

THEOREM 1 (Zassenhaus). *If U and V are subgroups of G, u and v normal subgroups of U and V, respectively, then the following three factor groups are isomorphic: $u(U \cap V)/u(U \cap v)$, $v(U \cap V)/v(u \cap V)$, $(U \cap V)/(u \cap V)(v \cap U)$.*

It is obvious that $U \cap v$ is a normal subgroup of $U \cap V$. Let f be the natural mapping of U on U/u. Call $f(U \cap V) = H$ and $f(U \cap v) = K$. Then $f^{-1}(H) = u(V \cap V)$ and $f^{-1}(K) = u(U \cap v)$ from which it follows that $u(U \cap V)/u(U \cap v)$ is isomorphic to H/K. If, however, we view f as defined only over $U \cap V$, then $f^{-1}(K) = [u \cap (U \cap V)](U \cap v) = (u \cap V)(V \cap v)$ so that $(U \cap V)/(u \cap V)(U \cap v)$ is also isomorphic to H/K. Thus the first and third of the above factor groups are isomorphic to each other. Similarly, the second and third factor groups are isomorphic.

COROLLARY 1. *If H is a subgroup and N a normal subgroup of the group G, then $H/H \cap N$ is isomorphic to HN/N, a subgroup of G/H.*

PROOF. Set $G = U$, $N = u$, $H = V$ and the identity $1 = v$ in Theorem 1. □

COROLLARY 2. *Under the conditions of Corollary 1, if G/N is abelian, so also is $H/H \cap N$.*

Let us call a group G *solvable* if it contains a sequence of subgroups $G = G_0 \supset G_1 \supset \cdots \supset G_s = 1$, each a normal subgroup of the preceding, and with G_{i-1}/G_i abelian.

THEOREM 2. *Any subgroup of a solvable group is solvable.*

For let H be a subgroup of G, and call $H_i = H \cap G_i$. Then that H_{i-1}/H_i is abelian follows from Corollary 2 above, where G_{i-1}, G_i and H_{i-1} play the role of G, N and H.

THEOREM 3. *The homomorph of a solvable group is solvable.*

Let $f(G) = G'$, and define $G'_i = f(G_i)$ where G_i belongs to a sequence exhibiting the solvability of G. Then by (c) there exists a homomorphism mapping G_{i-1}/G_i on G'_{i-1}/G'_i. But the homomorphic image of an abelian group is abelian so that the groups G'_i exhibit the solvability of G'.

B. Permutation Groups

Any one to one mapping of a set of n objects on itself is called a *permutation.* The iteration of two such mappings is called their *product.* It may be readily verified that the set of all such mappings forms a group in which the unit is the identity map. The group is called the *symmetric* group on n letters.

Let us for simplicity denote the set of n objects by the numbers $1, 2, \ldots, n$. The mapping S such that $S(i) = i + 1 \mod n$ will be denoted by $(123\ldots n)$ and more generally $(ij\ldots m)$ will denote the mapping T such that $T(i) = j, \ldots, T(m) = i$. If $(ij\ldots m)$ has k numbers, then it will be called a k cycle. It is clear that if $T = (ij\ldots s)$ then $T^{-1} = (s\ldots ji)$.

We now establish the

LEMMA. *If a subgroup U of the symmetric group on n letters $(n > 4)$ contains every 3-cycle, and if u is a normal subgroup of U such that U/u is abelian, then u contains every 3-cycle.*

PROOF. Let f be the natural homomorphism $f(U) = U/u$ and let $x = (ijk)$, $y = (krs)$ be two elements of U, where i, j, k, r, s are 5 numbers. Then, since U/u is abelian, setting $f(x) = x'$, $f(y) = y'$ we have $f(x^{-1}y^{-1}xy) = {x'}^{-1}{y'}^{-1}x'y' = 1$, so that $x^{-1}y^{-1}xy \in u$. But $x^{-1}y^{-1}xy = (kji)\cdot(srk)\cdot(ijk)\cdot(krs) = (kjs)$ and for each k, j, s we have $(kjs) \in u$. □

THEOREM 4. *The symmetric group G on n letters is not solvable for $n > 4$.*

If there were a sequence exhibiting the solvability, since G contains every 3-cycle, so would each succeeding group, and the sequence could not end with the unit.

C. Solution of Equations by Radicals

The extension field E over F is called an *extension by radicals* if there exist intermediate fields $B_1, B_2, \ldots, B_r = E$ and $B_i = B_{i-1}(\alpha_i)$ where each α_i is a root of an equation of the form $x^{n_i} - a_i = 0$, $a_i \in B_{i-1}$. A polynomial $f(x)$ in a field F is said to be *solvable by radicals* if its splitting field lies in an extension by radicals. We assume unless otherwise specified that the base field has characteristic 0 and that F contains as many roots of unity as are needed to make our subsequent statements valid.

Let us remark first that any extension of F by radicals can always be extended to an extension of F by radicals which is normal over F. Indeed B_1 is a normal extension of B_0 since it contains not only α_1, but $\varepsilon\alpha_1$, where ε is any n_1-root of unity, so that B_1 is the splitting field of $x^{n_1} - a_1$. If $f_1(x) = \prod_\sigma(x^{n_2} - \sigma(a_2))$, where σ takes all values in the group of automorphisms of B_1 over B_0, then f_1 is in B_0, and adjoining successively the roots of $x^{n_2} - \sigma(a_2)$ brings us to an extension of B_2 which is normal over F. Continuing in this way we arrive at an extension of E by radicals which will be normal over F. We now prove the

THEOREM 5. *The polynomial $f(x)$ is solvable by radicals if and only if its group is solvable.*

Suppose $f(x)$ is solvable by radicals. Let E be a normal extension of F by radicals containing the splitting field B of $f(x)$, and call G the group of E over F. Since for each i, B_i is a Kummer extension of B_{i-1},[1] the group of B_i over B_{i-1} is abelian. In the sequence of groups $G = G_{B_0} \supset G_{B_1} \supset \cdots \supset G_{B_r} = 1$ each is a normal subgroup of the preceding since $G_{B_{i-1}}$ is the group of E over B_{i-1} and B_i is a normal extension of B_{i-1}. But $G_{B_{i-1}}/G_{B_i}$ is the group of B_i over B_{i-1}, and hence is abelian. Thus G is solvable. However, G_B is a normal subgroup of G, and G/G_B is the group of B over F, and is therefore the group of the polynomial $f(x)$. But G/G_B is a homomorph of the solvable group G and hence is itself solvable.

On the other hand, suppose the group G of $f(x)$ to be solvable and let E be the splitting field. Let $G = G_0 \supset G_1 \supset \cdots \supset G_r = 1$ be a sequence with abelian factor groups. Call B_i the fixed field for G_i. Since G_{i-1} is the group of E over B_{i-1} and G_i is a normal subgroup of G_{i-1}, then B_i is normal over B_{i-1} and the group G_{i-1}/G_i is abelian. Thus B_i is a Kummer extension of B_{i-1}, hence is splitting field of a polynomial of the form $(x^n - a_1)(x^n - a_2)\cdots(x^n - a_s)$ so that by forming the successive splitting fields of the $x^n - a_k$ we see that B_i is an extension of B_{i-1} by radicals, from which it follows that E is an extension by radicals.

REMARK. The assumption that F contains roots of unity is not necessary in the above theorem. For if $f(x)$ has a solvable group G, then we may adjoin to F a primitive n^{th} root of unity, where n is, say, equal to the order of G. The group of $f(x)$ when considered as lying in F' is, by the theorem on Natural Irrationalities, a subgroup G' of G, and hence is solvable. Thus the splitting field over F' of $f(x)$ can be obtained by radicals. Conversely, if the splitting field E over F of $f(x)$ can be obtained by radicals, then by adjoining a suitable root of unity E is extended to E' which is still normal over F'. But E' could be obtained by adjoining first the root of unity, and then the radicals, to F; F would first be extended to F' and then F' would be extended to E'. Calling G the group of E' over F and G' the group of E' over F', we see that G' is solvable and G/G' is the group of F' over F and hence abelian. Thus G is solvable. The factor group G/G_E is the group of $f(x)$ and being a homomorph of a solvable group is also solvable.

D. The General Equation of Degree n

If F is a field, the collection of rational expressions in the variables $u_1, u_2, \ldots, u_n$ with coefficients in F is a field $F(u_1, u_2, \ldots, u_n)$. By the *general equation of degree n* we mean the equation

$$f(x) = x^n - u_1 x^{n-1} + u_2 x^{n-2} - \cdots + (-1)^n u_n. \tag{1}$$

Let E be the splitting field of $f(x)$ over $F(u_1, u_2, \ldots, u_n)$. If $v_1, v_2, \ldots, v_n$ are the roots of $f(x)$ in E, then $u_1 = v_1 + v_2 + \cdots + v_n$, $u_2 = v_1v_2 + v_1v_3 + \cdots + v_{n-1}v_n$, $\ldots$, $u_n = v_1 \cdot v_2 \cdot \ldots \cdot v_n$.

We shall prove that the group of E over $F(u_1, u_2, \ldots, u_n)$ is the symmetric group.

Let $F(x_1, x_2, \ldots, x_n)$ be the field generated from F by the variables $x_1, x_2, \ldots,$ x_n. Let $\alpha_1 = x_1 + x_2 + \cdots + x_n$, $\alpha_2 = x_1x_2 + x_1x_3 + \cdots + x_{n-1}x_n$, $\ldots$, $\alpha_n =$

[1] The reader should observe that the B_i are subfields of E, not of B. *Editor's note.*

$x_1 \cdot x_2 \cdot \ldots \cdot x_n$ be the elementary symmetric functions, i.e., $(x - x_1)(x - x_2)\cdots(x - x_n) = x^n - \alpha_1 x^{n-1} + \cdots + (-1)^n \alpha_n = f^*(x)$. If $g(\alpha_1, \alpha_2, \ldots, \alpha_n)$ is a polynomial in $\alpha_1, \alpha_2, \ldots, \alpha_n$, then $g(\alpha_1, \alpha_2, \ldots, \alpha_n) = 0$ only if g is the zero polynomial. For if $g(\sum x_i, \sum x_i x_k, \ldots) = 0$, then this relation would hold also if the x_i were replaced by the v_i. Thus, $g(\sum v_i, \sum v_i v_k, \ldots) = 0$ or $g(u_1, u_2, \ldots, u_n) = 0$ from which it follows that g is identically zero.

Between the subfield $F(\alpha_1, \ldots, \alpha_n)$ of $F(x_1, \ldots, x_n)$ and $F(u_1, u_2, \ldots, u_n)$ we set up the following correspondence: Let $f(u_1, \ldots, u_n)/g(u_1, \ldots, u_n)$ be an element of $F(u_1, \ldots, u_n)$. We make this correspond to $f(\alpha_1, \ldots, \alpha_n)/g(\alpha_1, \ldots, \alpha_n)$. This is clearly a mapping of $F(u_1, u_2, \ldots, u_n)$ on all of $F(\alpha_1, \ldots, \alpha_n)$. Moreover, if $f(\alpha_1, \alpha_2, \ldots, \alpha_n)/g(\alpha_1, \alpha_2, \ldots, \alpha_n) = f_1(\alpha_1, \alpha_2, \ldots, \alpha_n)/g_1(\alpha_1, \alpha_2, \ldots, \alpha_n)$, then $fg_1 - gf_1 = 0$. But this implies by the above that $f(u_1, \ldots, u_n) \cdot g_1(u_1, \ldots, u_n) - g(u_1, \ldots, u_n) \cdot f_1(u_1, \ldots, u_n) = 0$ so that $f(u_1, \ldots, u_n)/g(u_1, u_2, \ldots, u_n) = f_1(u_1, \ldots, u_n)/g_1(u_1, u_2, \ldots, u_n)$. It follows readily from this that the mapping of $F(u_1, u_2, \ldots, u_n)$ on $F(\alpha_1, \alpha_2, \ldots, \alpha_n)$ is an isomorphism. But under this correspondence $f(x)$ corresponds to $f^*(x)$. Since E and $F(x_1, x_2, \ldots, x_n)$ are respectively splitting fields of $f(x)$ and $f^*(x)$, by Theorem 10 the isomorphism can be extended to an isomorphism between E and $F(x_1, x_2, \ldots, x_n)$. Therefore, the group of E over $F(u_1, u_2, \ldots, u_n)$ is isomorphic to the group of $F(x_1, x_2, \ldots, x_n)$ over $F(\alpha_1, \alpha_2, \ldots, \alpha_n)$.

Each permutation of $x_1, x_2, \ldots, x_n$ leaves $\alpha_1, \alpha_2, \ldots, \alpha_n$ fixed and, therefore, induces an automorphism of $F(x_1, x_2, \ldots, x_n)$ which leaves $F(\alpha_1, \alpha_2, \ldots, \alpha_n)$ fixed. Conversely, each automorphism of $F(x_1, x_2, \ldots, x_n)$ which leaves $F(\alpha_1, \ldots, \alpha_n)$ fixed must permute the roots $x_1, x_2, \ldots, x_n$ of $f^*(x)$ and is completely determined by the permutation it effects on $x_1, x_2, \ldots, x_n$. Thus, the group of $F(x_1, x_2, \ldots, x_n)$ over $F(\alpha_1, \alpha_2, \ldots, \alpha_n)$ is the symmetric group on n letters. Because of the isomorphism between $F(x_1, \ldots, x_n)$ and E, the group for E over $F(u_1, u_2, \ldots, u_n)$ is also the symmetric group. If we remark that the symmetric group for $n > 4$ is not solvable, we obtain from the theorem on solvability of equations the famous theorem of Abel:

THEOREM 6. *The group of the general equation of degree n is the symmetric group on n letters. The general equation of degree n is not solvable by radicals if $n > 4$.*

E. Solvable Equations of Prime Degree

The group of an equation can always be considered as a permutation group. If $f(x)$ is a polynomial in a field F, let $\alpha_1, \alpha_2, \ldots, \alpha_n$ be the roots of $f(x)$ in the splitting field $E = F(\alpha_1, \ldots, \alpha_n)$. Then each automorphism of E over F maps each root of $f(x)$ into a root of $f(x)$, that is, permutes the roots. Since E is generated by the roots of $f(x)$, different automorphisms must effect distinct permutations. Thus, the group of E over F is a permutation group acting on the roots $\alpha_1, \alpha_2, \ldots, \alpha_n$ of $f(x)$.

For an irreducible equation this group is always *transitive.* For let α and α' be any two roots of $f(x)$, where $f(x)$ is assumed irreducible. $F(\alpha)$ and $F(\alpha')$ are isomorphic where the isomorphism is the identity on F, and this isomorphism can be extended to an automorphism of E (Theorem 10). Thus, there is an automorphism sending any given root into any other root, which establishes the "transitivity" of the group.

A permutation σ of the numbers $1, 2, \ldots, q$ is called a *linear substitution* modulo q if there exists a number $b \not\equiv 0$ modulo q such that $\sigma(i) \equiv bi + c \pmod{q}$, $i = 1, 2, \ldots, q$.

THEOREM 7. *Let $f(x)$ be an irreducible equation of prime degree q in a field F. The group G of $f(x)$ (which is a permutation group of the roots, or the numbers $1, 2, \ldots, q$) is solvable if and only if, after a suitable change in the numbering of the roots, G is a group of linear substitutions modulo q, and in the group G all the substitutions with $b = 1$, $\sigma(i) \equiv i+$ $(c = 1, 2 \ldots, q)$ occur.*

Let G be a transitive substitution group on the numbers $1, 2, \ldots, q$ and let G_1 be a normal subgroup of G. Let $1, 2, \ldots, k$ be the images of 1 under the permutations of G_1; we say $1, 2, \ldots, k$ is a *domain of transitivity* of G_1. If $i \leq q$ is a number not belonging to this domain of transitivity, there is a $\sigma \in G$ which maps 1 on i. Then $\sigma(1, 2, \ldots, k)$ is a domain of transitivity of $\sigma G_1 \sigma^{-1}$. Since G_1 is a normal subgroup of G, we have $G_1 = \sigma G_1 \sigma^{-1}$. Thus, $\sigma(1, 2, \ldots, k)$ is again a domain of transitivity of G_1 which contains the integer i and has k elements. Since i was arbitrary, the domains of transitivity of G_1 all contain k elements. Thus, the numbers $1, 2, \ldots, q$ are divided into a collection of mutually exclusive sets, each containing k elements, so that k is a divisor of q. Thus, in case q is a prime, either $k = 1$ (and then G_1 consists of the unit alone) or $k = q$ and G_1 is also transitive.

To prove the theorem, we consider the case in which G is solvable. Let $G = G_0 \supset G_1 \supset \cdots \supset G_{s+1} = 1$ be a sequence exhibiting the solvability. Since G_s is abelian, choosing a cyclic subgroup of it would permit us to assume the term before the last to be cyclic, i.e., G_s is cyclic. If σ is a generator of G_s, σ must consist of a cycle containing all q of the numbers $1, 2, \ldots, q$ since in any other case G_s would not be transitive [if $\sigma = (1ij \ldots m)(n \ldots p) \ldots$ then the powers of σ would map 1 only into $1, l, j \ldots m$, contradicting the transitivity of G_s]. By a change in the numbering of the permutation letters, we may assume

$$\sigma(i) \equiv i + 1 \pmod{q}$$

$$\sigma^c(i) \equiv i + c \pmod{q}$$

How let τ be any element of G_{s-1}. Since G_s is a normal subgroup of G_{s-1}, $\tau\sigma\tau^{-1}$ is an element of G_s, say $\tau\sigma\tau^{-1} = \sigma^b$. Let $\tau(i) = j$ or $\tau^{-1}(j) = i$, then $\tau\sigma\tau^{-1}(j) = \sigma^b(j) \equiv j+b \pmod{q}$. Therefore, $\tau\sigma(i) \equiv \tau(i)+b \pmod{q}$ or $\tau(i+1) \equiv \tau(i) + b$ for each i. Thus, setting $\tau(0) = c$, we have $\tau(1) \equiv c + b$, $\tau(2) \equiv \tau(1) + b = c + 2b$ and in general $\tau(i) \equiv c + ib \pmod{q}$. Thus, each substitution in G_{s-1} is a linear substitution. Moreover, the only elements of G_{s-1} which leave no element fixed belong to G_s, since for each $a \not\equiv 1$, there is an i such that $ai + b \equiv i \pmod{q}$ [take i such that $(a-1)i \equiv -b$] .

We prove by an induction that the elements of G are all linear substitutions, and that the only cycles of q letters belong to G_s. Suppose the assertion true of G_{s-n}. Let $\tau \in G_{s-n-1}$ and let σ be a cycle which belongs to G_s (hence also to G_{s-n}). Since the transform of a cycle is a cycle, $\tau^{-1}\sigma\tau$ is a cycle in G_{s-n} and hence belongs to G_s. Thus $\tau^{-1}\sigma\tau = \sigma^b$ for some b. By the argument in the preceding paragraph, τ is a linear substitution $bi + c$ and if τ itself does not belong to G_s, then τ leaves one integer fixed and hence is not a cycle of q elements.

We now prove the second half of the theorem. Suppose G is a group of linear substitutions which contains a subgroup N of the form $\sigma(i) = i + c$. Since the only linear substitutions which do not leave an integer fixed belong to N, and since the transform of a cycle of q elements is again a cycle of q elements, N is a normal subgroup of G. In each coset $N \cdot \tau$ where $\tau(i) \equiv bi + c$ the substitution $\sigma^{-1}\tau$ occurs, where $\sigma \equiv i + c$. But $\sigma^{-1}\tau(i) \equiv (bi + c) - c = bi$. Moreover, if $\tau(i) \equiv bi$ and $\tau'(i) \equiv b'i$ then $\tau\tau'(i) \equiv bb'i$. Thus, the factor group (G/N) is isomorphic to a multiplicative subgroup of the numbers $1, 2, \ldots, q-1 \mod q$ and is therefore abelian. Since (G/N) and N are both abelian, G is solvable.

COROLLARY 1. *If G is a solvable transitive substitution group on q letters (q prime), then the only substitution of G which leaves two or more letters fixed is the identity.*

This follows from the fact that each substitution is linear modulo q and $bi+c \equiv i \pmod{q}$ has either no solution ($b \equiv 1$, $c \not\equiv 0$) or exactly solution ($b \not\equiv 1$) unless $b \equiv 1$, $c \equiv 0$ in which case the substitution is the identity.

COROLLARY 2. *A solvable, irreducible equation of prime degree in a field which is a subset of the real numbers has either one real root or all its roots are real.*

The group of the equation is a solvable transitive substitution group on q (prime) letters. In the splitting field (contained in the field of complex numbers) the automorphism which maps a number into its complex conjugate would leave fixed all the real numbers. By Corollary 1, if two roots are left fixed, then all the roots are left fixed, so that if the equation has two real roots all its roots are real.

F. Ruler and Compass Constructions

Suppose there is given in the plane a finite number of elementary geometric figures, that is, points, straight lines and circles. We seek to construct others which satisfy certain conditions in terms of the given figures.

Permissible steps in the construction will entail the choice of an arbitrary point interior to a given region, drawing a line through two points and a circle with given center and radius, and finally intersecting pairs of lines, or circles, or a line and circle.

Since a straight line, or a line segment, or a circle is determined by two points, we can consider ruler and compass constructions as constructions of points from given points, subject to certain conditions.

If we are given two points we may join them by a line, erect a perpendicular to this line at, say, one of the points and, taking the distance between the two points to be the unit, we can with the compass lay off any integer n on each of the lines. Moreover, by the usual method, we can draw parallels and can construct m/n. Using the two lines as axes of a cartesian coordinate system, we can with ruler and compass construct all points with rational coordinates.

If $a, b, c, \ldots$ are numbers involved as coordinates of points which determine the figures given, then the sum, product, difference and quotient of any two of these numbers can be constructed. Thus, each element of the field $R(a, b, c, \ldots)$ which they generate out of the rational numbers can be constructed.

It is required that an arbitrary point is any point of a given region. If a construction by ruler and compass is possible, we can always choose our arbitrary points as points having rational coordinates. If we join two points with coefficients in $R(a, b, c, \ldots)$ by a line, its equation will have coefficients in $R(a, b, c, \ldots)$ and the intersection of two such lines will be a point with coordinates in $R(a, b, c, \ldots)$.

The equation of a circle will have coefficients in the field if the circle passes through three points whose coordinates are in the field or if its center and one point have coordinates in the field. However, the coordinates of the intersection of two such circles, or a straight line and circle, will involve square roots.

It follows that if a point can be constructed with a ruler and compass, its coordinates must be obtainable from $R(a, b, c, \dots)$ by a formula only involving square roots, that is, its coordinates will lie in a field $R_s \supset R_{s-1} \supset \cdots \supset R_1 = R(a, b, c, \dots)$ where each field R_i is splitting field over R_{i-1} of a quadratic equation $x^2 - \alpha = 0$. It follows (Theorem 6, p. 13) since either $R_i = R_{i-1}$ or $(R_i/R_{i-1}) = 2$, that (R_s/R_1) is a power of two. If x is the coordinate of a constructed point, then $(R_1(x)/R_1) \cdot (R_s/R_1(x)) = (R_s/R_1) = 2^\nu$ so that $R_1(x)/R_1$ must also be a power of two.

Conversely, if the coordinates of a point can be obtained from $R(a, b, c, \dots)$ by a formula involving square roots only, then the point can be constructed by ruler and compass. For, the field operations of addition, subtraction, multiplication and division may be performed by ruler and compass constructions and, also, square roots using $1 : r = r : r_1$ to obtain $r = \sqrt{r_1}$, may be performed by means of ruler and compass constructions,

As an illustration of these considerations, let us show that it is impossible to trisect an angle of 60°. Suppose we have drawn the unit circle with center at the vertex of the angle, and set up our coordinate system with X-axis as a side of the angle and origin at the vertex.

Trisection of the angle would be equivalent to the construction of the point $(\cos 20°, \sin 20°)$ on the unit circle. From the equation $\cos 3\theta = 4\cos^3\theta - 3\cos\theta$, the abscissa would satisfy $4x^3 - 3x = 1/2$. The reader may readily verify that this equation has no rational roots, and is therefore irreducible in the field of rational numbers. But since we may assume only a straight line and unit length given, and since the 60° angle can be constructed, we may take $R(a, b, c, \dots)$ to be the field R of rational numbers. A root α of the irreducible equation $8x^3 - 6x - 1 = 0$ is such that $(R(\alpha)/R) = 3$, and not a power of two.

THEORY OF ALGEBRAIC NUMBERS

Emil Artin

Notes by Gerhard Wurges from lectures held at the Mathematisches Institut, Göttingen, Germany in the Winter Semester, 1956/7. Translated by George Striker.

Notation

$A = \{a \mid \Delta_1, \Delta_2, \ldots\}$	The set of all objects a, with the properties $\Delta_1, \Delta_2, \ldots$
$a \in A$, $a \notin A$	a is, is not, an element of A, respectively.
$A \subset B$, $A \not\subset B$	A is, is not, contained in B, respectively.
$A \cup B$	$\{c \mid c \in A \text{ or } c \in B\}$ (The union of A and B)
$A \cap B$	$\{c \mid c \in A \text{ and } c \in B\}$ (The intersection of A and B)
$\varnothing$	The empty set.
$f\colon A \to B$, or $A \xrightarrow{f} B$	f maps A into B.
$f(C)$	$\{f(c) \mid c \in C\}$ (defined for $C \subset A$)
$f^{-1}(D)$	$\{a \mid a \in A,\ f(a) \in D\}$ (defined for $D \subset A$)
$f\|C$	Limitation of f onto C (defined for $C \subset A$)
$\Delta_1 \implies \Delta_2$	Δ_1 implies Δ_2.
$\Delta_1 \iff \Delta_2$	Δ_1 equivalent with Δ_2, Δ_1 if and only if Δ_2.

Furthermore, the following notation will be used throughout where no other explicity meaning is given for these symbols:

$\mathbb{Q}$: The field of rational numbers
$\mathbb{Z}$: The ring of rational integers
$\mathbb{R}$: The field of real numbers
$\mathbb{C}$: The field of complex numbers.

CHAPTER 1

SET-THEORETICAL PRELIMINARIES

1. Maps of Sets

A map

$$f\colon A \to B$$

of a set A into a set B is a function, defined on A, such that for every $a \in A$ (the *inverse image*), there exists a unique image, an element $b = f(a) \in B$.

If every element of B appears as the image of at least one element $a \in A$, then we say f is a map of A *onto* B; if every element of B is the image of at most one element of A, then we say f is a *one-to-one map* of A into B (or, as the case may be, of A onto B).

Two maps:

$$f\colon A \to B \qquad g\colon B \to C$$

may be combined into a single map, $h = gf$ of A into C, "*the product of f with g*" by the operation

$$h(a) = g(f(a)) \quad \text{for all } a \in A.$$

On a set A we can define an *equivalence relation* R: Among certain elements a relationship $a_1 \sim a_2$ (read a_1 is equivalent to a_2) exists, satisfying

(1) $a \sim a$ for all $a \in A$ (Reflexitivity)
(2) $a_1 \sim a_2 \implies a_2 \sim a_1$ (Symmetry)
(3) $a_1 \sim a_2$, $a_2 \sim a_3 \implies a_1 \sim a_3$ (Transitivity).

Obviously, the elements of A can now be separated into disjoint equivalence classes of elements equivalent modulo R. The totality of the classes is called the *quotient set* A/R of A modulo R.

The map of A onto A/R, which maps every element of A into the equivalence class to which it belongs is called the *canonical map* of A onto A/R.

Every map $f\colon A \to B$ of A into B can be split into three maps (*canonical maps*) by the following procedure:

In A, let an equivalence relation R be defined by

$$a_1 \sim a_2 \iff f(a_1) = f(a_2)$$

Let f_1 be the canonical map of A onto A/R

$$f_1(a) = \{a' \mid a' \in A,\ f(a') = f(a)\}$$

Let the map f_2 order to every class of A/R the element $f(a)$ in $f(A)$, where a is taken as any representative of this class. This definition does not depend on the choice of the representative, f is a one-to-one map of A/R onto $f(A)$. f_3 remains, finally, as the injection of $f(A)$ into B, i.e. the map of $f(A)$ into B by which each element of $f(A)$ is mapped onto itself. We then have

$$f = f_3 f_2 f_1 \quad A \xrightarrow{f_1} A/R \xrightarrow{f_2} f(A) \xrightarrow{f_3} B$$

The important part of the map is f_1; in many cases the maps f_2 and f_3 may be neglected completely.

2. Ordered Sets

Let a relation $a \leq b$ be defined for the elements of a set A, where the following axioms are satisfied.

(1) $a \leq b,\ b \leq a \iff a = b$
(2) $a \leq b,\ b \leq c \implies a \leq c$
(3) for any two elements $a, b \in A$, either $a \leq b$ or $b \leq a$.

Such a set A is termed *ordered* (completely ordered). If the ordering relation satisfies only axioms (1) and (2) the set A is *partially ordered.*

A *maximal element* of a partially ordered set A is an element $a' \in A$ such, that there exists no $a \in A$ for which $a' < a$ (i.e. $a' \leq a,\ a' \neq a$),

An *upper bound* of a subset B of the partially ordered set A is an element $a \in A$ with the property $b \leq a$ for all $b \in B$.

Maximal elements and upper bounds need not exist for every subset.

A partially ordered set A is called *inductively ordered* if every (completely) ordered subset of A has an upper bound in A.

We introduce the following axiom for sets (Zorn's Lemma);

AXIOM. Every non-empty inductively ordered set has a maximal element.

This axiom is equivalent to the Axiom of Choice or the Well-ordering Theorem of the theory of sets.

If x is an element of the inductively ordered set A, then the set

$$\{y \mid y \in A,\ x \leq y\}$$

is also inductively ordered. Thus the axiom introduced above implies the

COROLLARY. *In a non-empty inductively ordered set A, for every element $x \in A$ there exists a maximal element x' of A satisfying $x \leq x'$.*

It is in this form that we will mostly apply our axiom.

3. Ideals in Commutative Rings

We consider the set of ideals of a commutative ring, $\mathfrak{o}$. Letting S be a semigroup of elements of the ring $\mathfrak{o}$, i.e. a subset of $\mathfrak{o}$ closed under multiplication, we assume the existence of an ideal $\mathfrak{a}_0$ of $\mathfrak{o}$ so that

$$\mathfrak{a}_0 \cap S = \varnothing$$

(If $0 \notin S$ we can set $\mathfrak{a}_0 = (0)$.)

The set A of all ideals of $\mathfrak{o}$ satisfying

$$\mathfrak{a}_0 \cap S = \varnothing, \tag{1}$$

being non-empty can now be partially ordered by the set-theoretical containing relation "$\subset$", in fact to an inductive ordering.

This can be shown by letting $\{\mathfrak{a}_\alpha\}$ be any (completely) ordered set of ideals satisfying (1), so that for any two ideals $\mathfrak{a}_\alpha$ and $\mathfrak{a}_\beta$ either $\mathfrak{a}_\alpha \subset \mathfrak{a}_\beta$ or $\mathfrak{a}_\beta \subset \mathfrak{a}_\alpha$. Now, $\mathfrak{b} = \bigcup_\alpha \mathfrak{a}_\alpha$ is an ideal in $\mathfrak{o}$, and satisfies (1), hence $\mathfrak{b} \in A$. But $\mathfrak{b}$ is an upper bound for the subset $\{\mathfrak{a}_\alpha\}$ of A,

Let the ideal $\mathfrak{p}$ be a maximal element of the set A of ideals, the existence of which we have by the axiom. $\mathfrak{p}$ is a *prime ideal.*

Assume the contrary, i.e. there exist elements $a, b \in \mathfrak{o}$ with $a \notin \mathfrak{p}$, $b \notin \mathfrak{p}$, but $ab \in \mathfrak{p}$. Then $\mathfrak{p}$ is non-trivial subset of both $(\mathfrak{p}, a)$ and $(\mathfrak{p}, b)$, which implies neither of these sets satisfy (1), since $\mathfrak{p}$ is maximal. Hence there exist $s_1, s_2 \in S$

$$\left.\begin{aligned} s_1 &= p_1 + ac_1 + an_1 \\ s_2 &= p_2 + bc_2 + an_2 \end{aligned}\right\} \quad n_i \text{ are rational integers}, \quad p_i \in \mathfrak{p},\ c_i \in \mathfrak{o} (i = 1, 2)$$

Then $s_1 s_2 \in \mathfrak{p}$, but $s_1 s_2 \in S$, which contradicts (1).

Let R be the intersection of all prime ideals of $\mathfrak{o}$:

$$R = \bigcap_{\mathfrak{p}} \mathfrak{p}$$

R then consists precisely of all the nilpotent elements of the ring $\mathfrak{o}$, we say R is the radical of $\mathfrak{o}$.

PROOF. I. Let $a^n = 0$. Then $a^n \in \mathfrak{p}$ for every prime ideal $\mathfrak{p}$

$$\Longrightarrow\ a \in \mathfrak{p} \text{ for every prime ideal } \mathfrak{p} \ \Longrightarrow\ a \in R.$$

II. Let a be non-nilpotent, and set $S = \{a, a^2, a^3, \dots\}$. S is then a semigroup not containing 0.

$$\Longrightarrow \text{ there exists a prime ideal } \mathfrak{p} \text{ with } \mathfrak{p} \cap S = \varnothing$$

$$\Longrightarrow\ a \notin \mathfrak{p} \ \Longrightarrow\ a \notin R \qquad \square$$

If the ring $\mathfrak{o}$ contains a unity element 1 and $1 \neq 0$, we can set $S = \{1\}$, from which we see: There exist maximal ideals in $\mathfrak{o}$, i.e. ideals which are maximal in the set of all non-trivial ideals. Every ideal of $\mathfrak{o}$ is contained in a maximal ideal.

The set B of all prime ideals of $\mathfrak{o}$ can be inductively ordered (in the declining direction) by the set-theoretical containing relation: "$\supset$".

PROOF. Obviously we have a partial ordering. If $\{\mathfrak{p}_\alpha\}$ is a ordered subset of B, then $\mathfrak{p} = \bigcap_\alpha \mathfrak{p}_\alpha$ is an ideal; in fact a prime ideal because of the complete ordering of $\{\mathfrak{p}_\alpha\}$. $\mathfrak{p}$ is a lower bound of the subset $\{\mathfrak{p}_\alpha\}$. $\square$

Since $\mathfrak{o}$ itself is a prime ideal the set B cannot be empty, and it follows that: Every ring $\mathfrak{o}$ has minimal prime ideals. Every prime ideal of $\mathfrak{o}$ contains a minimal prime ideal.

Hence, we can represent the radical R of $\mathfrak{o}$ in the form:

$$R = \bigcap_{\mathfrak{p} \text{ min}} \mathfrak{p}$$

in which the intersection of only the minimal prime ideals is taken.

CHAPTER 2

PLACES OF A FIELD

1. Valuation Rings

In the following we always assume that $\mathfrak{o}$ is a *commutative ring with unity* 1. A *unit* e of $\mathfrak{o}$ is an element of $\mathfrak{o}$ for which there exists an *inverse* element e^{-1} in $\mathfrak{o}$, i.e. $e \cdot e^{-1} = 1$.

DEFINITION. $\mathfrak{o}$ is called a local ring if the non-units of $\mathfrak{o}$ form an ideal in $\mathfrak{o}$.

THEOREM. *A ring $\mathfrak{o}$ is a local ring if and only if there exists exactly one maximal ideal in $\mathfrak{o}$.*

PROOF. I. Let $\mathfrak{o}$ be a local ring. The ideal of non-units is then a maximal ideal, and the only such, since a non-trivial ideal cannot contain a unit.

II. Let the ring $\mathfrak{o}$ contain $\mathfrak{p}$ as only maximal ideal, and let α be a non-unit in $\mathfrak{o}$. Then we have

$$(\alpha) = \alpha\mathfrak{o} \neq \mathfrak{o} \implies (\alpha) \subset \mathfrak{p}$$

since every non-trivial ideal is contained in a maximal ideal. $\mathfrak{p}$, which therefore contains all the non-units, and cannot contain any other elements, is precisely the set of non-units of $\mathfrak{o}$, which is thereby an ideal. □

Given a ring $\mathfrak{o}$, we can always define the *quotient ring* $\mathfrak{o}_S$: Choosing a semigroup S in $\mathfrak{o}$, we set

$$\mathfrak{o}'_S = \{a/s \mid a \in \mathfrak{o},\ s \in S\}.$$

$\mathfrak{o}'_S$ is a ring, if multiplication and addition of the symbols a/s are defined in the usual manner:

$$\frac{a_1}{s_1} + \frac{a_2}{s_2} = \frac{a_1 s_2 + a_2 s_1}{s_1 s_2}, \qquad \frac{a_1}{s_1} \cdot \frac{a_2}{s_2} = \frac{a_1 a_2}{s_1 s_2}.$$

An equivalence relation, R, is introduced into $\mathfrak{o}'_S$:

$$\frac{a_1}{s_1} \sim \frac{a_2}{s_2} \iff (a_1 s_2 - a_2 s_1)s = 0 \quad \text{for some } s \in S.$$

The quotient set, $\mathfrak{o}_S = \mathfrak{o}'_S/R$ is a ring if addition and multiplication are defined for the classes by representatives. $\mathfrak{o}_S$ is called the quotient ring of $\mathfrak{o}$ by S.

The ring $\mathfrak{o}$ itself can be mapped into $\mathfrak{o}_S$ by the homomorphism f, given by

$$f(a) = \frac{as}{s} \mod R \quad \text{for all } a \in \mathfrak{o}, \text{ taking any } s \in S.$$

If S contains no divisors of 0 in $\mathfrak{o}$, then f is an isomorphism of $\mathfrak{o}$ into $\mathfrak{o}_S$.

By the definition of a non-trivial prime ideal $\mathfrak{p}$, the complementary set M of $\mathfrak{p}$ in $\mathfrak{o}$ is a semigroup in $\mathfrak{o}$. It can be easily seen that the quotient ring $\mathfrak{o}_M$ is a local ring.

Let $\mathfrak{o}$ now be a subring of a field k.

DEFINITION. $\mathfrak{o}$ is called a valuation ring in k if at least one of every pair of inverse elements of k, a or a^{-1}, is contained in $\mathfrak{o}$.

THEOREM. *A valuation ring, $\mathfrak{o}$, is a local ring, i.e. the non-units of $\mathfrak{o}$ form an ideal, $\mathfrak{p}$.*

PROOF. Let $\mathfrak{o}$ be a valuation ring in k.
I. Let $a, b \in \mathfrak{o}$ such that $a+b$ is a unit of $\mathfrak{o}$, i.e. $(a+b)$ and $(a+b)^{-1}$ in $\mathfrak{o}$. We prove that either a or b is unit of $\mathfrak{o}$. If either is zero this is trivial. We therefore let $a, b \neq 0$, and let $a/b \in \mathfrak{o}$ (otherwise $b/a \in \mathfrak{o}$).

$$\Longrightarrow\ 1 + a/b \in \mathfrak{o} \Longrightarrow \frac{a+b}{b} \in \mathfrak{o} \Longrightarrow \frac{1}{b} \in \mathfrak{o} \text{ (since } (a+b)^{-1} \in \mathfrak{o})$$
$$\Longrightarrow\ b \text{ is a unit of } \mathfrak{o}.$$

Thus, if a and b are non-units, $a \pm b$ is a non-unit.
II. Let $a, b \in \mathfrak{o}$, and ab be a unit of $\mathfrak{o}$, i.e. $a^{-1}b^{-1} \in \mathfrak{o}$.

$$\Longrightarrow\ aa^{-1}b^{-1} = b^{-1} \in \mathfrak{o},\ ba^{-1}b^{-1} = a^{-1} \in \mathfrak{o} \Longrightarrow a \text{ and } b \text{ are units of } \mathfrak{o}.$$

Hence, if a is a non-unit then ab is also a non-unit for any $b \in \mathfrak{o}$.
I and II are sufficient to prove the theorem. □

Since the ideal $\mathfrak{p}$ of non-units of a valuation ring $\mathfrak{o}$ is a maximal ideal, it follows that the residue class ring $\mathfrak{o}/\mathfrak{p}$ is a field.

2. Places

Let $\mathfrak{o}$ be a valuation ring of k and $\mathfrak{p}$ the corresponding maximal ideal of non-units of $\mathfrak{o}$. The canonical map

$$\varphi\colon \mathfrak{o} \to \mathfrak{o}/\mathfrak{p}$$

of $\mathfrak{o}$ onto the residue class field $\mathfrak{o}/\mathfrak{p}$ is given by:

$$\varphi(a) = a + \mathfrak{p} \quad \text{for all } a \in \mathfrak{o}.$$

We extend it to a map of k by setting

$$\varphi(a) = \infty \quad \text{for } a \in k,\ a \notin \mathfrak{o}.$$

It is thereby a map of the field k onto the set $\mathfrak{o}/\mathfrak{p} \cup \{\infty\}$

DEFINITION. Let F be a field. A map φ of the field k onto the set $F \cup \infty$ is called a *place* of k if

(1) $\mathfrak{o} = \varphi^{-1}(F)$ is a ring, and $\varphi|\mathfrak{o}$ is a homomorphism of $\mathfrak{o}$ into F.
(2) $\varphi(a) = \infty \Longrightarrow \varphi(a^{-1}) = 0$.

The map of k onto $\mathfrak{o}/\mathfrak{p} \cup \{\infty\}$ defined above is a place of k, as property (2) can here be verified as follows (property (1) coming directly from the definition):

$$a \in k,\ \varphi(a) = \infty \Longrightarrow a \notin \mathfrak{o} \Longrightarrow a^{-1} \in \mathfrak{o} \Longrightarrow a^{-1} \text{ is a non-unit in } \mathfrak{o}$$
$$\Longrightarrow a^{-1} \in \mathfrak{p} \Longrightarrow \varphi(a^{-1}) = 0,$$

since $\mathfrak{p}$ is the kernel of the homomorphism $\varphi|\mathfrak{o}$.
We have thus proven:

THEOREM. *To every valuation ring $\mathfrak{o}$ of k there exists a place φ of k.*

The converse also holds:

THEOREM. *To every place φ of k there corresponds a unique valuation ring of $\mathfrak{o}$.*

PROOF. Let φ be a place of k, in other words, a map of k into some set $F \cup \{\infty\}$, where F is a field and φ satisfies conditions (1) and (2). $\varphi^{-1}(F) = \mathfrak{o}$ is a ring,

$$a \in k,\ a \notin \mathfrak{o} \implies \varphi(a) = \infty \implies \varphi(a^{-1}) = 0 \implies a^{-1} \in \mathfrak{o}.$$

Hence, $\mathfrak{o}$ is a valuation ring; we will call it the valuation ring corresponding to the place φ. □

THEOREM. *The units of the valuation ring corresponding to the place φ are fully characterized by the properties:*

$$\varphi(a) \neq 0, \quad \varphi(a) \neq \infty;$$

the non-units being, therefore, characterized by

$$\varphi(a) = 0.$$

PROOF. I. Let $a \in k$, $\varphi(a) \neq 0$, $\varphi(a) \neq \infty$, then $\varphi(a) \in F$ and $a \in \varphi^{-1}(F) = \mathfrak{o}$ but if $a^{-1} \notin \mathfrak{o} \implies \varphi(a^{-1}) = \infty \implies \varphi(a) = 0$, contrary to hypothesis.

Hence, $a \in \mathfrak{o}$, $a^{-1} \in \mathfrak{o}$, i.e. a is a unit of $\mathfrak{o}$.

II. We now let a be a unit of $\mathfrak{o}$.

$$\implies \varphi(a) \in F,\ \varphi(a^{-1}) \in F \implies \varphi(a) \neq \infty,\ \varphi(a^{-1}) \neq \infty.$$

Because $\varphi(a) \cdot \varphi(a^{-1}) = \varphi(aa^{-1}) = \varphi(1) = 1 \neq 0$ we have:

$$\varphi(a) \neq 0, \quad \varphi(a^{-1}) \neq 0.$$ □

In addition and multiplication we can calculate with the symbol ∞ precisely as is done with this symbol in the theory of functions:

$$\varphi(a) = \infty,\ \varphi(b) \neq \infty \implies a \notin \mathfrak{o},\ b \in \mathfrak{o} \implies a \pm b \notin \mathfrak{o} \implies \varphi(a \pm b) = \infty$$

$$\varphi(a) = \infty,\ \varphi(b) \neq 0 \implies a^{-1} \in \mathfrak{p},\ b \notin \mathfrak{p} \implies ab \notin \mathfrak{p}$$

$$(\text{since otherwise } b \in a^{-1}\mathfrak{o} \subset \mathfrak{p}\mathfrak{o} \subset \mathfrak{p}) \implies \varphi(ab) = \infty.$$

As in the theory of functions, however, we can give the product $\infty \cdot 0$ and the sum $\infty \pm \infty$ no immediate meaning.

The homomorphism $\varphi|\mathfrak{o}$ of $\mathfrak{o} = \varphi^{-1}(F)$ into the field F defined by the place φ of k can be split. In the manner described earlier, into three canonical maps:

$$\mathfrak{o} \xrightarrow[\substack{\text{canonical}\\ \text{homomorphism}\\ \text{onto}}]{\varphi_1} \mathfrak{o}/\mathfrak{p} \xrightarrow[\substack{\text{isomorphism}\\ \text{onto}}]{\varphi_2} \varphi(\mathfrak{o}) \xrightarrow[\substack{\text{injection}\\ \text{into}}]{\varphi_3} F.$$

We can neglect the maps φ_2 and φ_3 and therefore denote two places of k as *equivalent places* if they differ only in their respective canonical maps φ_2 and φ_3. We see immediately that two places of k are equivalent if and only if their valuation rings coincide.

We make a remark here as to the intuitive meaning of places. The concept of a place is a generalization of function-theoretical concept.

In particular, if k is a field of analytic functions, of the complex variable z, then one may get the values of these functions at, say, the place 2 of the argument z (at $s = 2$) by substituting in each function $f(z)$ the complex number 2 for the argument z. It is easily seen that the map:

$$f(z) \xrightarrow{\varphi} f(2)$$

represents a place by the above definition. The field $\mathbb{C}$ of complex numbers here plays the part of the field F, the ring $\mathfrak{o}$ consists precisely of all functions regular at $z = 2$, and the ideal $\mathfrak{p}$ of those functions of k that have a zero at $z = 2$.

Generalizing this analogy to function theory, we call the place φ a zero of the element $a \in k$ if $\varphi(a) = 0$, and a pole of a if $\varphi(a) = \infty$. Later on we will also introduce the concept of the order of a pole or zero, as in the theory of functions.

As an example, we now consider the field $k = \mathbb{Q}$ of *rational numbers.*

To every prime number p there exists a place φ_p with valuation ring $\mathfrak{o}_p$, made up of all rational numbers whose denominator is prime to p. The maximal ideal $\mathfrak{p}_p$ of $\mathfrak{o}_p$ consists of those rational numbers whose numerator (in reduced form) is divisible by p.

The homomorphism $\varphi_p|\mathfrak{o}_p$ is the homomorphism of $\mathfrak{o}_p$ onto the residue class field $\mathfrak{o}_p/\mathfrak{p}_p$. Two rational numbers, a/b and a'/b', both in $\mathfrak{o}_p$, belong to the same residue class if $ab' - a'b$ is divisible by p.

That these places, together with the trivial place exhaust the places of this field, will be shown later.

3. The Extension Theorem

The following theorem is important in the theory of places:

EXTENSION THEOREM. *Let $\mathfrak{o}$ be any subring of a field, k; F an algebraically closed field; let there be given a homomorphism f of $\mathfrak{o}$ into F. Then there exists a place*

$$\varphi\colon k \to F \cup \{\infty\}$$

of k, such that $\varphi|\mathfrak{o} = f$.

PROOF. The idea of the proof is the succesive generation of the place φ, i.e. the homomorphism f is gradually extended to a larger and larger subring of k until finally a valuation ring is attained. Two sorts of extension are, particularly, used.

I. Let S be the semigroup of all $s \in \mathfrak{o}$ with $f(s) \neq 0$. The quotient ring $\mathfrak{o}_S$ of all quotients of the form a/s with $a \in \mathfrak{o}$, $s \in S$ can be formed within the field k, where addition, multiplication and equality of elements are already defined.

$$g(a/s) = \frac{f(a)}{f(s)}$$

defines a homomorphism g of $\mathfrak{o}_S$ into F. The definition is, indeed, meaningful, since

$$\frac{a}{s} = \frac{a'}{s'} \implies as' = a's \implies f(a)f(s') = f(a')f(s) \implies \frac{f(a)}{f(s)} = \frac{f(a')}{f(s')}.$$

The homomorphism property

$$g(a/s + a'/s') = g(a/s) + g(a'/s')$$

can be obtained from the corresponding property of f. Furthermore, $g|\mathfrak{o} = f$, since for a $a \in \mathfrak{o}$ we have

$$g(a) = g\left(\frac{as}{s}\right) = \frac{f(a)f(s)}{f(s)} = f(a)$$

Hence, g is a extension of f on $\mathfrak{o}_S$.

The image $g(\mathfrak{o}_S) = F_1$ is a subring of F, it is even a subfield of F. For this we need only show the existence of an inverse of every element other than 0:

$$\text{But,}\quad g\left(\frac{a}{s}\right) \implies \frac{f(a)}{f(s)} \neq 0 \implies f(a) \neq 0 \implies a \in S$$

$$\frac{s}{a} \in \mathfrak{o}_S \implies g\left(\frac{a}{s}\right) \in g(\mathfrak{o}_S) = F_1$$

Because $g\left(\frac{s}{a}\right) \cdot g\left(\frac{a}{s}\right) = g\left(\frac{s}{a} \cdot \frac{a}{s}\right) = g(1) = 1$, $g\left(\frac{a}{s}\right)$ is inverse $g\left(\frac{a}{s}\right)$.

It should be pointed out that the extension g of f on $\mathfrak{o}_S$ is not necessarily an extension on a larger ring; it is possible that $\mathfrak{o}_S = \mathfrak{o}$. This is the case precisely then when all the elements of S are units of $\mathfrak{o}$.

II. We now assume that $\mathfrak{o}_S = \mathfrak{o}$. In I we have proven that $f(\mathfrak{o})$ is a field $F_1 \subset F$. Letting $\alpha \neq 0$ be an element of k, we attempt to extend f to the subring

$$\mathfrak{o}[\alpha] = \{a_0 + a_1\alpha + \cdots + a_n\alpha^n \mid a_i \in \mathfrak{o}\}$$

of k, or, in the event that this proves impossible, on $\mathfrak{o}[\alpha^{-1}]$. We will prove that at least one of these attempts must be fruitful.

We choose some element $\xi \in F$ and experimentally set

$$g(a_0 + a_1\alpha + \cdots + a_n\alpha^n) = f(a_0) + f(a_1)\xi + \cdots + f(a_n)\xi^n.$$

For a simplified notation we write:

$$f(a) = \bar{a} \quad \text{for } a \in \mathfrak{o}, \quad f(\mathfrak{o}) = \bar{\mathfrak{o}} = F_1$$

$$\bar{h}(x) = \bar{a}_0 + \bar{a}_1x + \cdots + \bar{a}_nx^n, \quad \text{for } h(x) = a_0 + a_1x + \cdots + a_nx^n \in \mathfrak{o}[x].$$

The map:

$$h(x) \to \bar{h}(x)$$

is a homomorphism of $\mathfrak{o}[x]$ onto $\bar{\mathfrak{o}}[x] = F_1[x]$.

In this notation our experiment reads:

$$g(h(\alpha)) = \bar{h}(\xi).$$

If this definition is meaningful it obviously defines a homomorphism of $\mathfrak{o}[\alpha]$. Necessary and sufficient for the meaningfulness of the definition is the condition:

$$h_1(\alpha) = h_2(\alpha) \implies \bar{h}_1(\xi) = \bar{h}_2(\xi),$$

or, equivalently:

$$h(\alpha) = 0 \implies \bar{h}(\xi) = 0.$$

Let A now be the set of all polynomials $h(x) \in \mathfrak{o}[x]$, for which $h(\alpha) = 0$, and $\bar{A}$ the set of their images under the homomorphism of $\mathfrak{o}[x]$ onto $F_1[x]$. $\bar{A}$ is an ideal in $F_1[x]$:

In particular, if $\bar{h}_1, \bar{h}_2 \in \bar{A}$, $\bar{m} \in F_1[x]$, with inverse images $h_1, h_2 \in A$ and $m \in \mathfrak{o}[x]$, then $h_1 \pm h_2 \in A$, $mh_1 \in A$, implying $\bar{h}_1 \pm \bar{h}_2 \in \bar{A}$, $\overline{mh_1} \in \bar{A}$.

Since in $F_1[x]$ every ideal is a principal ideal, $\bar{A}$ consists exactly of all multiples of a certain polynomial $\bar{h}_0(x)$. The above condition is therefore equivalent to:

$$\bar{h}_0(\xi) = 0.$$

If $\bar{h}_0(x)$ is not a non-zero constant one can always find a ξ in the algebraically closed field F so that this equation is satisfied. If we use this ξ in the above indicated manner we arrive at a extension of f on $\mathfrak{o}[\alpha]$.

If, on the other hand, $\bar{h}_0(x)$ is a constant other than 0 the condition $\bar{h}_0(\xi)$ cannot be satisfied. Since F_1 is a field, the constant $\bar{h}_0(x)$ can be normalized to 1. All inverse images of $\bar{h}_0(x)$ then have the form:

$$(1) \qquad h_0(x) = 1 + \pi_0 + \pi_1 x + \cdots + \pi_n x^n, \quad \bar{\pi}_i = 0 \ (i = 0, 1, 2, \ldots, n).$$

The failure of our attempt to extend f to $\mathfrak{o}[\alpha]$ indicates the existence of an equation:

$$(2) \quad (1 + \pi_0) + \pi_1 \alpha + \pi_2 \alpha^2 + \cdots + \pi_n \alpha^n = 0, \quad \bar{\pi}_i = 0 \ (i = 0, 1, 2, \ldots, n).$$

Similarly it can be shown that f is not extendable to $\mathfrak{o}[\alpha^{-1}]$ only if a relation of the following form exists:

$$(3) \ (1 + \pi'_0) + \pi'_1 \alpha^{-1} + \pi'_2 \alpha^{-2} + \cdots + \pi'_m \alpha^{-m} = 0, \quad \bar{\pi}'_i = 0 \ (i = 0, 1, 2, \ldots, m).$$

To show that f is extendable either to $\mathfrak{o}[\alpha]$ or to $\mathfrak{o}[\alpha^{-1}]$, we must now only show that the equations (2) and (3) are incompatible.

Assume that they both hold. Among all such equations we choose those of the minimal degree n and m, respectively. Certainly, then $n \geq 1$, $m \geq 1$, since $\overline{1 + \pi_0} = 1 \neq 0$, $\overline{1 + \pi'_0} = 1 \neq 0$ implies $1 + \pi_0 \neq 0$, $1 + \pi'_0 \neq 0$. Say, $m \leq n$.

After multiplication with $\frac{\alpha^n}{1+\pi'_0}$, equation (3) gives:

$$(4) \qquad \alpha^n = \pi''_1 \alpha^{n-1} + \cdots + \pi''_m \alpha^{n-m}$$

with $\pi''_i = -\frac{\pi'_i}{1-\pi'_0} \in \mathfrak{o}_S = \mathfrak{o}$, $\bar{\pi}''_i = 0 \ (i = 1, 2, \ldots, m)$.

By substituting (4) into (2) one gets another equation of the form (2), but of degree $n - 1$, which is in contradiction to the minimal property of n.

The equations (2) and (3) are therefore incompatible.

III. After these two preliminary steps the proof of the extension theorem can be undertaken as follows:

Let B be the set of all maps g which act as non-trivial homomorphisms of a subring $\mathfrak{o}'$ of k into F. In B we define the following partial ordering: If g is a homomorphism of $\mathfrak{o}'_1$, g_2 a homomorphism of $\mathfrak{o}'_2$, then we let $g_1 \leq g_2$ if and only if $\mathfrak{o}'_1 \subset \mathfrak{o}'_2$ and $g_2 | \mathfrak{o}'_1 = g_1$ both hold. B is in fact inductively ordered:

Let $\{g_\nu\}$ be a completely ordered subset of B, $\{\mathfrak{o}'_\nu\}$ the set of corresponding rings. Then,

$$\mathfrak{o}' = \bigcup_\nu \mathfrak{o}'_\nu$$

is a ring, since for any pair of rings $\mathfrak{o}'_\mu, \mathfrak{o}'_\lambda$ one is contained in the other. For the elements $a \in \mathfrak{o}'$ we then define $g(a) = g_\nu(a)$ if $a \in \mathfrak{o}'_v$ (which must be the case for some $\mathfrak{o}'_\nu$). By the definition of the ordering relation and by the complete ordering of $\{\mathfrak{G}_\nu\}$ the above definition of g is meaningful, i.e. it does not depend on the choice of the ring $\mathfrak{o}'_\nu$ containing a.

g is a homomorphism of $\mathfrak{o}'$ into F, hence $g \in B$. Obviously g is an extension of every g_ν on the ring $\mathfrak{o}'$, and therefore $g \geq g_\nu$ for all ν. g is an upper bound of the subset $\{g_\nu\}$.

The originally suggested homomorphism f of $\mathfrak{o}$ into F belongs to the set B. There exists, therefore, a φ in B such that $\varphi \geq f$ with φ maximal. Let $\mathfrak{O}$ be the subring corresponding to φ; we have $\mathfrak{o} \subset \mathfrak{O}$.

Let

$$S = \{s \mid s \in \mathfrak{O}, \ \varphi(s) \neq 0\},$$

then $\mathfrak{O}_S = \mathfrak{O}$, because φ is maximal. Otherwise, in particular, φ would be extendable to $\mathfrak{O}_S$ by I, and not be maximal.

Thus, every $s \in S$ is a unit of $\mathfrak{O}$. Conversely, if s is a unit of $\mathfrak{O}$ then $s \in \mathfrak{O}$, $s^{-1} \in \mathfrak{O}$, and $\varphi(s) \cdot \varphi(s^{-1}) = \varphi(ss^{-1}) = \varphi(1) = 1$, hence $\varphi(s) \neq 0$ and $s \in S$. It now follows that the non-units of $\mathfrak{O}$ are precisely those $a \in \mathfrak{O}$ with $\varphi(a) = 0$, the kernel of φ. These form an ideal $\mathfrak{p}$ in $\mathfrak{O}$. $\mathfrak{O}$ is a local ring.

Let $\alpha \neq 0$ be an element of k. By II, φ can he extended to either $\mathfrak{O}[\alpha]$ or to $\mathfrak{O}[\alpha^{-1}]$. Since, however, φ is maximal, either α or α^{-1} must lie in $\mathfrak{O}$.

$\mathfrak{O}$ is a valuation ring and φ, except for ∞, the place of $\mathfrak{O}$. If we enlarge φ by setting

$$\varphi(\alpha) = \infty \quad \text{when } \alpha \in k,\ \alpha \notin \mathfrak{O},$$

it becomes a place of k with $\varphi|\mathfrak{o} = f$, whereby the extension theorem is proven. □

It should still be pointed out that the assumption that F is algebraically closed is not of primary importance, as any field can be closed algebraically, and we have defined two homomorphisms to be equivalent if they differ only in the injection into a larger field.

4. Places of the Rational Number Field

We again take up the example of the rational number field $\mathbb{Q}$. As base ring we choose the smallest ring containing 1, the ring $\mathbb{Z}$ of rational integers. Any non-trivial homomorphism f of $\mathbb{Z}$ into a field (e.g. the field $\mathbb{C}$ of complex numbers) can be continued, according to the last theorem, to a place φ of $\mathbb{Q}$.

To gain oversight over all the places of $\mathbb{Q}$ and their valuation rings we start with any such valuation ring $\mathfrak{O}$ of $\mathbb{Q}$. As $1 \in \mathfrak{O}$, $\mathbb{Z} \subset \mathfrak{O}$. Let φ be the place corresponding to $\mathfrak{O}$ and $\mathfrak{p}$ the maximal ideal in $\mathfrak{O}$. $\mathfrak{p} \cap \mathbb{Z}$ is then a prime ideal of $\mathbb{Z}$:

There are only two possibilities —

(1) $\mathfrak{p} \cap \mathbb{Z}$ is the zero ideal of $\mathbb{Z}$. All integers other than 0 are units of $\mathfrak{O}$, therefore the same applies for all non-zero rational numbers. Hence $\mathfrak{O} = \mathbb{Q}$. In this case we talk of the *trivial valuation ring* and the *trivial place* of $\mathbb{Q}$.

(2) $\mathfrak{p} \cap \mathbb{Z} = p\mathbb{Z}$, where p is a prime number. Integers prime to p are units, and with them all rational numbers with both denominator and numerator prime to p. $\mathfrak{O}$ consists of all rational numbers with denominator prime to p, $\mathfrak{p}$ of rational numbers with numerator divisible by p.

As $1 \notin \mathfrak{p}$, the case $\mathfrak{p} \cap \mathbb{Z} = \mathbb{Z}$ is excluded, and all possibilities are really taken care of, since $\mathbb{Z}$ has no further prime ideals.

$\mathbb{Q}$, therefore, has only the following places:

(1) the trivial place,

(2) places φ_p corresponding to the prime numbers p.

5. Theory of Integers

Consider the field k, and any subring $\mathfrak{o}$ containing unity, Let α be an element of k.

DEFINITION. α is called integral with respect to $\mathfrak{o}$ if it satisfies an equation

$$\alpha^n + a_1\alpha^{n-1} + a_2\alpha^{n-2} + \cdots + a_n = 0, \quad a_i \in \mathfrak{o}\ (i = 1, 2, \ldots, n)$$

with highest coefficient 1.

We want to characterize the integers of k with respect to $\mathfrak{o}$ place-theoretically: If the above equation is divided by α^{n-1} it gives

$$\alpha = -a_1 - a_2\alpha^{-1} - \cdots - a_n\alpha^{-(n-1)}.$$

We now have, therefore, the following equivalent statements:

$$\alpha \text{ integral with respect to } \mathfrak{o}$$

$$\alpha \in \mathfrak{o}[\alpha^{-1}]$$

$$\alpha \text{ unit in } \mathfrak{o}[\alpha^{-1}]$$

$$\alpha^{-1}\mathfrak{o}[\alpha^{-1}] = \mathfrak{o}[\alpha^{-1}]$$

Hence, α is non-integral with respect to $\mathfrak{o}$ if $\alpha^{-1}\mathfrak{o}[\alpha^{-1}] \subset \mathfrak{o}[\alpha^{-1}]$ but $\alpha^{-1}\mathfrak{o}[\alpha^{-1}] \neq \mathfrak{o}[\alpha^{-1}]$, or if the following statement holds:

In $\mathfrak{o}[\alpha^{-1}]$ there exist maximal prime ideals containing α^{-1}:

Let α be non-integral with respect to $\mathfrak{o}$, and let $\mathfrak{p}$ be a maximal prime ideal of $\mathfrak{o}[\alpha^{-1}]$ containing α^{-1}. The canonical homomorphism

$$f\colon \mathfrak{o}[\alpha^{-1}] \to \mathfrak{o}[\alpha^{-1}]/\mathfrak{p}$$

of $\mathfrak{o}[\alpha^{-1}]$ onto the field $\mathfrak{o}[\alpha^{-1}]/\mathfrak{p}$ is non-trivial. Thus, there exists a place φ with $\varphi|\mathfrak{o}[\alpha^{-1}] = f$.

For every such place φ, $\varphi|\mathfrak{o}$ is finite, since even $\varphi|\mathfrak{o}[\alpha^{-1}]$ is finite. Further,

$$\varphi(\alpha^{-1}) = f(\alpha^{-1}) \in f(\mathfrak{p}) = 0, \quad \text{and hence } \varphi(\alpha) = \infty$$

The kernel of φ in $\mathfrak{o}[\alpha^{-1}]$ is the maximal ideal $\mathfrak{p}$. The kernel of φ in $\mathfrak{o}$ is a prime ideal $\mathfrak{p}_0 = \mathfrak{o} \cap \mathfrak{p}$ of $\mathfrak{o}$. $\mathfrak{p}_0$ is, in fact, a maximal ideal in $\mathfrak{o}$, which implies that $\mathfrak{o}/\mathfrak{p}_0$ is a field:

For, letting $a \in \mathfrak{o}$, $a \not\equiv 0 \pmod{\mathfrak{p}_0}$

$$\Longrightarrow \varphi(a) \neq 0, \infty \Longrightarrow a \in \mathfrak{o}[\alpha^{-1}],\ a \not\equiv 0 \pmod{\mathfrak{p}}.$$

In $\mathfrak{o}[\alpha^{-1}]$ there exists a $\beta = b_0 + b_1\alpha^{-1} + \cdots + b_r a^{-r}$ with $b_i \in \mathfrak{o}$ and $\alpha\beta \equiv 1 \pmod{\mathfrak{p}}$. Because $\alpha^{-1} \equiv 0 \pmod{\mathfrak{p}}$ this implies $ab_0 \equiv 1 \pmod{\mathfrak{p}}$ and even further $ab_0 \equiv 1 \pmod{\mathfrak{p}_0}$.

Compiling our results we see:

THEOREM. *If α is non-integral with respect to $\mathfrak{o}$ there exists a place φ of k, such that $\varphi|\mathfrak{o}$ is finite, the kernel of $\varphi|\mathfrak{o}$ is a maximal ideal in $\mathfrak{o}$, and $\varphi(\alpha) = \infty$.*

There exists a converse to this theorem. To derive it we now let α be integral with respect to $\mathfrak{o}$, and φ be some place of k for which $\varphi|\mathfrak{o}$ is finite. We show that then $\varphi(\alpha)$ is finite:

Assume $\varphi(\alpha) = \infty$. This would imply $\varphi(\alpha^{-1}) = 0$. Applying

$$\alpha^n + a_1\alpha^{n-1} + \cdots + a_n = 0, \quad a_i \in \mathfrak{o}$$

we get for α

$$1 + a_1\alpha^{-1} + \cdots + a_n\alpha^{-n} = 0$$

and applying φ:

$$1 + 0 + \cdots + 0 = 0, \quad \text{a contradiction.}$$

We have thus proven the following theorem:

THEOREM. *α is integral with respect to $\mathfrak{o}$ if and only if for every place φ of k for which $\varphi|\mathfrak{o}$ is finite, $\varphi(\alpha)$ is also finite.*

Now let $\mathfrak{O}$ be the set of all elements of k integral with respect to $\mathfrak{o}$ (called the *integral closure* of $\mathfrak{o}$ in k). From the last theorem we then have: $\mathfrak{O}$ is a ring. The integral closure of $\mathfrak{O}$ is $\mathfrak{O}$ itself, since the places that are finite in $\mathfrak{O}$ are precisely the places finite in $\mathfrak{o}$. The ring $\mathfrak{O}$ is the intersection of all valuation rings of k which contain the ring $\mathfrak{o}$.

CHAPTER 3

VALUATIONS

1. Ordered Groups

In order to give a general definition of what is meant by a valuation of a field we first need the concept of ordered groups. Consider first the example of the multiplicative group of positive rational or real numbers. These groups have the following properties:

(1) A complete ordering is given on the set of elements by the relation "$<$".
(2) The subset $S = \{s \mid s < 1\}$ is a semigroup.
(3) $s \in S$ and a any group element imply: $as < a$, $sa < a$; Hence, S is an invariant semigroup: $aSa^{-1} \subset S$ for all group elements a (which, incidentally implies $aSa^{-1} = S$)
(4) $s < 1$ implies $1 < s^{-1}$.

Analogously we define:

DEFINITION. The group $\mathfrak{G}$ is an *ordered group* if it contains a semigroup S with the following properties:

(1) $aSa^{-1} \subset S$ for all $a \in \mathfrak{G}$.
(2) $\mathfrak{G} = S \cup \{1\} \cup S^{-1}$, the components being disjoint, where $S^{-1} = \{s^{-1} \mid s \in S\}$.

An ordered group is a completely ordered set by the order relation:

$$a < b \iff ab^{-1} \in S.$$

This definition is symmetric with respect to left and right multiplication, since

$$ab^{-1} \in S \implies b^{-1}a = b^{-1}ab^{-1}b \in b^{-1}Sb \subset S,$$

and similarly

$$b^{-1}a \in S \implies ab^{-1} \in S.$$

The transitivity of the relation can be shown:

$$a < b,\ b < c \implies ab^{-1} \in S,\ bc^{-1} \in S \implies ab^{-1}bc^{-1} = ac^{-1} \in S \implies a < c.$$

That this ordering is complete follows directly from the definition of the ordered group, according to which, for any pair of elements a, b always $ab^{-1} \in S$ or $ab^{-1} = 1$ or $ab^{-1} \in S^{-1}$ holds, from which follow $a < b$, $a = b$, $b < a$, respectively.

This complete ordering of an ordered group $\mathfrak{G}$ has the following properties:

(1) $a < b \implies ac < bc$, $ca < cb$ for any $c \in \mathfrak{G}$,
(2) $a < b$, $c < d \implies ac < bd$,
(3) $a < b \iff b^{-1} < a^{-1}$,
(4) $a < 1 \iff a \in S$.

By a (multiplicative) group $\mathfrak{G}$ with 0-element we understand a set with the property: the totality of elements other than 0 form, under multiplication, a group $\mathfrak{G}^*$ in the usual sense, and $0a = a0 = 0$ for any $a \in \mathfrak{G}$. For example, all the elements of a field k form, with respect to multiplication, a group with 0-element. An ordering of a group $\mathfrak{G}$ with 0-element is an ordering such that $\mathfrak{G}^*$ is an ordered group in the sense of our definition, and such that $0 < a$ for every $a \in \mathfrak{G}$.

2. Valuations of a Field

DEFINITION. A *valuation* of a field k is a map $|\ |$ of k into an ordered group $\mathfrak{G}$ with 0-element, which ascribes to every $a \in k$ an $|a| \in \mathfrak{G}$ so that the following axioms are satisfied:

(1) $|a| = 0$ if and only if $a = 0$,
(2) $|ab| = |a| \cdot |b|$,
(3) $|a| \leq 1 \implies |1 + a| \leq 1$.

$|a|$ is called the absolute value of a, or the value of a by the valuation $|\ |$.

The map

$$a \to |a| \quad \text{for } a \in k^*$$

is thus a homomorphism of the multiplicative group k^* of k into the multiplicative group $\mathfrak{G}^*$, and therefore $|1| = 1$. Because of this we can restate Axiom 3 in the equivalent

(3a) $|a + b| \leq \text{Max}(|a|, |b|)$.

(3) results from (3a) by setting $b = 1$. Conversely, (3) also implies (3a), for if a and b are non-zero elements of k (if either $a = 0$ or $b = 0$ (3a) is trivial), and, say, $|a| \geq |b|$

$$\implies |b/a| \leq 1 \implies |1 + b/a| \leq 1 \text{ by (3)} \implies |a + b| \leq |a| = \text{Max}(|a|, |b|).$$

To a given valuation $|\ |$ of k we define the following subset of k:

$$\mathfrak{o} = \{a \mid a \in k,\ |a| \leq 1\}.$$

From the valuation axioms it can be seen that $\mathfrak{o}$ is a ring. $\mathfrak{o}$ is, in fact, even a valuation ring, for

$$a \in k,\ a \notin \mathfrak{o} \implies |a| > 1 \implies |a^{-1}| < 1 \implies a^{-1} \in \mathfrak{o}.$$

A valuation determines, in this manner, a unique corresponding valuation ring. Two valuations of k with the same corresponding valuation ring $\mathfrak{o}$ are designated *equivalent valuations.* The group of units of $\mathfrak{o}$ is

$$\mathfrak{u} = \{a \mid |a| = 1\};$$

the maximal ideal $\mathfrak{p}$ of all non-units is

$$\mathfrak{p} = \{a \mid |a| < 1\}.$$

The homomorphism

$$k \xrightarrow{|\ |} \mathfrak{G}$$

of k into $\mathfrak{G}$ can be split into its three canonical maps

$$k \xrightarrow{\text{I}} k/\mathfrak{u} \xrightarrow{\text{II}} |k| \xrightarrow{\text{III}} \mathfrak{G}.$$

I is the homomorphism: $a \to a\mathfrak{u}$ of k onto $k/\mathfrak{u}$,
II is the isomorphism: $a\mathfrak{u} \to |a|$ of $k/\mathfrak{u}$ onto $|k|$,

III is the injection of $|k|$ into $\mathfrak{G}$,

Because of the isomorphism II, the order of the group $|k|$, which is, of course, a subgroup of $\mathfrak{G}$, can be carried over to the group $k/\mathfrak{u}$. In this ordering of $k/\mathfrak{u}$ the part of S is played by the semigroup $\{a\mathfrak{u} \mid a \in \mathfrak{p}\}$, since

$$a\mathfrak{u} < 1 \iff |a| < 1 \iff a \in \mathfrak{p}$$

If we substitute, for the valuation group $|k|$, the isomorphic group $k/\mathfrak{a}$, defined as ordered group by the semigroup $S = \{a\mathfrak{u} \mid a \in \mathfrak{p}\}$ then a new valuation of k,

$$|a| = a\mathfrak{u}$$

is defined, which is equivalent to the original valuation.

On the other hand, if we start with a valuation ring $\mathfrak{o}$ of k with maximal ideal $\mathfrak{p}$ and group of units $\mathfrak{u}$, and define $k/\mathfrak{u}$ as ordered group using the semigroup $S = \mathfrak{p}/\mathfrak{u} = \{a\mathfrak{u} \mid a \in \mathfrak{p}\}$, then

$$|a| = a\mathfrak{u}$$

is a valuation of k. For we have

$$k/\mathfrak{u} = \mathfrak{p}/\mathfrak{u} \cup \mathfrak{u}/\mathfrak{u} \cup \mathfrak{p}^{-1}/\mathfrak{u} \quad \text{where } \mathfrak{p}^{-1} = \{a \mid a^{-1} \in \mathfrak{p}\},$$

so that S indeed generates an ordering of $k/\mathfrak{u}$. The valuation axioms are also easily shown to hold; the first two being obvious, and the third from

$$\begin{aligned} |a| \leq 1 &\implies a\mathfrak{u} \in \mathfrak{p}/\mathfrak{u} \cup \mathfrak{u}/\mathfrak{u} = \mathfrak{o}/\mathfrak{u} \implies a \in \mathfrak{o} \\ &\implies 1 + a \in \mathfrak{o} \implies (1+a)\mathfrak{u} \in \mathfrak{o}/\mathfrak{u} \implies |1+a| \leq 1. \end{aligned}$$

It has thus been shown that every class of equivalent valuations has a unique corresponding valuation ring, and conversely that every valuation ring corresponds uniquely to a class of equivalent valuations. Consequently, there also corresponds to every class of equivalent valuations a place of k, and conversely.

Another notation is frequently used for valuations. The group $\mathfrak{G}^*$ is written as an additive group, while at the same time the inequality sign of the order relation for any two elements is reversed. The 0 is then to be replaced by taking in addition to the group $\mathfrak{G}^*$ a symbol which is greater than all the elements of $\mathfrak{G}^*$ — it is denoted by ∞. If, finally, $\operatorname{ord} a$ is written instead of $|a|$, then the valuation axioms take the form:

(1) $\operatorname{ord} a = \infty$ if and only if $a = 0$,
(2) $\operatorname{ord} ab = \operatorname{ord} a + \operatorname{ord} b$,
(3) $\operatorname{ord} a \geq 0 \implies \operatorname{ord}(1+a) \geq 0$.

This notation brings out the connection with the theory of functions particularly clearly, as the order here defined (ord) is the exact analogon of the order of a zero of a function. The third axiom, in terms of function theory, says that with a also $1 + a$ is regular.

The connection of the new notation with the earlier notation and with that of place theory is seen as follows:

$$\operatorname{ord} a > \operatorname{ord} b \iff |a| < |b| \iff |a/b| < 1 \iff a/b \in \mathfrak{p} \iff \varphi(a/b) = 0,$$

where φ is the place of k corresponding to the valuation ring $\mathfrak{o}$.

We include here the following remarks about valuations:

1.) If the valuation ring is k, hence the corresponding place φ a trivial place of k, then the valuation is called the *trivial valuation.* We then have

$$|a| \leq 1 \quad \text{for all } a \in k \implies |0| = 0, \ |a| = 1 \quad \text{for all } a \in k^*;$$

and thus, $|k| = \{0, 1\}$.

This is obviously an uninteresting case. It cannot, however, always be excluded, as, for example, a non-trivial valuation of a field can induce the trivial valuation upon a subfield.

2.) The third axiom in the form

(3a) $|a + b| \leq \operatorname{Max}(|a|, |b|)$

is in many cases of itself stronger. If, in particular, $|a| > |b|$, then

$$|a + b| \leq |a|,$$

but also

$$|a| = |(a + b) - b| \leq \operatorname{Max}(|a + b|, |b|).$$

This maximum is necessarily equal to $|a + b|$, since otherwise $|a| \leq |b|$, a contradiction. The two inequalities together now yield

$$|a + b| = |a| = \operatorname{Max}(|a|, |b|).$$

An easy inductive prove leads to the generalization

$$|a_1| > |a_i| \quad \text{for all } i = 2, 3, \ldots, n \implies |a_1 + a_2 + \cdots + a_n| = |a_1|.$$

This result is often applied in the following form:

Let $a_1 + a_2 + \cdots + a_n = 0$. Then there exist at least two different a_i with maximal absolute value among the a_j $(j = 1, 2, \ldots, n)$.

3.) To illustrate the concept of valuations, we again consider the example of the field $\mathbb{Q}$ of rational numbers. In this case it is easy to gain oversight of all possible valuations.

Let p be a fixed prime number. Every rational number $a \neq 0$ can be uniquely written in the form

$$a = p^\nu b,$$

where ν is an integer and b is a rational number prime to p (i.e. in reduced form both numerator and denominator are prime to p). The power ν of p is uniquely determined by the fundamental theorem of number theory.

We then define:

$$\operatorname{ord}_p a = \begin{cases} \infty & \text{for } a = 0, \\ \nu & \text{for } a = p^\nu b, \ b \text{ prime to } p. \end{cases}$$

This is a valuation of $\mathbb{Q}$ (in the second notation for valuations). Axioms (1) and (2) are again trivially fulfilled, and axiom (3) follows from:

$$\operatorname{ord}_p a \geq 0 \iff \text{denominator of } a \text{ prime to } p$$
$$\iff \text{denominator } (1 + a) \text{ prime to } p \iff \operatorname{ord}_p(1 + a) \geq 0.$$

The corresponding valuation ring $\mathfrak{o} = \mathfrak{o}_p$ consists of all rational numbers whose denominators are prime to p.

For every prime number of p one gets, in this way, a valuation of $\mathbb{Q}$. Since, as was previously recognized, no valuation rings other than the $\mathfrak{o}_p$ exist in $\mathbb{Q}$, the thus arrived at valuations are, up to equivalence, all possible valuations of $\mathbb{Q}$.

As an application of valuation theory, we can now easily prove the following theorem:

THEOREM. *Let* $\mathfrak{o}$ *be a subring of* k; $x_1, x_2, \ldots, x_n$ *be elements of* k, *and let*

$$x_i^{m_i} = f_i(x_1, x_2, \ldots, x_n) \quad (i = 1, 2, \ldots, n)$$

where $f_i(X_1, X_2, \ldots, X_n)$ *is a polynomial in* $X_1, X_2, \ldots, X_n$ *with coefficients in* $\mathfrak{o}$ *and of degree* $< m_i$ (*for* $i = 1, 2, \ldots, n$). *Then, the* $x_1, x_2, \ldots, x_n$ *are integral with respect to* $\mathfrak{o}$.

PROOF. If the assertion were wrong, there would exist a place φ of k, finite in $\mathfrak{o}$, such that

$$\varphi(x_i) = \infty \quad \text{for at least one } i.$$

Let $|\ |$ be the corresponding valuation. We then have

$$|x_i| > 1 \quad \text{for at least one } i.$$

Choosing an index j such that

$$|x_j| = \mathrm{Max}(|x_1|, |x_2|, \ldots, |x_n|) \quad (> 1),$$

our assumption yields

$$x_j^{m_j} = f_j(x_1, x_2, \ldots, x_n) = \sum_{\nu_1, \ldots, \nu_n} a_{\nu_1, \ldots, \nu_n} x_1^{\nu_1} x_2^{\nu_2} \ldots x_n^{\nu_n}$$

or

$$\begin{aligned} 1 &= \frac{f_j(x_1, x_2, \ldots, x_n)}{x_j^{m_j}} \\ &= \sum_{\nu_1, \ldots, \nu_n} a_{\nu_1, \ldots, \nu_n} \left(\frac{x_1}{x_j}\right)^{\nu_1} \left(\frac{x_2}{x_j}\right)^{\nu_2} \cdots \left(\frac{x_n}{x_j}\right)^{\nu_n} \left(\frac{1}{x_j}\right)^{m_j - \nu_1 - \cdots - \nu_n}. \end{aligned}$$

Applying φ here one gets, because of the maximal property of x_j, because $m_j - \nu_1 - \cdots - \nu_n > 0$, and because $\varphi\left(\frac{1}{x_j}\right) = 0$, the equation $1 = 0$, obviously a contradiction. $x_1, x_2, \ldots, x_n$ must in fact all be integral with respect to $\mathfrak{o}$. □

This theorem yields as a special case: If $\mathfrak{O}$ is any finite extension ring of $\mathfrak{o}$ in the field k, in other words, if

$$\mathfrak{O} = \mathfrak{o}x_1 + \mathfrak{o}x_2 + \cdots + \mathfrak{o}x_n, \quad x_i \in \mathfrak{O} \subset k,$$

then $\mathfrak{O}$ is integral with respect to $\mathfrak{o}$.

Because $x_i^2 \in \mathfrak{O}$, we have, indeed, the equations

$$x_i^2 = a_{i1}x_1 + a_{i2}x_2 + \cdots + a_{in}x_n, \quad a_{ij} \in \mathfrak{o}\ (j = 1, 2, \ldots, n).$$

3. Valuations of Rank 1

A special and particularly important type of valuation is the valuation of rank 1; valuations are said to be of this type if the valuation group $\mathfrak{G}^*$ can be injected order-isomorphically into the group $\mathbb{R}^+$ of positive real numbers. $\mathbb{R}^+$ (or, as the case may be, $\mathbb{R}^+ \cup \{0\}$) can then be taken as the valuation group.

In this case one mostly subjects the valuation concept to a slight generalization, by weakening to some extent the third axiom, and requiring of a valuation the following axioms:

(1) $|a| = 0 \iff a = 0$,

(2) $|ab| = |a| \cdot |b|$

(3) There exists a certain definite constant C in the valuation group $\mathbb{R}^+$ such that:

$$|a| \leq 1 \implies |1 + a| \leq C$$

Certainly $C \geq 1$, as is seen by letting $a = 0$.

If a valuation satisfies these axioms with $C = 1$, then it is a valuation by our earlier definition.

Let us assume that for a valuation (3) is satisfied taking $C \leq 2$. Then (3) certainly holds with $C = 2$. Let $a, b \neq 0$ be two elements of k, and let $|a| \geq |b|$. Then,

$$|a + b| = |a| \cdot \left|1 + \frac{b}{a}\right| \leq |a| \cdot 2 = 2|a|,$$

and in general

$$|a + b| \leq 2 \operatorname{Max}(|a|, |b|).$$

This inequality can be strengthened in the following manner:

Repeating the same argument we arrive at

$$|a_1 + a_2 + a_3 + a_4| \leq 2^2 \cdot \operatorname{Max}_i(|a_i|),$$

and induction yields

$$|a_1 + a_2 + \cdots + a_{2^r}| \leq 2^r \cdot \operatorname{Max}_i(|a_i|).$$

For any n we can complete the sum $a_1 + a_2 + \cdots + a_n$, by addition of a suitable number of zeros, until the next power of 2 larger than n is reached. Obviously then

$$|a_1 + a_2 + \cdots + a_n| \leq 2n \operatorname{Max}_i(|a_i|),$$

and in the special case $a_1 = a_2 = \cdots = a_n = 1$ we have:

$$|n| \leq 2n.$$

The originally arrived at inequality we put in the weaker form:

$$|a_1 + a_2 + \cdots + a_n| \leq 2n \cdot (|a_1| + |a_2| + \cdots + |a_n|).$$

By applying the binomial theorem and the inequality $|n| \leq 2n$ we arrive at

$$|a + b|^n = |(a + b)^n| \leq 2(n + 1) \sum_{\nu=0}^{n} \left|\binom{n}{\nu}\right| \cdot |a^{n-\nu}| \cdot |b^\nu|$$

$$\leq 4(n + 1) \sum_{\nu=0}^{n} \binom{n}{\nu} |a|^{n-\nu} |b|^\nu = 4(n + 1)(|a| + |b|)^n$$

$$\implies |a + b| \leq \sqrt[n]{4(n + 1)}(|a| + |b|) \quad \text{for every natural } n,$$

and, by letting n tend toward infinity we see

$$|a + b| \leq |a| + |b|.$$

The valuation satisfies the *triangle inequality*.

If, on the other hand, the valuation $|\ |$ satisfies the triangle inequality,

$$|a + 1| \leq |a| + |1|,$$

which shows that we can choose $C = 2$.

We have thus proven the

THEOREM. *A valuation of rank* 1 *satisfies the triangle inequality if, and only if, valuation axiom* (3) *is satisfied with the constant* $C = 2$.

THEOREM. *Let* $|\ |$ *be a rank* 1 *valuation of* K, *and let* k *be a subfield of* K. *The triangle inequality holds in* K *if and only if it holds in* k.

PROOF. We can write the constant C of axiom 3 in the form $C = 2^\beta$, with real $\beta \geq 0$. In K we then have

$$|a| \leq 1 \implies |1 + a| \leq 2^\beta$$

which implies

$$|a_1 + a_2| \leq 2^\beta \operatorname{Max}_\nu(|a_\nu|)$$

and

$$|a_1 + a_2 + a_3 + a_4| \leq 2^{2\beta} \operatorname{Max}_\nu(|a_\nu|)$$

and then, as in the argument of the previous theorem,

$$|a_1 + a_2 + \cdots + a_n| \leq (2n)^\beta \operatorname{Max}_\nu(|a_\nu|).$$

Now let the triangle inequality hold in k. Because $n = n \cdot 1 \in k$:

$$|n| \leq n \quad \text{for every "natural number" } n,$$

and thus, for any $a, b \in K$

$$\begin{aligned} |a+b|^n &= \left| a^n + \binom{n}{1} a^{n-1} b + \cdots + b^n \right| \\ &\leq (2(n+1))^\beta \cdot \operatorname{Max}_\nu \left(\binom{n}{\nu} |a|^{n-\nu} |b|^\nu \right) \\ &\leq (2(n+1))^\beta \sum_{\nu=0}^{n} \binom{n}{\nu} |a|^{n-\nu} |b|^\nu = (2(n+1))^\beta (|a| + |b|)^n \end{aligned}$$

$$\implies |a+b| \leq (\sqrt[n]{2(n+1)})^\beta \cdot (|a| + |b|),$$

and, letting n tend toward infinity,

$$|a+b| \leq |a| + |b|.$$

Hence, the triangle inequality holds in K if it holds in k; which proves the theorem, since the converse is trivial. □

For these generalized valuations of rank 1 we also require an equivalence relationship. To this end we first consider a group Γ (e.g. the multiplicative group of a field k), and the group $\mathbb{R}^+$ of positive real numbers. Let there be given two homomorphisms, $|\ |_1$ and $|\ |_2$ of Γ into $\mathbb{R}^+$, satisfying:

(1) $|\ |_1$ is not trivial, i.e. $|\ |_1$ does not map the entire group Γ onto 1.
(2) $|a|_1 < 1 \implies |a|_2 < 1$.

Substituting $1/a$ for a in (2) yields immediately

$$|a|_1 > 1 \implies |a|_2 > 1.$$

Now let $|a|_1 = 1$. By (1) there exist $b \in \Gamma$ with $|b|_1 < 1$. But then,

$$|a^n b|_1 < 1 \implies |a^n b|_2 < 1 \implies |a|_2 \leq \frac{1}{\sqrt[n]{|b|_2}} \quad \text{for any } n,$$

and, letting $n \to \infty$, $|a|_2 \leq 1$.

We have thus shown that

$$|a|_1 = 1 \implies |a|_2 = 1,$$

since the same considerations hold using $\frac{1}{a}$.

Under our assumptions (1) and (2), it can, therefore, be easily seen that

$$\begin{aligned} |a|_1 < |b|_1 &\implies |a|_2 < |b|_2, \\ |a|_1 = |b|_1 &\implies |a|_2 = |b|_2, \\ |a|_1 > |b|_1 &\implies |a|_2 > |b|_2. \end{aligned}$$

If we now choose a fixed $c \in \Gamma$ with $|c|_1 > 1$, which is possible by (1), we can, for any given element $a \in \Gamma$ find a real number γ so that

$$|a|_1 = |c|_1^{\gamma}.$$

γ can be generated by a Dedekind cut in the set of rational numbers. For any rational number $n/m > \gamma$ we have

$$|a|_1 < |c|_1^{n/m} \implies |a^m|_1 < |c^n|_1 \implies |a^m|_2 < |c^n|_2 \implies |a|_2 < |c|_2^{n/m}.$$

Similarly, $\gamma > n/m \implies |a|_2 > |c|_2^{n/m}$; so that we have proved

$$|a|_2 = |c|_2^{\gamma}.$$

Therefore,

$$\gamma = \frac{\log |a|_1}{\log |c|_1} = \frac{\log |a|_2}{\log |c|_2},$$

and $|a|_2 = |a|_1^{\alpha}$, with $\alpha = \frac{\log |c|_2}{\log |c|_1} > 0$. Since c was chosen fixed, and α does not depend upon a, this equation must hold for all $a \in \Gamma$ with the same α. The homomorphism $|\ |_2$ can thus be attained by applying first the homomorphism $|\ |_1$ and then raising to the α-th power (in $\mathbb{R}^+$).

Let us apply these results to our valuations of rank 1, by identifying Γ with the multiplicative group k^* of k. We then see: two such valuations of k, from which the same inequalities between the values of the field elements arise, are nothing but positive powers of each other. This leads to the

DEFINITION. Two valuations of rank 1, $|\ |_1, |\ |_2$, are called equivalent when

$$|\ |_1 = |\ |_2^{\alpha} \quad \text{with real } \alpha > 0.$$

If, on the other hand, we start with a valuation $|\ |$ and take a positive power, α, we again get a valuation; all that is necessary is to replace the constant C in axiom 3 by C^{α}. A class of equivalent valuations consists, therefore, of all positive powers of some fixed valuation.

For $C = 1$ — in this case we speak of a *non-archimedean valuation*, in contrast to an *archimedean valuation*, where C must be greater than 1 — the valuation is a valuation by our earlier definition. It is easily seen that in this case the new equivalence relation coincides with the one previously given.

The following theorem provides a criterium as to whether a valuation is non-archimedean, i.e., as to whether one can choose $C = 1$.

THEOREM. *A valuation $|\ |$ is non-archimedean if, and only if, the values of the "natural numbers" (the multiples of* 1) *are bounded.*

PROOF. In one direction the theorem is trivial, since for a non-archimedean valuation we already have:

$$|1+1+\cdots+1| \leq 1.$$

To prove the converse we assume $|m| \leq D$, for all natural numbers m. We may assume $C = 2$, since otherwise this can be attained by going over to a suitable positive power. The triangle inequality now holds, and because $\left|\binom{n}{\nu}\right| \leq D$:

$$|a+b|^n \leq D(|a|^n + |a|^{n-1}|b| + \cdots + |b|^n) \leq D(n+1)(\operatorname{Max}(|a|,|b|))^n.$$

Thus, $|a+b| \leq \sqrt[n]{D(n+1)} \cdot \operatorname{Max}(|a|,|b|)$ and letting $n \to \infty$, $|a+b| \leq \operatorname{Max}(|a|,|b|)$, which establishes the non-archimedean character of the valuation. □

4. Valuations of the Field of Rational Numbers

As a *valuation of an integral domain* $\mathfrak{o}$, we define a mapping of $\mathfrak{o}$ into $\mathbb{R}^+ \cup \{0\}$ with the properties:

(1) $|a| = 0 \iff a = 0$,
(2) $|ab| = |a| \cdot |b|$,
(3) $|a+b| \leq C \cdot \operatorname{Max}(|a|,|b|)$, for some fixed $C \in \mathbb{R}^+$.

It is then not difficult to show that such a valuation of $\mathfrak{o}$ can be uniquely continued to a valuation of the quotient field k of $\mathfrak{o}$.

This last fact can be used to determine all valuations of the field $\mathbb{Q}$ of rational numbers. We arrived at all non-archimedean valuations earlier; these were generated by the prime numbers. We need now, therefore, only seek archimedean valuations of $\mathbb{Q}$.

Let us assume that $| \ |$ is a archimedean valuation of $\mathbb{Q}$. We can, again, assume that the constant $C = 2$, and that therefore the triangle inequality holds.

Now let m and n be integers > 1. m can then be represented in the *n-adic cipher system*:

$$m = a_0 + a_1 n + \cdots + a_r n^r \quad \text{with } 0 \leq a_i < n,\ n^r \leq m \ (a_i \text{ integers}).$$

We have

$$r \leq \frac{\log m}{\log n}.$$

For natural q the triangle inequality yields $|q| \leq q$; and therefore $|a_i| < n$ for all i, and

$$|m| < n\left(\frac{\log m}{\log n} + 1\right)(\operatorname{Max}(1,|n|))^{\frac{\log m}{\log n}}.$$

Substituting for m the positive power m^s, it follows that:

$$|m| < \sqrt[s]{n\left(s \cdot \frac{\log m}{\log n} + 1\right)}(\operatorname{Max}(1,|n|))^{\frac{\log m}{\log n}}$$

and letting $s \to \infty$:

$$|m| \leq (\operatorname{Max}(1,|n|))^{\frac{\log m}{\log n}}$$

for all integers $m, n > 1$.

Two cases are now possible:

1) There exists an integer $n > 1$ with $|n| \leq 1$.

But then $|m| \leq 1$ for all integers m, and thereby $| \ |$ is a non-archimedean valuation, against the hypothesis.

2) $|n| > 1$ for every integer $n > 1$. Then

$$|m| \leq |n|^{\frac{\log m}{\log n}},$$

or

$$|m|^{\frac{1}{\log m}} \leq |n|^{\frac{1}{\log n}} \quad \text{for all integers } m, n > 1.$$

Exchanging m and n shows that the equality must hold, and hence

$$|n|^{\frac{1}{\log n}} = e^{\alpha} \quad (\alpha > 0),$$

where α is independent of n, and we have

$$|n| = e^{\alpha \log n} = n^{\alpha} \quad \text{for all integers } n > 1.$$

Since this equation holds for $n = 1$ and $n = 0$ as well, and because $|-1| = 1$, we have:

$$|n| = \begin{cases} n^{\alpha} & \text{for non-negative } n, \\ (-n)^{\alpha} & \text{for negative } n. \end{cases}$$

This valuation of the integral domain of rational integers can be extended to a valuation of $\mathbb{Q}$ in one and only one way, and since

$$|a| = \begin{cases} a^{\alpha} & \text{for non-negative } a, \\ (-a)^{\alpha} & \text{for negative } a \end{cases}$$

is a valuation of $\mathbb{Q}$, and, since it coincides with the above valuation on the rational integers, it must be the archimedean valuation of $\mathbb{Q}$ with which we started. Thus, we have proven:

THEOREM. *Up to equivalence (i.e. positive powers), the field $\mathbb{Q}$ of rational numbers has only one archimedean valuation: the usual absolute value.*

5. The Approximation Theorem. (Independence Th.)

The question to be considered is whether a connection can be found between the various inequivalent valuations of rank 1 of a field k.

Let $|\ |_1, |\ |_2, \ldots, |\ |_n$ be a (finite) system of non-trivial inequivalent valuations of rank 1 of k. Since we are here concerned with equivalence classes, we may immediately assume that the triangle inequality is satisfied.

We will show that these n valuations are, to a great degree, independent, in the sense that none of them can be expressed in terms of the others.

LEMMA. *There exists an element $a \in k$, such that*

$$|a|_1 > 1, \quad \textit{but} \quad |a|_i < 1 \quad \textit{for } i = 2, 3, \ldots, n.$$

The proof is by induction on the number n of valuations in the system.

For $n = 1$ the assertion is trivial, for $n = 2$ we can show it to be correct by the following argument: Because the valuations are, by assumption, inequivalent, there exists a $b \in k$ with $|b|_1 < 1$, $|b|_2 \geq 1$, and also a $c \in k$ with $|c|_1 \geq 1$, $|c|_2 < 1$.

The element $a = c/b \in k$ then has the required property,

$$|a|_1 > 1, \quad |a|_2 < 1.$$

Let us therefore now assume that $n \geq 3$, that the lemma is correct for systems of at most $n-1$ inequivalent non-trivial valuations of rank 1, and that $|\ |_1, |\ |_2, \ldots, |\ |_n$

is a system of n non-equivalent valuations of rank 1. There then exist elements b and c of k such that

$$|b|_1 > 1, \quad |b|_i < 1 \quad (i = 2, 3, \ldots, n-1),$$
$$|c|_1 > 1, \quad |c|_n < 1.$$

We distinguish two cases:

Case 1) $|b|_n \leq 1$. We then set $a = b^r c$ (r natural), which yields:

$$|a|_1 = |b|_1^r |c|_1 > 1,$$
$$|a|_i = |b|_i^r |c|_i < 1 \quad \text{for sufficiently large } r \ (i = 2, 3, \ldots, n-1),$$
$$|a|_n = |b|_n^r |c|_n < 1.$$

Case 2) Let $|b|_n > 1$. Here we set $a = \frac{b^r c}{b^r + 1}$, and, applying the triangle inequality in the form $|\alpha + \beta| \geq |\alpha| - |\beta|$, we see

$$|a|_1 \geq \frac{|b|_1^r \cdot |c|_1}{|b|_1^r + 1} \to |c|_1 > 1 \quad \text{with } r \to \infty,$$
$$|a|_i \leq \frac{|b|_i^r \cdot |c|_i}{1 - |b|_i^r} \to 0 \quad \text{as } r \to \infty \ (i = 2, 3, \ldots, n-1),$$
$$|a|_n \leq \frac{|b|_n^r \cdot |c|_n}{|b|_n^r - 1} \to |c|_n < 1 \quad \text{as } r \to \infty.$$

Thus, r can be chosen large enough that all the requirements of the lemma are fulfilled.

LEMMA. *For every real $\varepsilon > 0$ there exists an element $b \in k$ so that the following inequalities are satisfied*:

$$|b - 1|_1 < \varepsilon, \quad |b|_i < \varepsilon \quad (i = 2, 3, \ldots, n)$$

Expressed visually, the lemma states that k contains an element that lies close to 1 in the $|\ |_1$-topology, and close to 0 in all the other $|\ |_i$-topologies. This can be so stated, since a valuation of rank 1 induces a topology into k, defined by the neighborhoods of 0: $\{a \mid |a| < \varepsilon\}$.

The proof makes use of the first lemma, which assures the existence of an $a \in k$ with

$$|a|_1 > 1, \quad |a|_i < 1 \quad (i = 2, 3, \ldots, n).$$

For the element $b = \frac{a^r}{a^r + 1}$ we then have

$$|b - 1|_1 = \left| \frac{1}{a^r + 1} \right|_1 \leq \frac{1}{|a|_1^r - 1} < \varepsilon,$$
$$|b|_i = \left| \frac{a^r}{a^r + 1} \right|_i \leq \frac{|a|_i^r}{1 - |a|_i^r} < \varepsilon \quad (i = 2, 3, \ldots, n)$$

for sufficiently large r.

The following theorem is now easily proven:

APPROXIMATION THEOREM (Theorem of Independence). *Let there be given n elements, $\beta_1, \beta_2, \ldots, \beta_n \in k$ and any real number $\varepsilon > 0$. Then there exists an element $a \in k$, such that*

$$|a - \beta_i|_i < \varepsilon \quad \textit{for } i = 1, 2, \ldots, n.$$

In other words, the elements β_i can, each in their corresponding $|\ |_i$-topology, be approximated to any degree by a single element a.

For the proof, we first find, for every i an element b_i, such that

$$|b_i - 1|_i < \varepsilon', \quad |b_i|_j < \varepsilon' \quad (\text{for } j \neq i),$$

to a given $\varepsilon' > 0$. Using the triangle inequality, it is then easy to show that the element $a = \beta_1 b_1 + \beta_2 b_2 + \cdots + \beta_n b_n \in k$ satisfies the inequalities of the theorem, if a suitably small ε' is chosen.

This theorem implies the following important

COROLLARY. *A relation of the form*

$$\prod_{i=1}^{n} |a|_i^{c_i} = 1$$

with real c_i, can hold for all $a \in k$ only if all the $c_i = 0$ $(i = 1, 2, \ldots, n)$.

6. The Product Formula in Fields of Algebraic Numbers and Functions

The Approximation Theorem, and the Corollary derived from it, show clearly the mutual independence of any finite number of inequivalent non-trivial valuations of rank 1. If, however, we consider infinitely many valuations, this no longer holds true, as can be seen from the following two examples.

1) Let $k = \mathbb{Q}$ be the field of rational numbers. We have already set up a complete system of inequivalent valuations for this field. By, where necessary, taking positive powers of these valuations, we can consider our system normalized as follows:

$$|a|_p = \left(\frac{1}{p}\right)^{\operatorname{ord}_p a} \qquad (a \neq 0)$$

is the representative of the equivalence class of valuations defined by the prime number p, and

$$|a|_\infty = \begin{cases} -a & \text{for negative } a, \\ a & \text{for non-negative } a \end{cases}$$

is the representative of the archimedean valuations (so that we have, in effect ∞ as the "infinite prime number").

The product

$$\varphi(a) = \prod_p |a|_p^{c_p},$$

taken over all prime numbers (including infinity), is meaningful for any $a \in k^*$, since, because $|a|_p = 1$ for almost all p (except for a finite number of p), it is a finite product for any fixed $a \in k^*$.

Because of the multiplicative property of valuations

$$|ab|_p = |a|_p \cdot |b|_p,$$

$\varphi(a)$ has the functional equation

$$\varphi(ab) = \varphi(a) \cdot \varphi(b),$$

and since furthermore, $\varphi(-1) = 1$, it suffices to know $\varphi(p)$ for all prime numbers p $(p \neq \infty)$. For these, however,

$$|p|_q = 1 \quad \text{for } q \neq p,\ q \neq \infty,$$

and therefore,

$$\varphi(p) = |p|_\infty^{c_\infty} |p|_p^{c_p} = p^{c_\infty} \left(\frac{1}{p}\right)^{c_p} = p^{c_\infty - c_p}.$$

This yields the theorem:

THEOREM. *In the field* $\mathbb{Q}$ *of rational numbers the* product formula

$$\prod_p |a|_p = 1$$

is true for all $a \in \mathbb{Q}$, $a \neq 0$, *and it is, essentially (except for powers), the only such formula.*

2) Such a product formula also holds in the *field of rational functions of one variable*; it then yields a familiar theorem of the theory of functions. As we have not yet examined the valuations of this field, we must now do so.

Let k_0 be any field which we extend by adjunction of an element x, transcendental over k_0, to the field of rational functions in x over the field of constants k_0:

$$k = k_0(x).$$

We do not, however, consider all possible valuations of k, rather just those which are trivial on the field of constants k_0; it is these valuations that particularly bring out the character of the field k as a field of rational functions over k_0. These valuations are necessarily non-archimedean, since all multiples of 1 lie in k_0, and hence their absolute values are certainly bounded.

For any given non-trivial, non-archimedean valuation $|\ |$ of k, trivial on k_0, either $|x| \leq 1$ or $|x| > 1$ must be the case.

Case 1) $|x| \leq 1$.

This brings with it: $|f(x)| \leq 1$ for every $f(x) \in k_0[x]$. If $\mathfrak{o}$ is the corresponding valuation ring, and $\mathfrak{p}$ its maximum ideal, then

$$k_0[x] \in \mathfrak{o}.$$

$k_0[x] \cap \mathfrak{p}$ is a prime ideal in $k_0[x]$. The valuation is, by assumption, non-trivial, hence $k_0[x] \neq (0)$, and is therefore generated by an irreducible polynomial:

$$k_0[x] \cap \mathfrak{p} = (p(x)); \quad p(x) \in k_0[x], \text{ irreducible.}$$

As in the field of rational numbers, this leads to an additively written valuation ord_p, or to the multiplicatively written valuation:

$$|g(x)| = d^{\operatorname{ord}_p g(x)}$$

with any real number d such that $0 < d < 1$. On the other hand, it is easily seen that any irreducible polynomial $p(x) \in k_0[x]$ yields such a valuation.

In the usual theory of functions one would understand, for the order of the corresponding zero of $g(x)$ the product of $\operatorname{ord}_p g(x)$ with the degree g_p of $p(x)$. We therefore want to normalize our valuations in such a way that the degree g_p of $p(x)$ is taken care of by the valuation itself. We set

$$d = c^{g_p},$$

where c is any real number with $0 < c < 1$, which, however, is held fixed for all valuations. We thus arrive at the normalized valuations

$$|g(x)|_p = (c^{g_p})^{\operatorname{ord}_p g(x)}.$$

Case 2) $|x| > 1$.

Let $f(x) = a_n x^n + a_{n-1}x^{n-1} + \cdots + a_0 \in k_0[x]$, $a_n \neq 0$. Among the values $|a_i x^i| = |x|^i$ $(i = 0, 1, \ldots, n)$ $|a_n x^n|$ is now greater than any other.

$$\implies |f(x)| = |a_n x^n| = |x|^n = x^{\deg f}$$

$$\implies \left|\frac{f(x)}{g(x)}\right| = |x|^{\deg f - \deg g}.$$

If this is a valuation, it is certainly uniquely defined by the value of x, and hence, up to equivalence, is the only valuation in this case. On the other hand, if we replace x by $1/x$, then the considerations for the first case show that a valuation with $\left|\frac{1}{x}\right| < 1$ exists (in particular, the valuation generated by the polynomial $p\left(\frac{1}{x}\right) = \frac{1}{x}$) and that in this valuation $|x| > 1$. But, as we saw above, if a valuation with $|x| > 1$ exists, it must be the function computed above, which is now, therefore, a valuation.

This valuation is normalized, using the same real number c as in the first case, to

$$\left|\frac{f(x)}{g(x)}\right|_\infty = \left(\frac{1}{c}\right)^{\deg f - \deg g},$$

In the same manner as it was shown for the field of rational numbers, it can be shown here that the product formula

$$\prod_p |a|_p = 1 \quad \text{for all } a \in k,\ a \neq 0$$

is true, the product being taken over all irreducible polynomials $p(x)$ (including ∞).

If k_0 is taken as the field of complex numbers, then every irreducible polynomial is of the first degree, and the product formula yields the following theorem: A rational function of one variable has the same number of poles and zeros on the Riemann Sphere.

It will just be mentioned here that the product formula also holds for finite extensions of the field of rational numbers, and for finite extensions of a field of rational functions, but only so far as the existence of the formula; its uniqueness for this type of formula no longer remains.

CHAPTER 4

NORMED FIELDS AND VECTOR SPACES

1. The Theorem of Gelfand–Tornheim

At a later point, the question will be raised as to which fields admit an archimedean valuation. In this study a theorem will be needed, which will now be derived.

Let $\mathbb{R}$ be the field of real numbers, with the usual absolute valuation $|\ |$, and $\mathbb{C}$ the field of complex numbers. For fields which contain $\mathbb{R}$ as a subfield we want to replace the valuation concept by one somewhat more general.

DEFINITION. A field, k, containing $\mathbb{R}$, is called normed if there exists a map

$$a \to \|a\| \quad (\text{Norm of } a)$$

of k into $\mathbb{R}^+ \cup \{0\}$, with the following properties:

(1) $\|a\| = 0 \iff a = 0$,
(2) $\|ab\| \leq \|a\| \cdot \|b\|$,
(3) $\|a + b\| \leq \|a\| + \|b\|$,
(4) $\|ab\| = |a| \cdot \|b\|$, for $a \in \mathbb{R}$, $b \in k$.

COROLLARY. $a \in \mathbb{R} \implies \|a\| = \|a \cdot 1\| = |a| \cdot \|1\|$; *the norm of elements of $\mathbb{R}$ differs from their absolute value by a fixed real factor only. Hence, the same topology is induced onto $\mathbb{R}$ by the norm as by the usual absolute value.*

The following theorem then holds:

THEOREM (Gelfand–Tornheim). *A normed field, as here defined, must either be equal to the field $\mathbb{R}$ of real numbers or equal to the field $\mathbb{C}$ of complex numbers.*

This theorem is originally due to Gelfand (Sbornik 1941), however with the more special assumption that k contains $\mathbb{C}$. It was then carried over to the theory of normed rings, and ever since has been a strong tool of modern analysis (cf., e.g. Loomis, Abstract Analysis). The original proof uses function-theoretical methods. It was in our general form that Tornheim proved the theorem purely algebraically. Another proof, by Ostrowski, limits itself to valuations of k. We reproduce here Tornheim's proof.

First we adjoin to the field k a root i of the equation

$$x^2 + 1 = 0.$$

The resulting field, $k(i)$ can possibly be the field k itself; if this is not the case, though, we must note that the field $k(i)$ is no longer normed, since the norm is only defined on k.

Let us assume the theorem is wrong; then it is possible to choose a fixed element, y, such that

$$y \in k, \quad y \notin \mathbb{C} = \mathbb{R}(i).$$

This will be led to a contradiction.

By our choice of y, for any $c \in \mathbb{C}$, $y - c \neq 0$, which implies the existence of

$$\frac{1}{y-c} \quad \text{for all } c \in \mathbb{C}.$$

Setting $c = a + ib$, $\bar{c} = a - ib$ $(a, b \in \mathbb{R})$, we have

$$A = \frac{1}{y-c} + \frac{1}{y-\bar{c}} = \frac{2(y-a)}{(y-a)^2 + b^2},$$

an element of k for any choice of $c \in \mathbb{C}$, of which we can therefore take the norm

$$\|A\| = \left\| \frac{2(y-a)}{(y-a)^2 + b^2} \right\| = \left\| \frac{2(y-a)}{y^2 - 2ay + a^2 + b^2} \right\|.$$

We will now show that $\|A\|$ is a continuous function of a and b (and therefore of c), i.e. that the map

$$c = a + ib \to \|A\|$$

of $\mathbb{C}$ into $\mathbb{R}$ is continuous, where continuous is to be understood in the sense of the topology induced by the usual absolute value $|\ |$ in $\mathbb{C}$ and in $\mathbb{R}$. To show this we split the map $c \to \|A\|$ into two maps

1) $c \to A$ (map of $\mathbb{C}$ into k),
2) $A \to \|A\|$ (map of k into $\mathbb{R}$) ,

and demonstrate that both of these maps are continuous, using in k the topology induced by the norm $\|\ \|$.

That the map 2) is continuous follows directly from the easily arrived at form of the triangle inequality for the norm:

$$\big|\|x\| - \|x_0\|\big| \leq \|x - x_0\|.$$

To demonstrate that the map 1) is also continuous we first consider the single terms of the quotient A. The expressions $2y, y^2, 2a, 2ay$, and $a^2 + b^2$ are all continuous in a and b, as, for example,

$$\|2ay - 2a_0y\| = \|2(a - a_0)y\| = 2|a - a_0| \cdot \|y\|.$$

If $u, v \in k$, then the map

$$(u, v) \to u + v$$

of $k \times k$ into k is a continuous map. Since a continuous function of a continuous function is again continuous, it follows thar both numerator and denominator of A, separately, are continuous in a and b.

Because

$$A = 2(y-a) \cdot \frac{1}{y^2 - 2ay + a^2 + b^2},$$

it now suffices to verify the following two statements:

a) If x is a variable in k, then $\frac{1}{x}$ is continuous for $x \neq 0$,
b) The map $(u, v) \to uv$ of $k \times k$ into k is continuous.

The assertion b) is immediately attained from

$$\begin{aligned}
\|uv - u_0v_0\| &= \|\{u_0 + (u - u_0)\} \cdot \{v_0 + (v - v_0)\} - u_0v_0\| \\
&= \|u_0(v - v_0) + (u - u_0)v_0 + (u - u_0)(v - v_0)\| \\
&\leq \|u_0\| \cdot \|v - v_0\| + \|u - u_0\| \cdot \|v_0\| + \|u - u_0\| \cdot \|v - v_0\|.
\end{aligned}$$

To prove a) we majorize as follows:

$$\left\|\frac{1}{x}-\frac{1}{x_0}\right\| = \left\|\frac{x_0-x}{xx_0}\right\| \le \|x-x_0\|\cdot\left\|\frac{1}{x_0}\right\|\cdot\left\|\frac{1}{x}\right\|,$$

and therefore now only have to show that $\frac{1}{x}$ remains bounded in a neighborhood of $x_0 \neq 0$. If we now set

$$z = \frac{x-x_0}{x_0},$$

then,

$$\|z\| \le \|x-x_0\|\cdot\left\|\frac{1}{x_0}\right\|,$$

$$\left\|\frac{1}{x}\right\| = \left\|\frac{1}{x_0+(x-x_0)}\right\| \le \left\|\frac{1}{x_0}\right\|\cdot\left\|\frac{1}{1+z}\right\|,$$

and the boundedness of $\left\|\frac{1}{x}\right\|$ in a neighborhood of x_0 follows from the boundedness of $\left\|\frac{1}{1+z}\right\|$ in a neighborhood of 0, which in its turn can be proven for $\|z\| \le \frac{1}{2}$, where the inequality

$$\left\|\frac{1}{1+z}\right\| = \left\|1-\frac{z}{1+z}\right\| \le \|1\|+\|z\|\left\|\frac{1}{1+z}\right\| \le \|1\|+\frac{1}{2}\left\|\frac{1}{1+z}\right\|$$

holds true, and can be restated as

$$\left\|\frac{1}{1+z}\right\| \le 2\cdot\|1\|.$$

Thus, it has been shown that A, and therefore also $\|A\|$ is continuous in c.

We consider large values of $|c|$, and rewriting:

$$A = \frac{2y-(c+\bar{c})}{y^2-(c+\bar{c})y+c\bar{c}} = \frac{\frac{2y}{c\bar{c}}-\left(\frac{1}{c}+\frac{1}{\bar{c}}\right)}{\frac{y^2}{c\bar{c}}-\left(\frac{1}{c}+\frac{1}{\bar{c}}\right)y+1},$$

we have, for $|c|\to\infty$: $\frac{1}{c\bar{c}}\to 0$, $\frac{1}{c}+\frac{1}{\bar{c}}\to 0$, and then, because $\|A\|$ is continuous, as already proven,

$$\|A\| = \left\|\frac{1}{y-c}+\frac{1}{y-\bar{c}}\right\| \to 0.$$

Hence, the maximum

$$M = \mathrm{Max}_{c\in\mathbb{C}}\left\|\frac{1}{y-c}+\frac{1}{y-\bar{c}}\right\|$$

exists, and is assumed for some finite $c\in\mathbb{C}$. For $c=0$ the corresponding value

$$\|A\| = \frac{2}{y} \neq 0,$$

therefore $M\neq 0$; hence $M>0$.

We now let T be the set of all $c\in\mathbb{C}$ for which the corresponding $\|A\|$ takes on the maximum value M. Since T is the inverse image of the point M by a continuous map, T is closed, and since it is also bounded, it is compact. The contradiction will be brought about by showing that T is also open.

The idea is the following: the function $\frac{1}{y-c}+\frac{1}{y-\bar{c}}$ can be treated as the real component of the analytic function $\frac{2}{y-c}$, with values in k. The mean value of such a function on a circle is equal to the value of the function at the center of the circle;

if the function has a maximum at the center, then it must also have one almost everywhere on the circle. This yields immediately, that T is open.

This idea can be carried out purely algebraically: Let $c_0 \in \mathbb{C}$, and $c = c_0 + r\varepsilon$, where r is any real number and ε runs through the set of all the n-th roots of unity. We take the mean value

$$S_n = \frac{1}{n}\sum_{\varepsilon}\left(\frac{1}{y - c_0 - r\varepsilon} + \frac{1}{y - \bar{c}_0 - r\bar{\varepsilon}}\right),$$

or, if for simplification we write,

$$z = y - c_0, \quad \bar{z} = y - \bar{c}_0,$$

$$S_n = \frac{1}{n}\sum_{\varepsilon}\left(\frac{1}{z - r\varepsilon} + \frac{1}{\bar{z} - r\bar{\varepsilon}}\right).$$

¿From the identity

$$x^n - r^n = \prod_{\varepsilon}(x - r\varepsilon),$$

by differentiating logarithmically we get,

$$\frac{nx^{n-1}}{x^n - r^n} = \sum_{\varepsilon}\frac{1}{x - r\varepsilon}.$$

As ε runs through all of the n-th roots of unity, so does $\bar{\varepsilon}$, hence

$$S_n = \frac{z^{n-1}}{z^n - r^n} + \frac{\bar{z}^{n-1}}{\bar{z}^n - r^n}.$$

To study the behaviour, for small r, of S_n as $n \to \infty$, we rewrite as follows:

$$S_n = \frac{\frac{1}{z}}{1 - \frac{r^n}{z^n}} + \frac{\frac{1}{\bar{z}}}{1 - \frac{r^n}{\bar{z}^n}} = \frac{\left(\frac{1}{z} + \frac{1}{\bar{z}}\right) - \frac{r^n}{z\bar{z}}\left(\frac{1}{z^{n-1}} + \frac{1}{\bar{z}^{n-1}}\right)}{1 - r^n\left(\frac{1}{z^n} + \frac{1}{\bar{z}^n}\right) + \frac{r^{2n}}{(z\bar{z})^n}}.$$

If, for the moment, we set $\frac{1}{z} = u + iv$, $\frac{1}{\bar{z}} = u - iv$ $(u, v \in k)$, then

$$\begin{aligned}
\left\|\frac{1}{z^q} + \frac{1}{\bar{z}^q}\right\| &= \|(u + iv)^q + (u - iv)^q\| \\
&= \left\|2\sum_{\nu}\binom{q}{2\nu}u^{q-2\nu}v^{2\nu}\right\| \\
&\le 2\sum_{\nu}\binom{q}{2\nu}\|u\|^{q-2\nu}\|v\|^{2\nu} \\
&\le 2\sum_{\mu}\binom{q}{\mu}\|u\|^{q-\mu}\|v\|^{\mu} \\
&= 2(\|u\| + \|v\|)^q,
\end{aligned}$$

increases slower than the power of a fixed number as $q \to \infty$. The same is true of $\frac{1}{(z\bar{z})^n}$, because $\left\|\frac{1}{(z\bar{z})^n}\right\| \leq \left\|\frac{1}{z\bar{z}}\right\|^n$. For sufficiently small r we therefore have:

$$\left.\begin{aligned} r^n\left(\frac{1}{z^n}+\frac{1}{\bar{z}^n}\right) &\to 0 \\ r^n\left(\frac{1}{z^{n-1}}+\frac{1}{\bar{z}^{n-1}}\right) &\to 0 \\ \frac{r^{2n}}{(z\bar{z})^n} &\to 0 \end{aligned}\right\} \quad \text{for } n \to \infty$$

convergence being defined in the topology generated in k by $\| \ \|$.

Because of the already demonstrated continuity of a product,

$$\|S_n\| \to \left\|\frac{1}{z}+\frac{1}{\bar{z}}\right\| = \left\|\frac{1}{y-c_0}+\frac{1}{y-\bar{c}_0}\right\| \quad \text{for } n \to \infty$$

with r sufficiently small.

Now let c_0 be a boundary point of T. Then there exists a d near c_0, such that $d \notin T$. Through d we place a circle with center at c_0. If d was chosen near enough to c_0, then the above convergence of S_n takes effect for the circle radius r.

Since the complement of the closed set T is open, it contains an entire arc,

$$\zeta = c_0 + r\alpha,$$

where α runs through the closed arc L of the unit circle between two consecutively placed s-th roots of unity. Setting $n = ms$ for large m, the number of all the roots of unity on L is $m+1$. If we now let

$$M' = \mathrm{Max}_{\alpha \in L}\left\|\frac{1}{y-c_0-r\alpha}+\frac{1}{y-\bar{c}_0-r\bar{\alpha}}\right\|,$$

then $M' < M$, since the arc lies in the set complementary to T. This implies:

$$\|S_n\| \leq \frac{1}{n}(mM' + (n-m)M) = M - \frac{M-M'}{s}$$

$$\Longrightarrow \lim_{n\to\infty} S_n = \left\|\frac{1}{y-c_0}+\frac{1}{y-\bar{c}_0}\right\| \leq M - \frac{M-M'}{s} < M,$$

and therefore, c_0 is not contained in T. T is thus open, a contradiction, since we have already proven it to be closed and bounded.

The Gelfand–Tornheim Theorem is thus proven. The question as to how such a norm $\| \ \|$ of k looks, for the case $k = \mathbb{C}$, the field of complex numbers, still remains open. This question did not arise in Gelfand's investigations; the norm there was always defined, up to a constant factor, as the common absolute value. We will, indeed, later see that the norm of $\mathbb{C}$ is essentially, uniquely determined. (The "essentially" here means that, although various norms may be found, they all induce the same topology.)

2. The Perfect Completion of a Field with Valuation

To study this last problem in full generality, we want to introduce here the concept of the *perfect completion* of a field with valuation. It shall be constructed in the same manner as the real numbers are constructed from the field of rational numbers (with the usual absolute valuation), by the Cauchy sequence method of Cantor.

Our starting point is a field k with a rank-1 valuation.

A sequence of elements $a_1, a_2, a_3, \dots$ of k is called a *Cauchy sequence* (C.S.), if, to every real number $\varepsilon > 0$ there exists a natural number N such that for $\nu, \mu > N$

$$|a_\nu - a_\mu| < \varepsilon \quad \text{always holds.}$$

In the usual manner, the following can then be shown:

1.) The C.S. of k form a ring, $\mathfrak{o}$.

2.) The zero-sequences, i.e. the C.S. with limit 0, form an ideal $\mathfrak{p}$ in $\mathfrak{o}$.

3.) If $a_1, a_2, a_3, \dots$ is a C.S. not contained in $\mathfrak{p}$, and if all the $a_i \neq 0$ $(i = 1, 2, 3, \dots)$, then $1/a_1, 1/a_2, 1/a_3, \dots$ is also a C.S. in k.

4.) $\mathfrak{p}$ is a maximal ideal in $\mathfrak{o}$, i.e. the residue class ring $\mathfrak{o}/\mathfrak{p}$ is a field $\tilde{k}$.

5.) Let $\alpha \in \tilde{k}$, and let α be represented by the C.S. $a_1, a_2, a_3, \dots$. We then define:

$$|\alpha| = \lim_{i\to\infty} |a_i|.$$

This definition is meaningful, and yields a valuation of $\tilde{k}$.

6.) The field k can be mapped into the field $\tilde{k}$ by

$$a \to \bar{a},$$

where the residue class $\bar{a} \in k$ is represented by the C.S. $a, a, a, \dots$. This map is an isomorphism of k into $\tilde{k}$, and furthermore, $|a| = |\bar{a}|$.

It is customary, at this point, to identify the field k with its isomorphic image, $\bar{k}$ in $\tilde{k}$. This identification would, however, prove to be disadvantageous for our next steps, therefore we postpone it.

7.) If $\alpha \in \tilde{k}$ and α is represented by the C.S. $a_1, a_2, a_3, \dots$, then

$$\alpha = \lim_{i\to\infty} \bar{a}_i.$$

In particular, since, by definition.

$$|\alpha - \bar{a}_i| = \lim_{n\to\infty} |a_n - a_i|,$$

$\alpha - \bar{a}_i$ being represented by the C.S. $a_1 - a_i, a_2 - a_i, a_3 - a_i, \dots$, and since $a_1, a_2, a_3, \dots$ is a C.S., we have

$$\lim_{i\to\infty} |\alpha - \bar{a}_i| = \lim_{i\to\infty} \lim_{n\to\infty} |a_n - a_i| = 0.$$

Thus, we have shown that every $\alpha \in \tilde{k}$ can be approximated by elements of $\bar{k}$ to any degree of accuracy (with respect to the valuation defined). In topological terms we have: $\bar{k}$ is *dense* in $\tilde{k}$.

8.) $\tilde{k}$ is *complete* (*perfect*) with respect to the given valuation; i.e. every C.S. of $\tilde{k}$ has a limit in $\tilde{k}$.

For, let $\alpha_1, \alpha_2, \alpha_3, \dots$ be a C.S. in $\tilde{k}$. We can then find elements $a_i \in k$, such that

$$|\alpha_i - \bar{a}_i| < \frac{1}{i} \quad (i = 1, 2, 3, \dots).$$

$$\Longrightarrow \alpha_1 - \bar{a}_1,\ \alpha_2 - \bar{a}_2,\ \alpha_3 - \bar{a}_3, \dots \text{ is a C.S. of } \tilde{k}.$$

$$\Longrightarrow \bar{a}_1, \bar{a}_2, \bar{a}_3, \dots \text{ is a C.S. of } \tilde{k}$$

$$\Longrightarrow a_1, a_2, a_3, \dots \text{ is a C.S. of } k$$

This C.S. of k represents a certain element, $\beta \in \tilde{k}$, and then,

$$\lim_{i \to \infty} \alpha_i = \beta.$$

9.) Finally, we identify the field k with its isomorphic image, $\bar{k}$ in $\tilde{k}$. Then, $\tilde{k}$ is an overfield to k, the valuation defined for $\tilde{k}$ is an extension of the given valuation of k, and $\tilde{k}$ is complete with respect to this valuation. $\tilde{k}$ is called the *perfect completion* of k.

3. Normed Vector Spaces over Complete Fields

We now consider a field k with a non-trivial valuation of rank 1, $| \ |$, such that k is complete with respect to this valuation.

Let V be a *vector space* over k. V is called *normed* if to every element A of V there corresponds a real number $\|A\|$, the norm of A, such that:

(1) $\|A\| > 0$; $\|A\| = 0 \iff A = 0$,
(2) $\|A + B\| \le \|A\| + \|B\|$,
(3) $\|aA\| = |a| \cdot \|A\|$ for every $a \in K$.

Such a norm $\| \ \|$ induces in V a definite topology, in which the neighborhoods of zero are defined by the "spheres"; $\|X\| < c$ (c real).

An additional assumption is made for V:

$$n = \dim V < \infty.$$

Under this assumption the following theorem can be proven:

THEOREM. *Any two norms in V induce the same topology.*

Two topologies are here considered to be the same if any neighborhood of zero in one topology contains a neighborhood of zero to the other, and vice-versa.

To prove the theorem it suffices to show that the topology induced by any norm $\| \ \|$ is the same as the topology induced by a certain given norm $\| \ \|_0$.

Let $A_1, A_2, \ldots, A_n$ be a basis of V over k. Any vector X of V can be written

$$X = x_1 A_1 + x_2 A_2 + x_3 A_3 + \cdots + x_n A_n, \quad x_i \in k.$$

The norm $\| \ \|_0$ will be defined as:

$$\|X\|_0 = \text{Max}(|x_1|, |x_2|, \ldots, |x_n|).$$

The norm properties can immediately be derived from the valuation properties in k.

Let $\| \ \|$ be any norm in V. Then certainly

$$\|X\| \le |x_1| \cdot \|A_1\| + |x_2| \cdot \|A_2\| + \cdots + |x_n| \cdot \|A_n\| \implies \|X\| \le c \cdot \|X\|_0$$

with the fixed real number $c = \|A_1\| + \|A_2\| + \cdots + \|A_n\|$. In every neighborhood of zero, $\|X\| < r$, there is therefore contained a neighborhood $\|X\|_0 < \frac{r}{c}$.

The converse of this last fact is proven by induction on the dimension n of the vector space. For $n = 1$ we have

$$\|X\| = |x_1| \cdot \|A_1\| = c \cdot \|X\|_0, \quad c = \|A_1\|,$$

and the equality of the two topologies is evident.

Let us now assume the equality of the topologies to be proven for all vector spaces with dimension less than n. Let us consider the $n-1$ dimensional subspace U of V:

$$U = \{x_1 A_1 + x_2 A_2 + \cdots + x_{n-1} A_{n-1} \mid x_i \in k\},$$

and the two norms induced on U by the norms $\| \ \|$ and $\| \ \|_0$. By the induction assumption the two thus induced topologies of U are equal.

In U, let $B_1, B_2, \ldots$ be a Cauchy sequence by the norm $\| \ \|$; it is then also a Cauchy sequence by the norm $\| \ \|_0$.

If $B = x_1 A_1 + \cdots + x_{n-1} A_{n-1}$, and if $\|B\|_0$ is small, then all the components $x_1, x_2, \ldots, x_{n-1}$ are small by the valuation of k. Hence, the components of the C.S. $B_1, B_2, \ldots$ corresponding to a single fixed A_ν, $x_{1\nu}, x_{2\nu}, x_{3\nu}, \ldots$ themselves are a C.S. in k, and converge in k, since it is a complete field. But then the sequence $B_1, B_2, \ldots$ also converges in U.

Since this holds for every Cauchy sequence in U, U is a closed set. The map

$$X \to A_n + X, \quad X \in V,$$

is a one-to-one topology preserving map of V onto itself (in both topologies). The image set

$$A_n + U$$

of the closed set U is therefore closed in both topologies, and its complement is open.

The zero vector 0 is not contained in $A_n + U$, since a representation of the form

$$0 = x_1 A_1 + x_2 A_2 + \cdots + x_{n-1} A_{n-1} + A_n, \quad x_i \in k,$$

would contradict the linear independence of the A_i. Hence there exists an entire neighborhood of 0 in the $\| \ \|$-topology that does not intersect with $A_n + U$. We can therefore find a real number $d > 0$ with the property:

$$X = x_1 A_1 + \cdots + x_n A_n, \ \|X\| < d \implies X \notin A_n + U, \text{ i.e. } x_n \neq 1.$$

If, now, for the same X, $|x_n| \geq 1$, then $\left\| \frac{X}{x_n} \right\| = \frac{\|X\|}{|x_n|} < d$, which is impossible, since $\frac{X}{x_n} \in A_n + U$. We have thus shown

$$\|X\| < d \implies |x_n| < 1.$$

The outstanding place of A_n in the preceding considerations is completely arbitrary. Such a real number $d > 0$ can be found for each A_i $(i = 1, 2, \ldots, n)$. If we replace all these numbers d by the smallest of them, say e, we have: There exists a real number $e > 0$ with the property:

$$\|X\| < e \implies \|X\|_0 < 1.$$

To any given real $\varepsilon > 0$ there exists an element $a \in k$ with $a \neq 0$, $|a| < \varepsilon$. Then,

$$\|X\| < e \cdot |a| < e\varepsilon \implies \left\| \frac{X}{a} \right\| < e \implies \left\| \frac{X}{a} \right\|_0 < 1 \implies \|X\|_0 < |a| < \varepsilon.$$

In every neighborhood $\|X\|_0 < |a|$ there exists a neighborhood $\|X\| < e|a|$.

Together with the already proven converse, this shows the equality of the two topologies.

CHAPTER 5

EXTENSIONS OF VALUATIONS

1. Extendability of a General Valuation

In the following the important question as to whether a valuation can be extended onto the extension of a field will be taken up. For the time we want to limit ourselves to non-archimedean valuations (not necessarily valuations of rank 1). We can, here, immediately eliminate the case of the trivial valuation, which can naturally be extended by the trivial valuation of the overfield.

Let k, therefore, be a field with a non-trivial, non-archimedean valuation $|\ |$, that maps k onto an ordered group $\mathfrak{G}$ with zero element.

$$k \xrightarrow{|\ |} \mathfrak{G}.$$

Let K be any field extension of k. Sought is a valuation of K, such that the valuation group contains the group $\mathfrak{G}$, and that coincides on k with the given valuation.

By the extension theorem we already know that a given place of k can be extended to a place of the overfield K. In particular, if

$$\varphi\colon k \to F \cup \{\infty\}$$

is a non-trivial place of k with the corresponding valuation ring $\mathfrak{o}$, then

$$\varphi|\mathfrak{o}\colon \mathfrak{o} \to F$$

is a non-trivial homomorphism of $\mathfrak{o}$. As $\mathfrak{o}$ is also a subring of K, the extension theorem shows the existence of a place ϕ of K with

$$\phi|\mathfrak{o} = \varphi|\mathfrak{o}.$$

But further,

$$\begin{aligned} a \in k,\ a \notin \mathfrak{o} &\implies \varphi(a) = \infty \implies \varphi(a^{-1}) = 0 \\ &\implies \phi(a^{-1}) = 0 \implies \phi(a) = \infty \implies \phi|k = \varphi|k, \end{aligned}$$

and ϕ extends the place φ onto K.

It therefore suffices to show the following: If φ is the place of k corresponding to the valuation $|\ |$, and if ϕ is a place of K which extends φ, then the valuation $|\ |_K$, generated by ϕ on K is, by suitable choice of the valuation group (which is, of course, only determined up to isomorphism), is an extension of the valuation $|\ |$ of k.

The valuation ring $\mathfrak{O}$ corresponding to ϕ is characterized as

$$\begin{aligned} \mathfrak{O} &= \{\alpha \mid \alpha \in K,\ \phi(\alpha) \text{ finite}\} \\ &= \{\alpha \mid \alpha \in K,\ |\alpha|_K \le 1\}, \end{aligned}$$

while the valuation ring $\mathfrak{o}$ of φ, because $\phi|k = \varphi|k$, can be characterized as

$$\begin{aligned}\mathfrak{o} &= \{a \mid a \in k,\ \phi(a) \text{ finite}\}\\ &= \{a \mid a \in k,\ |a|_K \leq 1\}.\end{aligned}$$

The two following statements therefore hold:

1.) ϕ extends $\varphi \implies \mathfrak{o} = \mathfrak{O} \cap k$,
2.) $|\ |_K$ extends $|\ | \implies \mathfrak{o} = \mathfrak{O} \cap k$.

The extendability of valuations would be immediately proven if it were demonstrated that the converse of these statements are true (by suitable choice of the image field F of ϕ and the valuation group of $|\ |_K$). Extension of the valuation would then be equivalent to extension of the place.

To prove this converse we start out with the two valuation rings $\mathfrak{O}$ and $\mathfrak{o}$ of K and k respectively, where $\mathfrak{o} = \mathfrak{O} \cap k$. Let $\mathfrak{P}$ and $\mathfrak{p}$ be the respective prime ideals, U_K and U_k the corresponding groups of units. Then

$$\mathfrak{P}^{-1} = \{\alpha^{-1} \mid \alpha \in \mathfrak{P}\}$$

is the complement of $\mathfrak{O}$ in K, and

$$\mathfrak{p}^{-1} = \{a^{-1} \mid a \in \mathfrak{p}\}$$

is the complement of $\mathfrak{o}$ in k. Thus,

$$\mathfrak{o} = \mathfrak{O} \cap k \implies \mathfrak{p}^{-1} = \mathfrak{P}^{-1} \cap k \implies \mathfrak{p} = \mathfrak{P} \cap k \implies U_k = U_K \cap k,$$

U_k is a subgroup of U_K.

I. As the valuations corresponding to $\mathfrak{O}$ and $\mathfrak{o}$, we define $|\ |_K$ and $|\ |_k$, respectively, as follows:

In K: $|\alpha|_K = \alpha U_K$, with the ordering relation $|\alpha|_K < 1 \implies \alpha \in \mathfrak{P}$.
In k: $|a|_k = aU_K$, with the ordering relation $|a|_K < 1 \implies a \in \mathfrak{p}$.

The map

$$aU_k \to aU_K$$

of the group of values aU_k into the group of values aU_K is well-defined, as is seen by $aU_K = (aU_k)U_K$, and is a homomorphism. It is even an isomorphism, since the kernel of the map consists of the unity of U_k only:

$$\begin{aligned}a \in k,\ aU_K = U_K &\iff a \in k,\ a \in U_K\\ &\iff a \in U_K \cap k = U_k\\ &\iff aU_k = U_k.\end{aligned}$$

The ordering is also invariant under this isomorphism:

$$\begin{aligned}a \in k,\ aU_k < 1 &\iff a \in \mathfrak{p} = \mathfrak{P} \cap k\\ &\iff a \in k,\ a \in \mathfrak{P}\\ &\iff a \in k,\ aU_K < 1.\end{aligned}$$

If the aU_k are identified with their images aU_K, then $|\ |_K$ is indeed an extension of $|\ |_k$.

II. As places ϕ and φ, corresponding to $\mathfrak{O}$ and $\mathfrak{o}$, respectively, we choose:

$$\text{In } K\colon \quad \phi(\alpha) = \begin{cases} \infty, & \alpha \notin \mathfrak{O}, \\ \alpha + \mathfrak{P}, & \alpha \in \mathfrak{O}. \end{cases}$$

$$\text{In } k\colon \quad \phi(a) = \begin{cases} \infty, & a \notin \mathfrak{o}, \\ a + \mathfrak{p}, & a \in \mathfrak{o}. \end{cases}$$

We map $\varphi(k)$ into $\phi(K)$:

$$\infty \to \infty,$$
$$a + \mathfrak{p} \to a + \mathfrak{P}.$$

¿From $a + \mathfrak{P} = (a + \mathfrak{p}) + \mathfrak{P}$ it follows that the map is well-defined, furthermore, its limitation on $\varphi(\mathfrak{o}) = \mathfrak{o}/\mathfrak{p}$ is an isomorphism. The residue class field $\mathfrak{o}/\mathfrak{p}$ can be isomorphically injected into the residue class field $\mathfrak{O}/\mathfrak{P}$. Let us again identify the inverse image with its isomorphic image; the place ϕ is then an extension in K of the place φ of k. We summarize our results in the

THEOREM. *Any non-archimedean valuation of a field k can be extended onto any extension field K of k.*

2. Ramification and Residue Class Degree

Now that we have determined the extendability of a valuation of a field to an overfield, the problem presents itself as to whether an extension of a valuation of rank 1 is also a valuation of this type. In general this is not the case; we will, however, show that it is always true for the case of a finite field extension.

The case of a trivial valuation will be handled straight away. Let $|\ |$ be the trivial valuation of k. An extension of $|\ |$ on some overfield of k can then very well be non-trivial, as our example of a field of algebraic functions showed.

If we assume, however, that K is an algebraic extension of k (not necessarily finite), we can show that only the trivial valuation of K extends the trivial valuation of k. For, if we had $\alpha \in K$, $|\alpha| > 1$, in some valuation $|\ |$ of K which is trivial on k, and if α satisfied the equation

$$\alpha^n + a_1\alpha^{n-1} + \cdots + a_n = 0, \quad a_i \in k,$$

this would imply

$$|\alpha^n + a_1\alpha^{n-1} + \cdots + a_n| = |\alpha|^n > 1,$$

since $|a_i| \leq 1$, and therefore

$$|a_i\alpha^{n-i}| = |a_i||\alpha|^{n-i} \leq |\alpha|^{n-i} \leq |\alpha|^n \quad \text{for } i = 1, 2, 3, \ldots, n.$$

This is in contradiction to the algebraic equation for α, and thus

$$|\alpha| \leq 1 \quad \text{for all } \alpha \in K,$$

which implies

$$|\alpha| = 1 \quad \text{for all } \alpha \in K,\ \alpha \neq 0.$$

Now let $|\ |$ be a non-trivial, non-archimedean valuation of K (not necessarily a valuation of rank 1), ϕ the corresponding place, and k some subfield of K. The limitation of the valuation $|\ |$ and the place ϕ onto k yields a valuation $|\ |$ and a place φ, respectively, of k. We will introduce two invariants that serve to describe the behaviour of an extension of this type.

The group of values $|K^*|$ contains the group of values $|k^*|$ (the latter being a subgroup of the former). The index

$$e = (|K^*| : |k^*|)$$

is called the *ramification*. It is a measure of how many new values are introduced by the extension, and corresponds exactly to what is understood by the same term in the theory of functions.

Similarly, the image field $\phi(\mathfrak{O})$ ($\cong \mathfrak{O}/\mathfrak{P}$) contains the image field $\varphi(\mathfrak{o})$ ($\cong \mathfrak{o}/\mathfrak{p}$) as a subfield. The degree

$$f = [\phi(\mathfrak{O}) : \varphi(\mathfrak{o})]$$

is called the *residue class degree.*

Both e and f can be infinite. Some clarity is introduced, though, by the

THEOREM. *If K is a finite extension of k, then both e and f are finite, and*

$$e \cdot f \leq n = [K : k].$$

PROOF. Let $\pi_1, \pi_2, \ldots, \pi_r$ be elements of K^*, so that the cosets $|\pi_1||k^*|$, $|\pi_2||k^*|$, $\ldots$, $|\pi_r||k^*|$ are all different from each other, and let $\omega_1, \omega_2, \ldots, \omega_s$ be elements of $\mathfrak{O}$, so that the images $\phi(\omega_1), \phi(\omega_2), \ldots, \phi(\omega_s)$ are linearly independent over $\phi(\mathfrak{o})$.

We will then show that the rs elements $\omega_i\pi_j$ ($j = 1, 2, \ldots, r$; $i = 1, 2, \ldots, s$) of K are linearly independent over k. This implies immediately that $rs \leq n$, and hence, $ef \leq n$.

Let $a_1, a_2, \ldots, a_s$ be elements of k, not all 0, and let $|a_m| = \mathrm{Max}_i\, |a_i|$. Then,

$$\begin{aligned} &|a_1\omega_1 + \cdots + a_s\omega_s| = |a_m| \cdot |x_1\omega_1 + \cdots + x_s\omega_s|, \quad \text{with } x_i = \frac{a_i}{a_m} \\ &\implies |x_i| \leq 1,\ |x_m| = 1 \\ &\implies x_1\omega_1 + \cdots + x_s\omega_s \in \mathfrak{O} \\ &\implies |x_1\omega_1 + \cdots + x_s\omega_s| \leq 1. \end{aligned}$$

We can even show that the equality holds here, since

$$\begin{aligned} \phi(x_1\omega_1 + \cdots + x_s\omega_s) &= \phi(x_1)\phi(\omega_1) + \cdots + \phi(x_s)\phi(\omega_s) \\ &= \varphi(x_1)\phi(\omega_1) + \cdots + \varphi(x_1)\phi(\omega_1) \neq 0 \end{aligned}$$

because $\varphi(x_m) = \varphi(1) = 1$ and because of the assumed linear independence of the $\phi(\omega_1), \ldots, \phi(\omega_s)$ over $\varphi(\mathfrak{o})$. This proves that

$$|a_1\omega_1 + \cdots + a_s\omega_s| = \mathrm{Max}_i\, |a_i| \in |k^*|.$$

Let us now consider a relation

$$\sum_{i,j} a_{ij}\omega_i\pi_j = 0, \quad a_{ij} \in k,$$

or, in slightly different form,

$$\sum_j (a_{1j}\omega_1 + a_{2j}\omega_2 + \cdots + a_{sj}\omega_s)\pi_j = 0,$$

so that we have

$$|a_{1j}\omega_1 + a_{2j}\omega_2 + \cdots + a_{sj}\omega_s| = \mathrm{Max}_i\, |a_{ij}| \in |k^*|,$$

if not all the a_{ij} are equal to zero. The absolute values of the separate non-zero summands (in the last sum, over j) are all different from each other, since they lie in different cosets of $|K^*|/|k^*|$. This yields

$$\begin{aligned}\left|\sum_{i,j} a_{ij}\omega_i\pi_j\right| = 0 &\implies \mathrm{Max}_j(|a_{1j}\omega_1 + \cdots + a_{sj}\omega_s| \cdot |\pi_j|) = 0\\ &\implies \mathrm{Max}_j\,|a_{1j}\omega_1 + \cdots + a_{sj}\omega_s| = 0\\ &\implies \mathrm{Max}_{i,j}\,|a_{ij}| = 0\\ &\implies a_{ij} = 0, \quad \text{for all } i, j.\end{aligned}$$ □

The $\omega_i\pi_j$ ($i = 1, 2, \ldots, s$; $j = 1, 2, \ldots, r$) are linearly independent over k. This last theorem can be generalized in such a manner as to consider not only one way of extending a given valuation, but immediately all possible extensions.

THEOREM. *If k is a field with a non-trivial, non-archimedean valuation of rank 1, $|\ |$, and K is a finite extension of k, with the unequal valuations $|\ |_1, |\ |_2, \ldots$, all of which are extensions of $|\ |$, with ramification $e_1, e_2, \ldots$, and the residue class degree $f_1, f_2, \ldots$, respectively, then*

$$\sum_\nu e_\nu f_\nu \leq n = [K : k].$$

It should be remarked that the theorem also holds true without the limitation to valuations of rank 1. The proof is then, however, considerably more difficult, as the Approximation Theorem loses its validity for valuations not of rank 1. The above theorem suffices for our applications.

PROOF. No two of the valuations are equivalent. For, if $|\ |_1^\alpha = |\ |_2$, then $\alpha = 1$, as can be seen by letting $a \in k$, $|a| \neq 1$, $|a| \neq 0$, and the valuations were assumed to be at least unequal. We can also assume immediately that the valuations are all of rank 1; it will later be shown that this follows from the last theorem. We can, therefore, apply the Approximation Theorem to the set $|\ |_1, |\ |_2, \ldots$ of valuations.

We choose elements $\omega_i^{(\nu)}$ ($i = 1, 2, \ldots, e_\nu$) and $\pi_j^{(\nu)}$ ($j = 1, 2, \cdots f_\nu$) to each valuation $|\ |_\nu$, as they were chosen in the proof of the last theorem. According to the Approximation Theorem we can now find elements $\bar{\pi}_j^{(\nu)}$ and $\bar{\omega}_i^{(\nu)}$ for every ν, i and j which satisfy

$$\begin{aligned}&|\bar{\pi}_j^{(\nu)} - \pi_j^{(\nu)}|_\nu < |\pi_j^{(\nu)}|, &&|\bar{\pi}_j^{(\nu)}|_\mu < \mathrm{Min}_{\rho,s}\,|\pi_s^{(\rho)}| \quad \text{for all } \mu \neq \nu,\\ &|\bar{\omega}_i^{(\nu)} - \omega_i^{(\nu)}|_\nu < 1, &&|\bar{\omega}_i^{(\nu)}|_\mu < 1 \quad \text{for all } \mu \neq \nu.\end{aligned}$$

Because $|\bar{\pi}_j^{(\nu)}|_\nu = |\pi_j^{(\nu)} - (\pi_j^{(\nu)} - \bar{\pi}_j^{(\nu)})|_\nu = |\pi_j^{(\nu)}|_\nu$ it follows that all the absolute values $|\bar{\pi}_j^{(\nu)}|_\nu$ ($j = 1, 2, \ldots, e_\nu$) represent different cosets of $|K^*|_\nu/|k^*|$. Similarly, for a given ν, all $\phi_\nu(\bar{\omega}_i^{(\nu)})$ ($i = 1, 2, \ldots, f_\nu$) are linearly independent over $\varphi(\mathfrak{o})$, for

$$\begin{aligned}|\bar{\omega}_i^{(\nu)} - \omega_i^{(\nu)}|_\nu < 1 &\implies \phi_\nu(\bar{\omega}_i^{(\nu)} - \omega_i^{(\nu)}) = 0\\ &\implies \phi_\nu(\bar{\omega}_i^{(\nu)}) = \phi_\nu(\omega_i^{(\nu)}).\end{aligned}$$

We must now show that the $\sum_\nu e_\nu f_\nu$ elements $\bar\omega_i^{(\nu)} \cdot \bar\pi_j^{(\nu)}$ of K are linearly independent over k. Let

$$\sum_{i,j,\nu} a_{ij}^{(\nu)} \bar\omega_i^{(\nu)} \bar\pi_j^{(\nu)} = 0, \quad a_{ij}^{(\nu)} \in k$$

and let, say $|a_{11}^{(1)}|$ be maximal among all the $|a_{ij}^{(\nu)}|$. If we now write this relation in the form

$$\sum_j \left(\sum_i a_{ij}^{(1)} \bar\omega_i^{(1)} \right) \bar\pi_j^{(1)} + \sum_{\substack{j,i \\ \nu \geq 2}} a_{ij}^{(\nu)} \bar\omega_i^{(\nu)} \bar\pi_j^{(\nu)} = 0$$

and take the absolute value $|\ |_1$, then, as in the proof of the last theorem, we have

$$\begin{aligned} \left| \left(\sum_i a_{i1}^{(1)} \bar\omega_i^{(1)} \right) \bar\pi_1^{(1)} \right|_1 &= \mathrm{Max}_i \, |a_{i1}^{(1)}| \cdot |\bar\pi_1^{(1)}|_1 \\ &= |a_{11}^{(1)}| \cdot |\bar\pi_1^{(1)}|_1, \end{aligned}$$

and therefore certainly

$$\mathrm{Max}_j \left| \left(\sum_i a_{ij}^{(1)} \bar\omega_i^{(1)} \right) \bar\pi_j^{(1)} \right|_1 \geq |a_{11}^{(1)}| \cdot |\bar\pi_1^{(1)}|_1.$$

For $\nu \geq 2$ we have, however,

$$|a_{ij}^{(\nu)}| \leq |a_{11}^{(1)}|, \quad |\bar\omega_i^{(\nu)}|_1 < 1, \quad |\bar\pi_i^{(\nu)}|_1 < |\pi_1^{(1)}|_1$$

and hence,

$$|a_{ij}^{(\nu)}| \cdot |\bar\omega_i^{(\nu)}|_1 \cdot |\bar\pi_j^{(\nu)}|_1 < |a_{11}^{(1)}| \cdot |\pi_1^{(1)}|_1,$$

so that the second part of the sum in the equation above has, after taking the absolute value $|\ |_1$, no effect on the left side. One then has

$$\left| \sum_j \left(\sum_i a_{ij}^{(1)} \bar\omega_i^{(1)} \right) \bar\pi_j^{(1)} \right|_1 = 0$$

which implies, as in the previous theorem:

$$\begin{aligned} |a_{11}^{(1)}| = 0, \quad |a_{ij}^{(\nu)}| &= 0 \quad \text{for all } i, j, \nu, \\ a_{ij}^{(\nu)} &= 0 \quad \text{for all } i, j, \nu. \end{aligned}$$

□

Corollary. *A given (non-archimedean) valuation of a field k has only finitely many extensions on a finite extension field K (at most $n = [K : k]$). If $|\ |_1, |\ |_2, \ldots, |\ |_r$ are all the possible extensions, with respective ramifications and residue class degrees $e_1, e_2, \ldots, e_r$ and $f_1, f_2, \ldots, f_r$, then*

$$\sum_{\nu=1}^{r} e_\nu f_\nu \leq n.$$

(Later it will be shown that the above inequality increases, in most cases, to equality. This holds, for example, in all fields of algebraic numbers and functions.)

The first of the two above theorems allows us to prove the result indicated earlier, that the property of being a valuation of rank 1 is not lost by extension to a finite overfield. This, in the form of the

THEOREM. *Let the field K be finite over the subfield k, the valuation $|\ |_K$ an extension of the valuation $|\ |$ of k. The group of values $|K^*|$ can then be mapped order-isomorphically into the group $\mathbb{R}^+$ of positive real numbers in such a manner as to leave each element of $|k^*|$ fixed.*

PROOF. Let $|K^*|_K = \mathfrak{G}$. We then have

$$|k^*| \subset \mathfrak{G}, \quad |k^*| \subset \mathbb{R}^+.$$

The group $\mathfrak{G}$ is abelian, and

$$(\mathfrak{G} : |k^*|) = e$$

is finite. Letting $\mathfrak{G}^e$ be the set of all e-th powers of elements of $\mathfrak{G}$,

$$\mathfrak{G}^e \subset |k^*| \subset \mathbb{R}^+.$$

We also define a modified absolute value $\|\ \|$, by setting

$$\|\alpha\| = (|\alpha|_K^e)^{\frac{1}{e}},$$

where the e-th root is extracted within $\mathbb{R}^+$. This is a valuation, the first two axioms being trivially satisfied, the third following from

$$\|\alpha\| \leq 1 \implies |\alpha| \leq 1 \implies |1+\alpha| \leq 1 \implies \|1+\alpha\| \leq 1.$$

For $a \in k$, $\|a\| = |a|$. The valuations $|\ |_K$ and $\|\ \|$ are equivalent, as the map

$$|\alpha| \to \|\alpha\|$$

is an isomorphism of the ordered group $|K^*|_K$ onto the ordered group $\|K^*\| \subset \mathbb{R}^+$. For, in an ordered group the raising of all elements to the e-th power is an isomorphism, as is also the extraction of e-th roots within $\mathbb{R}^+$. □

3. Extension of Valuations of Rank 1 of Complete Fields

We will here study the possible extensions of a valuation of rank 1, in the case that the basic field is complete.

Let $|\ |$ be a rank 1 valuation (archimedean or non-archimedean) of a field K, of finite degree, n, over a subfield k. Let the restriction of the valuation to k be non-trivial, and let k be complete with respect to this valuation. We can, by choosing an equivalent valuation if necessary, assume the triangle inequality to hold in k, and hence in K.

K can be treated as an n-dimensional vector space over k, and the valuation as a norm of this vector space. If $\omega_1, \omega_2, \ldots, \omega_n$ form a basis of K over k, then the general element ξ of K takes the form

$$\xi = x_1\omega_1 + x_2\omega_2 + \cdots + x_n\omega_n \quad \text{with } x_i \in k.$$

We showed earlier that the norm $|\xi|$ in K generates the same topology as $\|\xi\| = \operatorname{Max}_i |x_i|$.

We let $\xi_i = x_{i1}\omega_1 + \cdots + x_{in}\omega_n$ $(i = 1, 2, \ldots)$ be a Cauchy sequence of K

$$\begin{aligned} &\implies |\xi_i - \xi_j| \text{ is small for large } i, j \\ &\implies \|\xi_i - \xi_j\| \text{ is small for large } i, j \\ &\implies |x_{ir} - x_{jr}| \text{ is small for large } i, j \ (r = 1, 2, \ldots, n) \end{aligned}$$

Thus, for all r, the coefficient sequences $x_{1r}, x_{2r}, \ldots$ are Cauchy sequences, the limits lying in k: $\lim_{i\to\infty} x_{ir} = y_r \in k$.

We set $\eta = y_1\omega_1 + y_2\omega_2 + \cdots + y_n\omega_n$;

$$\begin{aligned} &\Longrightarrow \|\eta - \xi_i\| \text{ small for large } i \\ &\Longrightarrow |\eta - \xi_i| \text{ small for large } i \\ &\Longrightarrow \eta = \lim_{i\to\infty} \xi_i. \end{aligned}$$

Hence, K is complete.

THEOREM. *A valuation of rank* 1 *of a complete field k permits at most one extension onto a finite extension K of the field k. (That this one extension always exists is shown next.)*

In particular, the extension can be given in a formula using only the absolute values of elements of k.

As is usual, we denote with $N\xi$ the norm of the element ξ relative to k, in other words, the product of the conjugates of ξ (in the inseparable case taking each with a certain multiplicity, c.f., e.g. van der Waerden, Algebra I.) Then,

$$\begin{aligned} &N\xi \in k \\ &N(\xi\eta) = N\xi \cdot N\eta \\ &Na^n = a^n \quad \text{for all } a \in k. \end{aligned}$$

Let $\xi = x_1\omega_1 + x_2\omega_2 + \cdots + x_n\omega_n$. $N\xi$ can then be expressed as an homogenous polynomial of degree n in $x_1, x_2, \ldots, x_n$, with fixed coefficients in K.

Let α be some element of K with $|\alpha| < 1$. Hence, $|\alpha|^i \to 0$ and also $\alpha^i \to 0$, with $i \to \infty$. Let $\alpha^i = x_{i1}\omega_1 + x_{i2}\omega_2 + \cdots + x_{in}\omega_n$,

$$\begin{aligned} &\Longrightarrow x_{ir} \to 0, \quad \text{with } i \to \infty \ (r = 1, 2, \ldots, n) \\ &\Longrightarrow |N\alpha|^i \to 0, \quad \text{with } i \to \infty \\ &\Longrightarrow |N\alpha| < 1. \end{aligned}$$

Together with the same considerations, using $\frac{1}{\alpha}$, for $|\alpha| > 1$, this yields

$$\begin{aligned} |\alpha| < 1 &\Longrightarrow |N\alpha| < 1, \\ |\alpha| > 1 &\Longrightarrow |N\alpha| > 1, \end{aligned}$$

and therefore also

$$|N\alpha| = 1 \Longrightarrow |\alpha| = 1.$$

Take any element of K, $\beta \neq 0$, and set $\alpha = \frac{\beta^n}{N\beta}$. Then,

$$N\alpha = \frac{(N\beta)^n}{N(N\beta)} = \frac{(N\beta)^n}{(N\beta)^n} = 1$$

$$\Longrightarrow |N\alpha| = 1 \Longrightarrow |\alpha| = 1 \Longrightarrow |\beta|^n = |N\beta|$$

$$\Longrightarrow |\beta| = \sqrt[n]{|N\beta|}.$$

The valuation, on the assumption it exists, is uniquely determined by its limitation on k.

If our starting point is a non-archimedean valuation of the complete field k, then the last formula does, in fact, give a valuation of the extension K, the existence of which was already proven. Thus, in this case, there is always one and only one extension of the valuation.

In the case of an archimedean valuation we have not yet considered the existence of an extension on an overfield. We now do this.

Let k be a field with an archimedean valuation $|\ |$. Certainly k has the characteristic 0, for otherwise the absolute values of the integers would be bounded, k therefore contains as a subfield the field $\mathbb{Q}$ of rational numbers (or an isomorphic field).

The valuation $|\ |$ is equivalent, within $\mathbb{Q}$, to the usual absolute valuation. By going over to a positive power of $|\ |$ if necessary, we can, in fact, assume that $|\ |$ is the usual absolute valuation in $\mathbb{Q}$. The triangle inequality, which holds in $\mathbb{Q}$, therefore also holds true in k.

We go over to the completion $\tilde{k}$ of k;

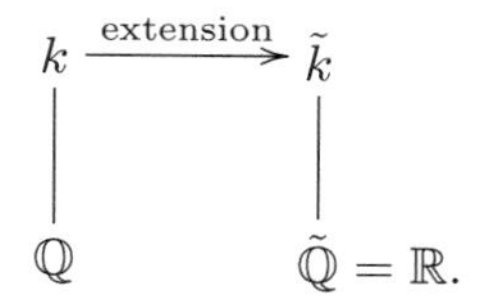

The equivalence classes of k representable by Cauchy sequences of elements of $\mathbb{Q}$ form a subfield $\tilde{\mathbb{Q}}$ of $\tilde{k}$, which is isomorphic to the field $\mathbb{R}$ of real numbers, by definition of the latter. We identify the subfield $\tilde{\mathbb{Q}}$ with $\mathbb{R}$. The valuation $|\ |$ of $\tilde{k}$ is, within $\mathbb{R} = \tilde{\mathbb{Q}}$ equal to the usual absolute valuation.

This valuation of $\tilde{k}$ is, therefore, a norm in the sense of Gelfand, and $\tilde{k}$ must therefore be equal either to $\mathbb{R}$, or to the field of complex numbers, $\mathbb{C}$. In the latter case our uniqueness theorem also shows that the valuation is the usual absolute valuation. Remembering that k is a subfield of $\tilde{k}$, we have proven the

THEOREM. *A field with archimedean valuation is always isomorphic to a subfield of the field $\mathbb{C}$ of complex numbers, the valuation being the usual absolute valuation (or a positive power of the same).*

Altogether, we have proven that a valuation of k is extendable onto a finite extension K of k, also in the archimedean case. For, with k, K is also isomorphic to a subfield of $\mathbb{C}$ ($\mathbb{C}$ being algebraically closed), and every possible embedding of K into the field $\mathbb{C}$ which leaves the embedding of k into $\mathbb{C}$, defined by the valuation of k, invariant, yields an extension of the valuation. Since the number of such embeddings is finite, so is the number of valuation extensions.

We have thus seen that only a very special class of fields allow archimedean valuations.

CHAPTER 6

COMPLETE NON-ARCHIMEDEAN FIELDS

1. The Transition to the Completion

Let $|\ |$ be a non-archimedean valuation of the field k. $\tilde{k}$ will denote the completion of k, and W_k and $W_{\tilde{k}}$ the respective groups of values.

LEMMA 1. $W_k = W_{\tilde{k}}$.

PROOF. Let $\alpha \in \tilde{k}$, $\alpha \neq 0$. We can then find an element $a \in k$, such that

$$|a - \alpha| < |\alpha|,$$

for k is dense in $\tilde{k}$. This, however implies that

$$|a| = |\alpha + (a - \alpha)| = |\alpha|.$$

All values of elements of $\tilde{k}$ appear, therefore, as values of elements of k. □

LEMMA 2. *Let φ be the place corresponding to the valuation $|\ |$ of k. The place corresponding to the valuation of $\tilde{k}$, is, with the valuation, an extension of the place φ, and will therefore also be denoted by φ. Furthermore, $\varphi(\tilde{k}) = \varphi(k)$.*

PROOF. For $\alpha \in \tilde{k}$, $\varphi(\alpha)$ finite, there exists an $a \in k$ with $|a - \alpha| < 1$. Then,

$$\varphi(a - \alpha) = 0, \quad \text{or} \quad \varphi(a) = \varphi(\alpha).$$

No new elements can appear by the transition from the residue class field of k to that of $\tilde{k}$. □

Let K be a finite extension of the field k, and $|\ |$ any valuation of K. In the completion $\tilde{K}$ of K, the set of elements representable by Cauchy sequences of elements of k forms a subfield, which is isomorphic to the completion $\tilde{k}$ of k. By identifying these isomorphic fields we can consider $\tilde{k}$ to be a subfield of $\tilde{K}$:

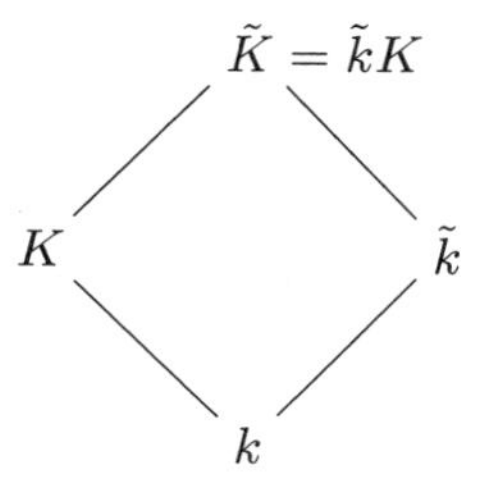

The composed field $\tilde{k}K$ of $\tilde{k}$ and K (the minimal field in $\tilde{K}$ containing both K and k), is contained in $\tilde{K}$ and is finite over $\tilde{k}$, for K is finite over k. Since every finite extension of a complete field is again complete (as was shown more generally for norms in vector spaces over complete fields), the field $\tilde{k}K$ is complete, and thus contains, with K, also $\tilde{K}$.

$$\tilde{K} = \tilde{k}K.$$

If we had started with a valuation of the base field k instead of immediately assuming a certain valuation of the extension K, certain difficulties would have arisen, as the following example will show.

Let $\mathbb{Q}$ be the field of rational numbers, with the usual absolute valuation, and consider the extension $\mathbb{Q}(\theta)$, generated by adjoining the symbol θ satisfying the equation $\theta^3 = 2$. We have $\tilde{\mathbb{Q}} = \mathbb{R}$. The relation corresponding to the above result, $\widetilde{\mathbb{Q}(\theta)} = \widetilde{\mathbb{R}(\theta)}$, is quite meaningless, since the polynomial $x^3 - 2$ is reducible in $\mathbb{R}$. If θ is interpreted as the real root of the equation we get the field of real numbers; taking θ, on the other hand, as an imaginary root, leads to the field of complex numbers. This indefiniteness has its cause in the existence of various extensions of the valuation of k onto K.

For a non-archimedean valuation of a field k, extended to a finite extension field K, the above considerations yield: the transition to the completions $\tilde{k}$ and $\tilde{K}$ leaves the ramification e and residue class degree f invariant.

2. Convergent Series in Complete Fields

Consider a complete field k. Every Cauchy sequence in k has a limit in k, and therefore the usual analysis can be carried over to k from the field of real numbers. In particular, the concept of a convergent series and the Cauchy convergence criterium can be taken over.

In the case of a non-archimedean valuation, the following remarkable property can be attained:

THEOREM. *A series* $\sum_{n=1} a_n$ *in a non-archimedean complete field converges if, and only if,* $\lim_{n\to\infty} a_n = 0$.

Necessity is trivial, the sufficiency follows from the proven inequality:

$$|a_n + a_{n+1} + \cdots + a_{n+p}| \leq \mathrm{Max}_{n\leq i\leq n+p} |a_i|.$$

The concept of unconditional convergence is unnecessary in the non-archimedean case; the terms of a convergent series can always be permuted at will. The concept of absolute convergence would also be inconvenient here, since it would lead out of the field.

As an example we consider a field k, complete by a non-archimedean valuation, where k contains the field $\mathbb{Q}$ of rational numbers, and the valuation in $\mathbb{Q}$ is one generated by the prime number p.

k then also contains as a subfield the completion $\tilde{\mathbb{Q}}$ of $\mathbb{Q}$, which is known as the *field of p-adic numbers.* The p-adic numbers are, with respect to the p-adic valuation, the exact analogon of the real numbers with respect to the usual absolute valuation. $\mathbb{Q}$ being dense in $\tilde{\mathbb{Q}}$, every p-adic number can be approximated to any degree by a rational number.

Examples of series in p-adic fields (or their extensions):

(1) $p = 2$ (dyadic numbers):

$$1 + 2 + 2^2 + 2^3 + \cdots = -1,$$

since the sum of the first n terms,

$$\frac{2^n - 1}{2 - 1} = 2^n - 1$$

converges to -1 as $n \to \infty$.

(2) $p = 5$:

$$\sqrt{-1} = \frac{1}{2}\sqrt{1-5} = \frac{1}{2}\sum_{n=0}^{\infty}(-1)^n\binom{\frac{1}{2}}{n}5^n.$$

To prove the convergence of this series, one demonstrates the terms converge to 0 as $n \to \infty$. The binomial theorem then shows the series converges to $\sqrt{-1}$. (Remark: $\sqrt{-1}$ here is a different object from the complex number i.)

(3) p any prime number:

We consider the convergence of the exponential series:

$$e^x = \sum_{n=0}^{\infty}\frac{x^n}{n!} \quad \text{in the field of } p\text{-adic numbers.}$$

Letting $n = a_0 + a_1 p + \cdots + a_r p^r$, $0 \le a_i \le p-1$, be the p-adic representation of n, we compute:[1]

$$\operatorname{ord} n! = \left[\frac{n}{p}\right] + \left[\frac{n}{p^2}\right] + \left[\frac{n}{p^3}\right]$$

$$\left[\frac{n}{p}\right] = a_1 + a_2 p + a_3 p^2 + \cdots + a_r p^{r-1}$$

$$\left[\frac{n}{p^2}\right] = \qquad a_2 + a_3 p + \cdots + a_r p^{r-2}$$

$$\cdots\cdots\cdots\cdots\cdots\cdots\cdots\cdots$$

$$\left[\frac{n}{p^r}\right] = \qquad\qquad a_r.$$

$$\operatorname{ord} n! = a_0\frac{p^0-1}{p-1} + a_1\frac{p^1-1}{p-1} + a_2\frac{p^2-1}{p-1} + \cdots + a_r\frac{p^r-1}{p-1} = \frac{n-s_n}{p-1},$$

where $s_n = a_0 + a_1 + a_2 + \cdots + a_r$ (the "diagonal sum" of n)

$$\Longrightarrow \operatorname{ord}\left(\frac{x^n}{n!}\right) = n\left(\operatorname{ord} x - \frac{1}{p-1}\right) + \frac{s_n}{p-1}.$$

Thus, the series converges then, and only then, when $\operatorname{ord} x > \frac{1}{p-1}$, for, $\frac{s_n}{p-1} \le \frac{\log n}{\log p}$. (Remark: $\operatorname{ord} x$ can be a rational fraction in an extension k of $\tilde{\mathbb{Q}}$.) The usual considerations lead, in the domain of convergence to: $e^x e^y = e^{x+y}$. The logarithmic series, handled similarly, yields a better convergence.

3. Some Topological Properties of a Complete Field

In the following k will be a complete field by a non-archimedean valuation, and K its algebraic closure.

Every element α of K lies in a finite extension, $k(\alpha)$ of k, hence an extension of the valuation onto K would be possible in at most one way:

$$|\alpha| = |N\alpha|^{\frac{1}{n}}, \quad n = [k(\alpha) : k] \quad (N = \text{norm in } k(\alpha)/k).$$

[1] For a real number α, $[\alpha]$ denotes the largest integer $m \le \alpha$.

The definition of the norm shows that one gets the same $|\alpha|$, even if one substitutes for $k(\alpha)$ any finite extension K_0 of k containing α, and thus this definition maps each α onto a unique $|\alpha|$. This map is obviously a valuation of K, since the axioms to be satisfied always involve at most three elements of K, for which a finite extension containing them can always be found.

The field K will not, as in the case of finite extensions, always be complete. It can be shown, however, that its completion K remains algebraically closed.

K can, like any field with valuation, be taken as a metric space with distance function $|\alpha - \beta|$. Also, for $\alpha \in K$, let

$$f = \operatorname{Irr}(\alpha, k)$$

be the normalized (highest coefficient 1) irreducible polynomial with coefficients in k, with root α. f is uniquely determined by α and k.

Let $\alpha = \alpha_1, \alpha_2, \ldots, \alpha_s$ $(s \geq 2)$ be the complete set of roots of $f(x) = \operatorname{Irr}(\alpha, k)$. Consider the points within the "circle" around α cutting the root nearest α, i.e. all $\beta \in K$, such that

$$|\beta - \alpha[< r = \operatorname{Min}_{i>1} |\alpha_i - \alpha|.$$

Let β be any such element of K and let $g(x) = \operatorname{Irr}(\alpha, k(\beta))$. α is the only root of $g(x)$.

For, certainly $g(x) \mid f(x)$. Let α_i be any root of $g(x)$, then the elements $\alpha - \beta$ and $\alpha_i - \beta$ are conjugates over $k(\beta) = k_1$, and therefore the field $k_1(\alpha - \beta)$ and $k_1(\alpha_i - \beta)$ are isomorphic.

$$\begin{aligned} &\Longrightarrow |\alpha_i - \beta| = |\alpha - \beta| < r \\ &\Longrightarrow |\alpha_i - \alpha| = |\alpha_i - \beta - (\alpha - \beta)| < r \\ &\Longrightarrow \alpha_i = \alpha \quad \text{(by our choice of } r). \end{aligned}$$

COROLLARY. *Let α be a separable element of K, $\alpha = \alpha_1, \alpha_2, \ldots, \alpha_n$ be the conjugates of α over k, and β any element of K satisfying*

$$|\beta - \alpha| < r = \operatorname{Min}_{i>1} |\alpha_i - \alpha|$$

(for $n \geq 2$). Then $\alpha \in k(\beta)$.

Now let $f(x)$ be any normalized polynomial in $k[x]$ of degree n whose coefficients have absolute values $\leq A$; we write

$$|f| \leq A.$$

If $\alpha \in K$, and $f(\alpha) = 0$, then $|\alpha| \leq A$.

For, let $f(\alpha) = \alpha^n + a_1\alpha^{n-1} + \cdots + a_n = 0$. If, now $|\alpha| > A$, then the term α^n would have higher absolute value than any other term, which is impossible.

Let $g(x)$ be another normalized polynomial out of $k[x]$, of degree n, and with

$$|f(x) - g(x)| \leq \varepsilon \leq A.$$

$$\Longrightarrow |g(x)| = |f(x) - (f(x) - g(x))| \leq A.$$

If, now, $f(x) = (x - \alpha_1)(x - \alpha_2)\ldots(x - \alpha_n)$ is the decomposition of f, and β is a root of g: $g(\beta) = 0$, then $|\beta| \leq A$, and

$$|f(\beta) = |f(\beta) - g(\beta)| \leq \varepsilon \cdot A^n.$$

$$\Longrightarrow |\beta - \alpha_1| \cdot |\beta - \alpha_2| \cdots |\beta - \alpha_n| \leq \varepsilon \cdot A^n$$

and therefore, for at least one α_i, say α_1, we have

$$|\beta - \alpha_1| \leq A\sqrt[n]{\varepsilon}.$$

If we also now take the multiplicity of the roots α_i under consideration, and write:

$$f(x) = (x-\alpha_1)^{\nu_1}(x-\alpha_2)^{\nu_2}\cdots(x-\alpha_s)^{\nu_s}, \quad \alpha_i \neq \alpha_j \text{ for } i \neq j,$$

it follows: For sufficiently small $\varepsilon > 0$, the inequality

$$|\beta - \alpha_1| \leq A\sqrt[n]{\varepsilon}$$

holds for ν_1 roots β of $g(x)$. The analogous statement holds true for the other roots $\alpha_2, \alpha_3, \ldots, \alpha_s$ of $f(x)$.

PROOF. Were the statement false, then to every $\varepsilon > 0$ there would still exist polynomials $g(x)$ for which a number, $\neq \nu_1$, of roots would lie near α_1. Since this number is bounded (lying between 0 and n), we can, from this set of exception-polynomials, choose a sequence with μ_1 roots near α_1, μ_2 roots near α_2, …, μ_s roots near α_s, and, say $\mu_1 \neq \nu_1$. The limit of the sequence would be $f(x)$, and then

$$f(x) = (x-\alpha_1)^{\mu_1}(x-\alpha_2)^{\mu_2}\cdots(x-\alpha_s)^{\mu_s},$$

in contradiction to the unique decomposition of polynomials. □

REMARK. A similar argument yields: The normalized irreducible polynomials of a given degree n form an open set under the topology defined by the valuation of the ring $k[x]$:

$$|f(x)| = |a_0x^n + a_1x^{n-1} + \cdots + a_n| = \mathrm{Max}_i\, |a_i|.$$

COROLLARY. *Let $f = (x-\alpha_1)(x-\alpha_2)\ldots(x-\alpha_n)$, $\alpha_i \in K$, be a separable and irreducible polynomial over k, and $g(x)$ be a normalized polynomial of degree n out of $k[x]$, lying sufficiently close to $f(x)$. If $\beta \in K$ is a root of $g(x)$, and, say, α_1 is the root of $f(x)$ near which β lies, then*

$$k(\alpha_1) = k(\beta).$$

We already know that $k(\alpha_1)$ is a subfield of $k(\beta)$, for $g(x)$ sufficiently near $f(x)$. The degrees of the fields imply the irreducibility of $g(x)$, and then from the equality of these degrees, we obtain the result.

4. Extensions of the Residue Class Field

The residue class field of k, i.e. the image of the valuation ring $\mathfrak{o}$ under the homomorphism $\varphi|\mathfrak{o}$, will be denoted by $\bar{k}$. $\bar{k} \cong \mathfrak{o}/\mathfrak{p}$ ($\mathfrak{p}$ maximal ideal in $\mathfrak{o}$). Similarly, let $\bar{K}$ be the residue class field of the algebraic closure K of k.

THEOREM. *$\bar{K}$ is the algebraic closure of $\bar{k}$.*

PROOF. If $\alpha \in K$, $|\alpha| \leq 1$, then let $\bar{\alpha}$ be the image of α by the transition to the residue class field. Similarly, for a polynomial,

$$f(x) = \alpha_0x^n + \alpha_1x^{n-1} + \cdots + \alpha_n \in K[x], \quad |\alpha_i| \leq 1,$$

let the image $\bar{f}(x)$ be defined by

$$\bar{f}(x) = \bar{\alpha}_0x^n + \bar{\alpha}_1x^{n-1} + \cdots + \bar{\alpha}_n \in \bar{K}[x].$$

1.) Let $\bar{\alpha} \in \bar{K}$, i.e. $\bar{\alpha}$ image of an element $\alpha \in K$ with $|\alpha| \le 1$. Further, let

$$f(x) = \operatorname{Irr}(\alpha, k),$$

and $\alpha = \alpha_1, \alpha_2, \dots, \alpha_n$ be the roots of $f(x)$:

$$f(x) = (x - \alpha_1)(x - \alpha_2) \dots (x - \alpha_n).$$

Since the α_i are conjugates over k, we have $|\alpha_i| = |\alpha| = 1$ for all $i = 1, 2, \dots, n$. Thus, the coefficients of $f(x)$ have absolute values ≤ 1. Also, since $f(\alpha) = 0$, $\bar{f}(\bar{\alpha}) = 0$; i.e. $\bar{\alpha}$ is algebraic over $\bar{k}$.

2.) Let $\bar{f}(x)$ be a normalized polynomial in $\bar{k}[x]$, and $f(x) = (x - \alpha_1)(x - \alpha_2) \dots (x - \alpha_n)$, $\alpha_i \in K$, be some inverse image of $\bar{f}(x)$. Since the coefficients of $f(x)$ all have absolute values ≤ 1, so do the roots α_i, $i = 1, 2, \dots, n$. Hence,

$$\bar{f}(x) = (x - \bar{\alpha}_1)(x - \bar{\alpha}_2) \cdots (x - \bar{\alpha}_n), \quad \bar{\alpha}_i \in \bar{K}.$$

By 1.) and 2.) the theorem is proved. □

REMARK. Under the assumptions and notation of 2.) we assume further that $\bar{f}(x)$ is irreducible in $\bar{k}$. We prove that then $f(x)$ is also irreducible, and $\overline{k(\alpha_1)} = \bar{k}(\bar{\alpha}_1)$.

For, let $F = k(\alpha_1)$, then $\bar{F} \supset \bar{k}(\bar{\alpha}_1)$

$$\Longrightarrow [\bar{F} : \bar{k}] = f \ge n.$$

In the inequality:

$$ef \le [F : k] \le n$$

we therefore have equality on both sides:

$$e = 1, \quad f = n, \quad [F : k] = n.$$

Hence, $f(x)$ is irreducible in k and $\bar{F} = \bar{k}(\bar{\alpha}_1)$.

We illustrate our results with the following diagram:

$$\begin{array}{ccc} k(\alpha_1) & \longrightarrow & \bar{k}(\bar{\alpha}_1) \\ {\scriptstyle e=1}\Big| {\scriptstyle f=n}^{\,n} & & \Big| {\scriptstyle n} \\ k & \longrightarrow & \bar{k} \end{array}$$

Applying the result twice yields

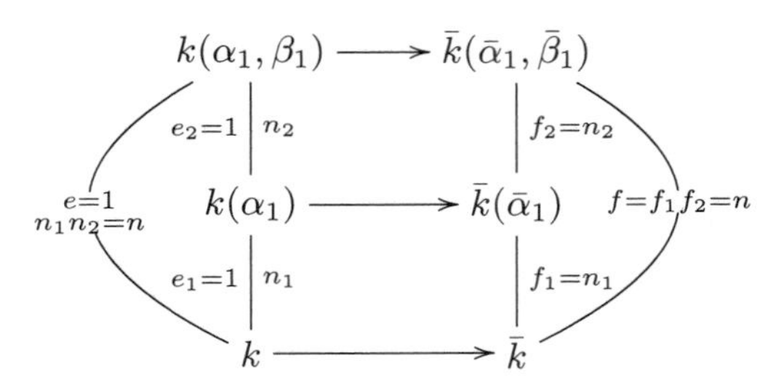

Finally, it is seen, that to any extension $\bar{F}$ of degree n over $\bar{k}$, we can find an extension F of degree n over k such that $\bar{F}$ is the residue class field of F:

$$\begin{array}{ccc} F & \longrightarrow & \bar{F} \\ {\scriptstyle e=1 \atop f=n} \Big| {\scriptstyle n} & & \Big| {\scriptstyle n} \\ k & \longrightarrow & \bar{k} \end{array}$$

REMARK. Still using the same notation, and under the same assumptions, assume further that $\bar{f}(x)$ is separable:

$$\bar{f}(x) = (x - \bar{\alpha}_1)(x - \bar{\alpha}_2)\cdots(x - \bar{\alpha}_n), \quad \bar{\alpha}_i \neq \bar{\alpha}_j \text{ for } i \neq j,\ n > 1.$$

For all α_i we have $|\alpha_i| \leq 1$, and because of the irreducibility of $\bar{f}(x)$ even

$$|\alpha_i| = 1 \quad \text{for } i = 1, 2, \ldots, n,$$

for $|\alpha_i| <1$ would imply $\bar{\alpha}_i = 0$. Similarly, because $\bar{f}(x)$ is separable,

$$|\alpha_i - \alpha_j| = 1 \quad \text{for } i \neq j.$$

Now, let β be an element of K with $\bar{\beta} = \bar{\alpha}_1$, then $|\beta - \alpha_1| < 1$: smaller than the distance between the roots. This yields

$$k(\alpha_1) \subset k(\beta).$$

Let F be an overfield of k, and $\bar{k}(\bar{\alpha}_1) \subset \bar{F}$. Then there exists a $\beta \in F$ with $\bar{\beta} = \bar{\alpha}_1$:

$$k(\alpha_1) \subset k(\beta) \subset F.$$

$k(\alpha_1)$ is contained in every extension F of k whose residue class field $\bar{F}$ contains the residue class field $\bar{k}(\bar{\alpha}_1)$ of $k(\alpha_1)$.

DEFINITION. A finite extension $E|k$ is called unramified, if:

(1) $f = n$, i.e. $[\bar{E} : \bar{k}] = [E : k]$; and,
(2) $\bar{E}$ is separable over $\bar{k}$.

The first property obviously implies that $e = 1$. The field considered in the last remark is unramified.

Let $E|k$ be unramified, so that $\bar{E}|\bar{k}$ is separable.

$$\Longrightarrow\ \bar{E} = \bar{k}(\bar{\alpha}_1),\ \bar{\alpha}_1 \text{ separable over } \bar{k},\ \alpha_1 \in E$$
$$\Longrightarrow\ k(\alpha_1) \subset E \ \Longrightarrow\ k(\alpha_1) = E,$$

because of the equality of field degrees. A residue class field $\bar{E}$ separable over $\bar{k}$ determines uniquely the corresponding unramified field E over k.

The above arguments also contain the existence property: To every such field $\bar{E}$ there exists an unramified field E with $\bar{E}$ as residue class field. There thus exists a one-to-one correspondence between separable extensions of $\bar{k}$ and unramified extensions of k.

The following sets up a criterium for unramified extensions:

THEOREM. *Let $f(x) = (x-\alpha_1)(x-\alpha_2)\cdots(x-\alpha_n)$, $\alpha_i \in K$, be a, not necessarily irreducible, element of $k[x]$. Let $|\alpha_i| \leq 1$ for all i, and $|\alpha_i - \alpha_j| = 1$ for $i \neq j$. Then, $k(\alpha_1)|k$ is unramified.*

PROOF. We can assume $f(x)$ to be irreducible. We show that then $\bar{f}(x)$ is also irreducible. If it were not we could find an irreducible $\bar{g}(x) \in k[x]$, with

$$\bar{f}(x) = \bar{g}(x)\bar{h}(x).$$

We choose some inverse image of $\bar{g}(x)$:

$$g(x) = (x - \beta_1)(x - \beta_2) \cdots (x - \beta_s), \quad \beta_i \in K, \ |\beta_i| \leq 1,$$

$$\Longrightarrow \bar{g}(x) = (x - \bar{\beta}_1)(x - \bar{\beta}_2) \cdots (x - \bar{\beta}_s), \quad \beta_i \in \bar{K},$$

and we can choose, by renumbering, $\bar{\beta}_1 = \bar{\alpha}_1, \ldots, \bar{\beta}_s = \bar{\alpha}_s$. We then have: $|\beta_i - \beta_j| = |\alpha_i - \alpha_j| = 1$ for $i \neq j$, and $|\alpha_1 - \beta_1| < 1$.

By applying our earlier considerations, first on $f(x)$ and then again on $g(x)$:

$$k(\alpha_1) \subset k(\beta_1), \quad k(\beta_1) \subset k(\alpha_1)$$

$$\Longrightarrow k(\alpha_1) = k(\beta_1)$$

$$\text{degree } f(x) = \text{degree } g(x)$$

$$\text{degree } \bar{f}(x) = \text{degree } \bar{g}(x)$$

$$\Longrightarrow \bar{f}(x) \text{ is irreducible.}$$

Thus, this case is identical with that of our first remark, and

$$f = n, \quad \overline{k(\alpha_1)} = \bar{k}(\bar{\alpha}_1),$$

which, together with the separability of $\bar{f}(x)$ proves the theorem. □

THEOREM. *If $E|k$ is unramified, and Ω is a finite extension of k, then the composed field $E\Omega$ is unramified over Ω.*

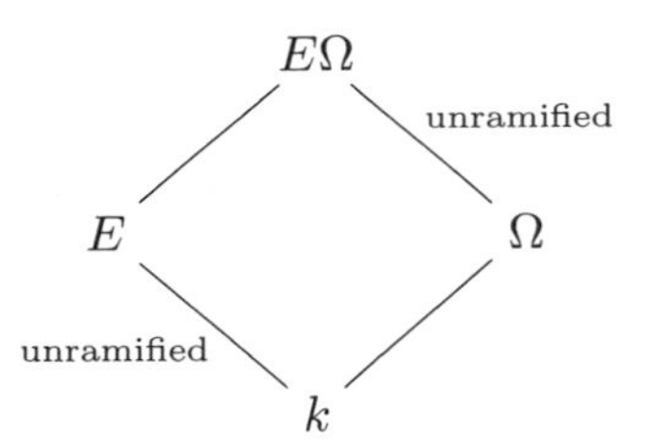

PROOF. Because of our assumptions, $E = k(\alpha_1)$, where $f(x) = \operatorname{Irr}(\alpha_1, k)$ has the same properties as $f(x)$ in the previous theorem. But then, $E\Omega = \Omega(\alpha_1)$ and α_1 is a root of $f(x)$. Hence $E\Omega|\Omega$ is unramified.

We can immediately compute the residue class fields:

$$\overline{E\Omega} = \bar{\Omega}(\bar{\alpha}_1) = \bar{\Omega}\bar{E}.$$ □

5. Finite Residue Class Fields

In most applications the residue class field is a field with a finite number of elements (*Galois Field*). We want to list, here, several properties of such fields (cf., e.g. van der Waerden, Algebra I).

A Galois field, $\bar{k}$, always is of non-zero characteristic, p, and the number of elements is a power of p, say $q = p^r$. The multiplicative group has $q - 1$ elements, hence

$$x^{q-1} = 1, \quad \text{for all } x \neq 0, \text{ or}$$

$$x^q = x \quad \text{for all } x \in \bar{k}.$$

The multiplicative group is cyclic.

An extension of the n-th degree of $\bar{k}$ is again a Galois field; it has q^n elements. Thus, every element of the extension field $\bar{E}$ satisfies the equation

$$x^{q^n} - x = 0$$

and we have

$$x^{q^n} - x = \prod_{\alpha \in \bar{E}} (x - \alpha).$$

$\bar{E}$ is the splitting field of this polynomial, and hence uniquely determined by the degree n of the extension. Since any Galois field with characteristic p must contain the prime field of characteristic p, it follows that the structure of a given Galois field is uniquely determined by the number of elements it contains. For, to a number, p^n, there is exactly one Galois field that contains this number of elements; it is the splitting field of the polynomial

$$x^{p^n} - x$$

over the prime field of characteristic p.

A finite extension of a Galois field $\bar{k}$ is always separable over $\bar{k}$, and is a normal field. All non-zero elements are powers of a single element.

Let k again be a complete field by a non-archimedean valuation, with algebraic closure K, but assume that the residue class field $\bar{k}$ is a finite field. It follows that to any degree n there exists, one, and only one, unramified extension E of k contained in K.

Thus, in the case of dyadic numbers ($p = 2$), although there are seven quadratic extension fields, only one can be unramified. Similarly, for uneven prime numbers p, only one of the three quadratic extension fields which always exist, is unramified.

The single unramified extension E of n-th degree can be found quite easily. Let the residue class field $\bar{k}$ have q elements. Then,

$$\bar{E} = \bar{k}(\bar{\xi}),$$

where $\bar{\xi}$ is any primitive root of the equation

$$x^{q^n - 1} - 1 = 0,$$

in $\bar{K}$. In $\bar{E}$ this polynomial then has the factor decomposition

$$x^{q^n - 1} - 1 = \prod_{\nu = 0}^{q^n - 2} (x - \bar{\xi}^{\nu}).$$

The polynomial $x^{q^n - 1} - 1$ is separable over k, since it is even separable over $\bar{k}$. If ξ is a primitive $(q^n - 1)$-th root of unity in K, the decomposition

$$x^{q^n - 1} - 1 = \prod_{\nu = 0}^{q^n - 2} (x - \xi^{\nu})$$

holds in K. ξ can thus be chosen such that precisely $\bar{\xi}$ is its image in $\bar{K}$, as is seen by taking the image of the last equation.

Thus, $k(\xi)$ is unramified, and

$$\overline{k(\xi)} = \bar{k}(\bar{\xi}) = \bar{E} \quad \text{and} \quad E = k(\xi)$$

since there can be only one unramified extension field of degree n, and $k(\xi)$ is one. Thus, the unramified quadratic extension of the p-adic number field can be generated by adjunction of a (p^2-1)-th primitive root of unity.

6. Discrete Valuations

Let the field k have a non-archimedean valuation with group of values W_k.

DEFINITION. The valuation of k is called discrete if the group of values W_k is cyclic.

Using additive notation for the group of values W_k of a discrete valuation, we can assume this group to be the group $\mathbb{Z}$ of positive integers:

$$W_k = \mathbb{Z}, \qquad \operatorname{ord} a \in \mathbb{Z} \text{ for } a \neq 0,$$

for, if this is not already the case, it can be attained by an order-isomorphic deformation.

By the above definition, there then exists an element $\pi \in k$, such that

$$\operatorname{ord} \pi = 1.$$

$|\pi|$, or in our present notation $\operatorname{ord} \pi$ generates the group of values W_k.

Thus, for any $a \in k$, $\operatorname{ord} a = n$, we have $\operatorname{ord} \frac{a}{\pi^n} = 0$, hence

$$\frac{a}{\pi^n} = \varepsilon, \quad \text{a unit of the valuation ring.}$$

We can, therefore, for any element a in the valuation ring find a unit ε with

$$a = \varepsilon \pi^n \quad (n = 0, 1, 2, \dots).$$

Arithmetic methods will, in general, yield no more than this about valuation rings, since arithmetic investigations give no further information about the units of a ring. The arithmetic in our valuation ring is of a quite trivial nature; it is generated by the "prime number" π. In rational number fields, for valuations generated by the natural prime number p, one can choose $\pi = p$.

THEOREM. *If K is a finite extension of k, then an extension of a discrete valuation of k onto K is a discrete valuation of K.*

PROOF. The index $(W_K : W_k)$ is finite, i.e. the quotient group W_K/W_k is finite. In our additive notation:

$$eW_K \subset W_k = \mathbb{Z}.$$

The group eW_K is an additive subgroup (modul) of $\mathbb{Z}$, it does not consist of 0 alone, and since any such subgroup is a principal ideal in $\mathbb{Z}$, it consists precisely of all multiples of some certain number: it is cyclic. But then W_K is also cyclic, for in the additive group of real numbers multiplication by $\frac{1}{e}$ is an isomorphism. □

In general, however, we no longer have $W_K = \mathbb{Z}$. In order to achieve this another deformation would be necessary, which would then make the value group W_K an additive subgroup of $\mathbb{Z}$; in particular, so that

$$W_k = eW_K = e\mathbb{Z},$$

as can be seen by: $(W_K : W_k) = (\mathbb{Z} : W_k) = e$.

The order of an element $a \in k$ is therefore not a uniquely defined concept, but depends upon the field to which the group of values is adapted. To indicate which order is meant, we write:

$$\mathrm{ord}_k \quad \text{or} \quad \mathrm{ord}_K, \quad \text{respectively,}$$

which correspond to each other by the formula

$$\mathrm{ord}_K\, a = e \cdot \mathrm{ord}_k\, a \quad \text{for any } a \in k.$$

(In case there are several ways of extending the valuation of k onto K, new ambiguities may arise; this can not happen, though, if the field k is complete).

The arithmetic in K can be deduced from the following: There exists an element $\Pi \in K$ such that

$$\mathrm{ord}_K\, \Pi = 1.$$

We then have $\mathrm{ord}_K\, \pi = e$; thus $\pi = \varepsilon \Pi^e$, Where ε is a unit of K.

EXAMPLE. Consider the field of rational numbers $\mathbb{Q}$ with the valuation determined by $p = 2$. In the quadratic number field $\mathbb{Q}(i)$, where $i^2 + 1 = 0$,

$$2 = (-i)(1+i)^2.$$

$-i$ is a unit, $1+i$ is integral — it must in fact be prime, since $\sum_{i=1}^{r} e_i f_i \leq 2$. Thus it follows that there exists only one extension of the valuation, and that for it $e = 2$, $f = 1$. In the Gauss number field $\mathbb{Q}(i) = K$, we therefore have:

$$\mathrm{ord}_K\, a = 2\,\mathrm{ord}_{\mathbb{Q}}\, a \quad \text{for every rational } a.$$

7. Complete Fields by Discrete Valuations

We now assume that the field k is complete by a discrete valuation. The valuation can then be uniquely extended to a valuation of an extension field K of degree n over k. Let e be the ramification and f be the residue class degree of the valuation extension. We can here replace our earlier inequality $ef \leq n$ by the stronger statement:

$$ef = n.$$

PROOF. It will suffice to prove that $ef \geq n$.

To every rational integer i choose an element $\Pi_i \in K$, with

$$\mathrm{ord}\,\Pi_i = \mathrm{ord}_K\, \Pi_i = i,$$

(e.g., Π^i). Choose, as before, elements $\omega_1, \omega_2, \ldots, \omega_f$ of the valuation ring of K, so that their images by the place form a basis of the residue class field of K over that of k (a *residue class basis*). Any element of the residue class field of K is an image of a linear combination

$$A = a_1\omega_1 + a_2\omega_2 + \cdots + a_f\omega_f,$$

a_i $(i = 1, 2, \ldots, f)$ elements of the valuation ring of k, and the image of such a linear combination is an element of the residue class field.

Let α be an element of K with $\mathrm{ord}\,\alpha \geq i$. $\frac{\alpha}{\Pi_i}$ is integral, since $\mathrm{ord}\,\frac{\alpha}{\Pi_i} \geq 0$. Thus, there exists a linear combination A whose image in the residue class field of

K is the same as that of $\frac{\alpha}{\Pi_i}$, and $\frac{\alpha}{\Pi_i} - A$ has as image 0.

$$\Longrightarrow \operatorname{ord}\left(\frac{\alpha}{\Pi_i} - A\right) \geq 1$$
$$\Longrightarrow \operatorname{ord}(\alpha - A\Pi_i) \geq i + 1.$$

Now let α be an integer in K: $\operatorname{ord} \alpha \geq 0$. By iteration of the above process: There exists a linear combination:

A_0 with: $\operatorname{ord}(\alpha - A_0\Pi_0) \geq 1$

A_1 with: $\operatorname{ord}(\alpha - A_0\Pi_0 - A_1\Pi_1) \geq 2.$

A_i $(i = 1, 2, \ldots, n)$ with: $\operatorname{ord}\left(\alpha - \sum_{i=0}^{n} A_i\Pi_i\right) \geq n + 1.$

This implies that α can be represented by a series

$$\sum_{i=0}^{\infty} A_i\Pi_i,$$

where the A_i are linear combinations of the ω_j $(j = 1, \ldots, f)$, with coefficients in the valuation ring of k, so that

$$\alpha = \sum_{i=0}^{\infty}(a_{i1}\omega_1 + a_{i2}\omega_2 + \cdots + a_{if}\omega_f)\Pi_i,$$

with a_{ij} integral elements of k.

We are now free to choose convenient Π_i: Let

$$\Pi = \Pi^j\pi^r, \quad \text{where } i = j + er,\ 0 \leq j \leq e - 1,\ r \in \mathbb{Z}.$$

Then,

$$\alpha = \sum_{r=0}^{\infty}\sum_{j=0}^{e-1}(a_{i1}\omega_1 + \cdots + a_{if}\omega_f)\pi^r\Pi^j$$
$$= \sum_{j=0}^{e-1}\left(\left(\sum_{r=0}^{\infty} a_{i1}\pi^r\right)\omega_1 + \cdots + \left(\sum_{r=0}^{\infty} a_{if}\pi^r\right)\omega_f\right)\Pi^j.$$

The convergence of the individual series $\sum_{r=0}^{\infty} a_{is}\pi^r$, which permitted the above interchange of summations, is, since k is complete, a convergence to elements b_{js} of k, which even lie in the valuation ring. Thus,

$$\alpha = \sum_{j=0}^{e-1}(b_{j1}\omega_1 + b_{j2}\omega_2 + \cdots + b_{jf}\omega_f)\Pi^j, \quad b_{js} \text{ integral in } k.$$

We can now state our result: Every integral element $\alpha \in K$ is a linear combination of the ef elements $\omega_i\Pi^j$ $(i = 1, \ldots, f;\ j = 0, \ldots, e - 1)$ with integral coefficients in k.

In k we now choose an element $a \neq 0$ with $|a| < 1$ (e.g. π). For any $\alpha \in K$ with $|\alpha| > 1$ there exists an $r \in \mathbb{Z}$ such that

$$|\alpha a^r| \leq 1,$$

or

$$\operatorname{ord}(\alpha a^r) \geq 0.$$

Thus αa^r, and therefore also α is a linear combination of the $\omega_i \Pi^j$ with coefficients in k (which, however, are no longer integral).

It now follows that $n = [K : k] \leq ef$, and hence

$$n = ef.$$

□

The elements $\omega_i \Pi^j$ constructed above are linearly independent over k, and form a basis of K over k.

We have also proved that, for the valuation rings $\mathfrak{O}$ and $\mathfrak{o}$ of K and k,

$$\mathfrak{O} = \sum_{i,j} \mathfrak{o} \cdot \omega_i \Pi^j.$$

An element of K is integral if, and only if, its representation in terms of the basis $\omega_i \Pi^j$ has all integral coefficients.

A basis of $K|k$ with this last property is called a minimal basis. We have proven that if the base field k is complete, such a minimal basis always exists, and have in fact, shown how to construct one, given a prime element of K and a residue class basis of $K|k$. A minimal basis need not exist if the base field is not complete.

Special Cases. 1.) $e = n$, $K|k$ is then called fully ramified. $\pi = \varepsilon \cdot \Pi^n$, $f = 1$. A minimal basis of $K|k$ is formed by the elements: $1, \Pi, \Pi^2, \ldots, \Pi^{e-1}$.

2.) $f = n$. Then $e = 1$ and a minimal basis of $K|k$ is formed by the elements

$$\omega_1, \omega_2, \ldots, \omega_f.$$

If, further, $\bar{K}$ is separable over $\bar{k}$ and thus $K|k$ unramified, then there exists an element $\alpha \in K$ such that $1, \alpha, \alpha^2, \ldots, \alpha^{n-1}$ is a minimal basis.

3.) $\bar{K}$ separable over $\bar{k}$ (e not necessarily 1).

It is possible, also in this case, to find a minimal basis in the form

$$1, \alpha, \alpha^2, \ldots, \alpha^{n-1}.$$

For, let α be an element which generates the residue class field extension corresponding to $K|k$. The image of α satisfies an equation of degree f, and therefore α itself must satisfy a congruence

$$\phi(\alpha) \equiv 0 \quad \text{of degree } f.$$

But the residue class extension is separable: $\phi'(\alpha) \neq 0$, and $\operatorname{ord} \phi'(\alpha) = 0$.

We may assume that $\phi(\alpha)$ is a prime element of K. For, the above congruence implies $\operatorname{ord} \phi(\alpha) \geq 1$. Should $\operatorname{ord} \phi(\alpha) \geq 2$, then we can replace α by $\alpha + \Pi$, where Π is an element of K with $\operatorname{ord} \Pi = 1$. (Then Π and $\alpha + \Pi$ have the same images.) $\alpha + \Pi$ satisfies the same congruence, and

$$0 \equiv \phi(\alpha + \Pi) = \phi(\alpha) + \Pi \phi'(\alpha) + \Pi^2 R(\alpha).$$

Since both $\phi(\alpha)$ and the remainder term $\Pi^2 R(\alpha)$ have orders ≥ 2, we have

$$\operatorname{ord} \phi(\alpha + \Pi) = 1.$$

Assuming then that $\operatorname{ord} \phi(\alpha) = 1$ and $\phi(\alpha)$ is a prime element, we see that each element of $\mathfrak{O}$ can be written as a polynomial in α with coefficients in $\mathfrak{o}$:

$$\mathfrak{O} = \mathfrak{o} + \mathfrak{o}\alpha + \mathfrak{o}\alpha^2 + \cdots + \mathfrak{o}\alpha^{n-1}.$$

For, a linear combination of elements $\alpha^i(\phi(\alpha))^j$ is a polynomial in α, and any higher powers can be replaced by linear combinations of $1, \alpha, \alpha^2, \ldots, \alpha^{n-1}$, since α satisfies an equation of degree n with integral coefficients in k.

CHAPTER 7

THE DIFFERENT

1. Complementary Sets

First we want to derive a formula which dates back to Euler.

Let k be any field, and $f(x)$ a separable polynomial out of $k[x]$, which is decomposed, in the splitting field to

$$f(x) = (x - \alpha_1)(x - \alpha_2)\cdots(x - \alpha_n), \quad \alpha_i \neq \alpha_j \text{ for } i \neq j.$$

The Lagrange Interpolation Formula then yields

$$\sum_{i=1}^{n} \frac{f(x)}{x - \alpha_i} \frac{\alpha_i^r}{f'(\alpha_i)} = x^r \quad (r = 0, 1, \ldots, n-1).$$

To prove the formula we note that both sides of the equation are polynomials of degree $\leq n-1$, and that it therefore suffices to show that these polynomials coincide for n different values of the argument, x. We choose for these values of x the n roots: $\alpha_1, \ldots, \alpha_n$, and because of the symmetry of the formula in these, we need only prove the coincidence of the values of the functions for one, α_1. But,

$$\left(\frac{f(x)}{x - \alpha_i}\right)_{x=\alpha_1} = \begin{cases} 0 & \text{for } i \neq 1, \\ (\alpha_1 - \alpha_2)(\alpha_1 - \alpha_3)\cdots(\alpha_1 - \alpha_n) = f'(\alpha_1) & \text{for } i = 1, \end{cases}$$

proving the formula for $x = \alpha_1$, and thus for all x.

Now let $K = k(\alpha)$ where $f(x) = \operatorname{Irr}(\alpha, k)$ is separable. The coefficient of $\frac{f(x)}{x-\alpha}$ lie in K. By the trace, $S(g(x))$ of a polynomial $g(x) \in K[x]$, we mean the polynomial in $k[x]$ arrived at by substituting for the coefficients a_i of $g(x)$ their respective traces $S(a_i) = S_{K|k}(a_i)$. We can then write our formula in the form:

$$S\left(\frac{f(x)}{x - \alpha} \cdot \frac{\alpha^r}{f'(\alpha)}\right) = x^r \quad (r = 0, 1, \ldots, n-1).$$

Now, let K be a separable extension of the field k, with the basis

$$\omega_1, \omega_2, \ldots, \omega_n.$$

THEOREM. *Given any n elements $a_1, a_2, \ldots, a_n \in k$, there exists a unique $\xi \in K$ with*

$$S(\xi\omega_i) = a_i \quad (i = 1, 2, \ldots, n).$$

PROOF. Set

$$\xi = x_1\omega_1 + x_2\omega_2 + \cdots + x_n\omega_n, \quad x_i \in k,$$

then the conditions for ξ can be written:

$$x_1 S(\omega_1\omega_i) + x_2 S(\omega_2\omega_i) + \cdots + x_n S(\omega_n\omega_i) = a_i \quad (i = 1, 2, \ldots, n).$$

This system of n equations in the n unknowns $x_1, \dots, x_n$, with coefficients in k, is uniquely solvable if the corresponding homogenous equations have only the trivial solution.

Let, therefore, $x_1, x_2, \dots, x_n$ be a solution of the homogenous equations, then

$$S(\xi \cdot \omega_i) = 0 \quad (i = 1, 2, \dots, n),$$

where $\xi = x_1\omega_1 + x_2\omega_2 + \cdots + x_n\omega_n$. By multiplying the i-th equation by an arbitrary $y_i \in k$ and then summing over all i, we can then attain

$$S(\xi \cdot \eta) = 0 \quad \text{for all } \eta \in K,$$

i.e.

$$S(\xi \cdot K) = 0.$$

If $\xi \neq 0$, then $\xi \cdot K = K$, and therefore also $S(K) = 0$, which contradicts the separability of K. Any solution of the homogenous equations is trivial. □

By this theorem we can choose $\omega'_1, \omega'_2, \dots, \omega'_n$, $\omega'_j \in K$, with

$$S(\omega'_j\omega_i) = \delta_{ij} = \begin{cases} 1 & \text{for } i = j \\ 0 & \text{for } i \neq j. \end{cases}$$

The elements $\omega'_1, \omega'_2, \dots, \omega'_n$, are then also independent over k, and form a basis of $K|k$, as can be seen by multiplying any possible relation

$$y_1\omega'_1 + y_2\omega'_2 + \cdots + y_n\omega'_n = 0, \quad y_i \in k,$$

by ω_i and taking the trace, thus showing all the y_i to be 0. This basis, $\omega'_1, \omega'_2, \dots, \omega'_n$ of $K|k$ is called the *complementary basis* to the basis $\omega_1, \omega_2, \dots, \omega_n$.

Let, in the following, all rings considered contain unity. Let $\mathfrak{o}$ be a subring of k having k for its quotient field. Let T be some subset of a separable extension field K of finite degree n over k. Dedekind introduced the following notion of a *complementary* set T' of T relative to $\mathfrak{o}$:

$$T' = \{t' \mid t' \in K,\ S(t'T) \subset \mathfrak{o}\}.$$

It is immediately apparent that these complementary sets satisfy

$$T_1 \subset T_2 \implies T'_1 \supset T'_2.$$

Special Cases. 1.) Let $\omega_1, \omega_2, \dots, \omega_n$ be a basis of $K|k$ and $T = \omega_1\mathfrak{o} + \omega_2\mathfrak{o} + \cdots + \omega_n\mathfrak{o}$. Then

$$T' = \{t' \mid t' \in K,\ S(t'\omega_i) \in \mathfrak{o}\} \quad \text{for all } i = 1, 2, \dots, n.$$

Now represent some $t' \in T'$ in the basis $\omega'_1, \omega'_2, \dots, \omega'_n$, complementary to our ω_i:

$$t' = y_1\omega'_1 + y_2\omega'_2 + \cdots + y_n\omega'_n, \quad y_i \in k.$$

Now, $S(t'\omega_i) \in \mathfrak{o}$ precisely then, when $y_i \in \mathfrak{o}$. We thus have

$$T' = \omega'_1\mathfrak{o} + \omega'_2\mathfrak{o} + \cdots + \omega'_n\mathfrak{o},$$

and $T'' = (T')' = T$ for this special case.

2.) We specialize the last case further, by letting K be generated by a single separable element α, and selecting the basis $1, \alpha, \alpha^2, \dots, \alpha^{n-1}$. Then

$$T = \mathfrak{o} + \alpha\mathfrak{o} + \alpha^2\mathfrak{o} + \cdots + \alpha^{n-1}\mathfrak{o}.$$

Using the derived interpolation formula we can calculate the complementary basis.

Let $f(x) = \operatorname{Irr}(\alpha, k) = a_0 + a_1 x + a_2 x^2 + \dots,\ a_i \in k,\ a_n = 1,\ a_{n+1} = a_{n+2} = \dots = 0.$

Then

$$\frac{f(x)}{x-\alpha} = \frac{f(x)-f(\alpha)}{x-\alpha} = \sum_{i\geq 1} a_i \frac{x^i - \alpha^i}{x-\alpha} = \sum_{r,s\geq 0} a_{r+s+1} x^r \alpha^s$$

$$= b_0 + b_1 x + \dots + b_{n-1}x^{n-1}, \quad \text{where } b_r = \sum_{s\geq 0} a_{r+s+1}\alpha^s = \left(\left[\frac{f(x)}{x^{r+1}}\right]\right)_{x=\alpha}$$

and the symbol $\left[\frac{f(x)}{x^{r+1}}\right]$ denotes the integral component of $\frac{f(x)}{x^{r+1}}$. We compute the b_r's:

$$\begin{aligned} b_{n-1} &= 1 \\ b_{n-2} &= a_{n-1} + \alpha \\ b_{n-3} &= a_{n-2} + a_{n-1}\alpha + \alpha^2 \\ &\cdots\cdots\cdots\cdots \\ b_0 &= a_1 + a_2\alpha + a_3\alpha^2 + \dots + \alpha^{n-1}. \end{aligned}$$

The interpolation formula now reads

$$S\left(\frac{b_0 + b_1 x + b_2 x^2 + \dots + b_{n-1}x^{n-1}}{f'(\alpha)}\alpha^r\right) = x^r \quad (0 \leq r \leq n-1),$$

and comparison of coefficients yields:

$$S\left(\frac{b_i}{f'(\alpha)}\alpha^r\right) = \begin{cases} 1 & \text{for } i = r \\ 0 & \text{for } i \neq r. \end{cases}$$

We have thus found the complementary basis to $1, \alpha, \dots, \alpha^{n-1}$. It is

$$\frac{b_0}{f'(\alpha)}, \frac{b_1}{f'(\alpha)}, \frac{b_2}{f'(\alpha)}, \dots, \frac{b_{n-1}}{f'(\alpha)}$$

so that $T' = \frac{b_0}{f'(\alpha)}\mathfrak{o} + \frac{b_1}{f'(\alpha)}\mathfrak{o} + \dots + \frac{b_{n-1}}{f'(\alpha)}\mathfrak{o}$.

3.) Assume, even more specialized, that α is integral, with respect to $\mathfrak{o}$, i.e. $a_i \in \mathfrak{o}$ $(i = 1, 2, \dots, n)$. Then

$$\frac{b_{n-1}}{f'(\alpha)} = \frac{1}{f'(\alpha)} \in T' \quad \text{and} \quad \frac{b_{n-2}}{f'(\alpha)} = \frac{a_{n-1}+\alpha}{f'(\alpha)} \in T'.$$

Further, since $\frac{1}{f'(\alpha)} \in T'$ and $a_{n-1} \in \mathfrak{o}$, we have $\frac{a_{n-1}}{f'(\alpha)} \in T'$ and then because of the linearity of the trace function, $\frac{\alpha}{f'(\alpha)} \in T'$. Proceeding in the same manner, we see that T' contains all the elements

$$\frac{1}{f'(\alpha)}, \frac{\alpha}{f'(\alpha)}, \frac{\alpha^2}{f'(\alpha)}, \dots, \frac{\alpha^{n-1}}{f'(\alpha)}$$

and hence $\frac{\mathfrak{o}+\alpha\mathfrak{o}+\alpha^2\mathfrak{o}+\dots+\alpha^{n-1}\mathfrak{o}}{f'(\alpha)} \subset T'$. But the inverse inclusion must also hold, since all the b_r can be expressed as polynomials in α of degree $\leq n-1$ with coefficients in $\mathfrak{o}$. Hence, our result:

$$T' = \frac{T}{f'(\alpha)}.$$

2. The Different of a Field Extension

The usual notion of an ideal, used until now, will be generalized:

DEFINITION. A (fractional) *ideal* $\mathfrak{a}$ with respect to $\mathfrak{o}$ is a subset of the quotient field of $\mathfrak{o}$, with the usual properties of ideals, i.e. closure under subtraction within itself, and under multiplication by an element of the ring, and for which there exists an element $d \neq 0$ in k, such that

$$d\mathfrak{a} \subset \mathfrak{o} \quad (d \text{ common denominator of elements of } \mathfrak{a}).$$

The usual ideals are those that are subsets of $\mathfrak{o}$:

$$\mathfrak{a} \subset \mathfrak{o}.$$

We now call them *integral ideals.*

The sum of fractional ideals $\mathfrak{a}$ and $\mathfrak{b}$ is defined like that of integral ideals:

$$\mathfrak{a} + \mathfrak{b} = \{a + b \mid a \in \mathfrak{a},\ b \in \mathfrak{b}\};$$

it is the smallest ideal containing both $\mathfrak{a}$ and $\mathfrak{b}$ similarly the product

$$\mathfrak{a} \cdot \mathfrak{b} = \left\{ \sum_{i=1}^{n} a_i b_i \mid a_i \in \mathfrak{a},\ b_i \in \mathfrak{b} \right\}.$$

We now assume that $\mathfrak{o}$ is integrally closed in its quotient field k:

$$\begin{array}{ccc} K & \supset & \mathfrak{O} \\ {\scriptstyle n}\Big| & & \Big| \\ k & \supset & \mathfrak{o} \end{array}$$

where K is an extension field of degree n over k, $\mathfrak{O}$ being the integral closure of $\mathfrak{o}$ in K. To every finite extension of the field k there corresponds, in this manner, a ring $\mathfrak{O}$.

If K is separable over k we can form the complementary set $\mathfrak{O}'$ of $\mathfrak{O}$; we denote this set as the *inverse different* of $K|k$. Using our earlier notation, the inverse different is defined by the condition $S(\mathfrak{O}'\mathfrak{O}) \subset \mathfrak{o}$.

We choose an $\alpha \in \mathfrak{O}$ which generates the field K. Such an α exists, for, since $K|k$ is separable, $K = k(\beta)$, for some $\beta \in K$, which satisfies an irreducible equation $a_0x^n + a_1x^{n-1} + \cdots + a_n = 0$, $a_i \in \mathfrak{o}$. By multiplying this equation by a_0^{n-1}, we see that the element $\alpha = a_0\beta$ satisfied the irreducible equation

$$x^n + a_1x^{n-1} + a_2a_0x^{n-2} + \cdots + a_na_0^{n-1} = 0,$$

and therefore is both an element of $\mathfrak{O}$ and a generating element of K.

Let $f(x) = \mathrm{Irr}(\alpha, k)$. Since $\mathfrak{o} \subset \mathfrak{O}$, $\alpha \in \mathfrak{O}$, we have

$$\mathfrak{o}[\alpha] = \mathfrak{o} + \alpha\mathfrak{o} + \alpha^2\mathfrak{o} + \cdots + \alpha^{n-1}\mathfrak{o} \subset \mathfrak{O}.$$

By what we proved in the last section we have

$$\frac{\mathfrak{o}[\alpha]}{f'(\alpha)} \supset \mathfrak{O}',$$

and furthermore, $\mathfrak{O} \subset \mathfrak{O}'$, because $S(\mathfrak{O}\mathfrak{O}) = S(\mathfrak{O}) \subset \mathfrak{o}$. Compiling, we see

$$\mathfrak{o}[\alpha] \subset \mathfrak{O} \subset \mathfrak{O}' \subset \frac{\mathfrak{o}[\alpha]}{f'(\alpha)} \subset \frac{\mathfrak{O}}{f'(\alpha)}.$$

The condition $\mathfrak{o}[\alpha] \subset \mathfrak{O} \subset \frac{\mathfrak{o}[\alpha]}{f'(\alpha)}$ serves to approximate $\mathfrak{O}$. To arrive at the exact integral closure $\mathfrak{O}$ of $\mathfrak{o}$ it suffices to check the elements of $\frac{\mathfrak{o}[\alpha]}{f'(\alpha)}$ for integrity with respect to $\mathfrak{o}$.

Now, the relation $f'(\alpha)\mathfrak{O}' \subset \mathfrak{O}$ shows that there exists a common denominator for all the elements of $\mathfrak{O}'$; the other conditions for $\mathfrak{O}'$ to be an ideal with respect to $\mathfrak{O}$ are also satisfied, the first being trivial and the second following from

$$S(\mathfrak{O}'\mathfrak{O}) \subset \mathfrak{o} \implies S(\mathfrak{O}'\mathfrak{O}\mathfrak{O}) \subset \mathfrak{o} \implies \mathfrak{O}\mathfrak{O}' \subset \mathfrak{O}'.$$

For the rings $\mathfrak{o}$ and $\mathfrak{O}$ we now make the following, limiting

ASSUMPTION. The rings $\mathfrak{o}$ and $\mathfrak{O}$ be such that their ideals (other than (0)) form a group by the operation of ideal multiplication.

The assumption amounts to the postulation of an inverse for each ideal, the other group axioms being obviously always satisfied (unit element: the ring itself).

We then take Dedekind's definition of the *different*:

$$\mathfrak{d}_{K|k} = (\mathfrak{O}')^{-1}.$$

If now, $\mathfrak{A}$ is any ideal ($\neq (0)$) of K, then the complementary set $\mathfrak{A}'$ has the usual properties of ideals, as is seen by $S(\mathfrak{A}'\mathfrak{A}) = S(\mathfrak{A}'\mathfrak{O}\mathfrak{A}) \subset \mathfrak{o}$, and the linearity of the trace function. $\mathfrak{A}'$ is defined as the largest set such that $S(\mathfrak{A}'\mathfrak{A}) \subset \mathfrak{o} \iff S(\mathfrak{A}'\mathfrak{A}\mathfrak{O}) \subset \mathfrak{o} \iff \mathfrak{A}'\mathfrak{A} \subset \mathfrak{O}' \iff \mathfrak{A}'\mathfrak{O} \subset \mathfrak{O}'\mathfrak{A}^{-1} \iff \mathfrak{A}' \subset \mathfrak{O}\mathfrak{A}^{-1}$; i.e. $\mathfrak{A}' = \mathfrak{O}'\mathfrak{A}^{-1}$. Then $\mathfrak{A}'$ is, as the product of two ideals, itself an ideal, and we have

$$\mathfrak{A}' = \mathfrak{d}_{K|k}^{-1}\mathfrak{A}^{-1} = \frac{1}{\mathfrak{d}_{K|k}\mathfrak{A}}.$$

An important property of the different is transitivity.

$$\begin{array}{ccc} K & \supset & \mathfrak{O}_K \\ | & & | \\ E & \supset & \mathfrak{O}_E \\ | & & | \\ k & \supset & \mathfrak{o} \end{array}$$

Let E be a separable extension of k, and K a separable extension of E, the corresponding ring being $\mathfrak{O}_E$ and $\mathfrak{O}_K$. We then have

$$\mathfrak{d}_{K|k} = \mathfrak{d}_{E|k} \cdot \mathfrak{d}_{K|E}.$$

PROOF. The trace function is transitive, i.e. $S_{K|k}(\alpha) = S_{E|k}(S_{K|E}(\alpha))$.

The different $\mathfrak{d}_{K|k}$ is thus symbolically defined by the condition

$$\begin{gathered} S_{E|k}(S_{K|E}(\mathfrak{d}_{K|k}^{-1}\mathfrak{O}_K)) \subset \mathfrak{o} \\ \iff S_{E|k}(S_{K|E}(\mathfrak{d}_{K|k}^{-1}\mathfrak{O}_K)\mathfrak{O}_E) \subset \mathfrak{o} \quad \text{since } \mathfrak{O}_E \subset E,\ \mathfrak{O}_E \subset \mathfrak{O}_K \\ \iff S_{K|E}(\mathfrak{d}_{K|k}^{-1}\mathfrak{O}_K) \subset \mathfrak{d}_{E|k}^{-1} \iff S_{K|E}(\mathfrak{d}_{K|k}^{-1}\mathfrak{d}_{E|k}\mathfrak{O}_K) \subset \mathfrak{O}_E \subset \mathfrak{O}_E \\ \iff \mathfrak{d}_{K|k}^{-1}\mathfrak{d}_{E|k} \subset \mathfrak{d}_{K|E}^{-1} \iff \mathfrak{d}_{K|k}^{-1} \subset \mathfrak{d}_{K|E}^{-1}\mathfrak{d}_{E|k}^{-1}. \end{gathered}$$

Since this can be taken as the defining equation for $\mathfrak{d}_{K|k}$, we have

$$\mathfrak{d}_{K|k}^{-1} = \mathfrak{d}_{K|E}^{-1}\mathfrak{d}_{E|k}^{-1},$$

which immediately proves the desired. □

3. The Different in Complete Fields

Let k be a complete field by a discrete valuation with valuation ring $\mathfrak{o}$.

THEOREM. *$\mathfrak{o}$ is a principal ideal ring (i.e. all ideals respective $\mathfrak{o}$ are principal ideals).*

PROOF. Let $\mathfrak{a} \neq (0)$ be an ideal with respect to $\mathfrak{o}$. There exists an element $b \in k$, $b \neq 0$, such that $b\mathfrak{a} \subset \mathfrak{o}$. Hence $\operatorname{ord} b\mathfrak{a} \geq 0$ and $\operatorname{ord} a > -\operatorname{ord} b$ for every $a \in \mathfrak{a}$. It is therefore that we can find an element of $\mathfrak{a}$ with minimal order, say a. Then,

$$a_1 \in \mathfrak{a} \implies \operatorname{ord} a_1 \geq \operatorname{ord} a \implies \operatorname{ord} \frac{a_1}{a} \geq 0 \implies c = \frac{a_1}{a} \in \mathfrak{o}$$
$$\implies a_1 = ac, \quad c \in \mathfrak{o}.$$

Since, obviously, $ac \in \mathfrak{a}$ for every $c \in \mathfrak{o}$, $\mathfrak{a}$ is the principal ideal

$$\mathfrak{a} = a\mathfrak{o}.$$ □

Since it follows from $\operatorname{ord} a_1 \geq \operatorname{ord} a$ that $a_1 = a \cdot \frac{a_1}{a} = ac \in \mathfrak{a}$, the elements a_1 of $\mathfrak{a}$ are precisely characterized by the condition

$$\operatorname{ord} a_1 \geq \operatorname{ord} a.$$

We have, however, shown, that every element $a \in k$ can be written in the form

$$a = \varepsilon \pi^r$$

where ε is a unit of $\mathfrak{o}$, and π is some given prime element, i.e. $\operatorname{ord} \pi = 1$.

Hence

$$a\mathfrak{o} = \pi^r \varepsilon \mathfrak{o} = \pi^r \mathfrak{o};$$

in k there can be no ideals (respective $\mathfrak{o}$) other than the principal ideals $\pi^r\mathfrak{o}$ generated by the powers of π.

In a principal ideal ring the ideals other than (0) obviously are a group, the inverse ideal to $a\mathfrak{o}$ being $a^{-1}\mathfrak{o}$, since $a\mathfrak{o}a^{-1}\mathfrak{o} = aa^{-1}\mathfrak{o} = \mathfrak{o}$. The unity element of the group, $\mathfrak{o}$, is also denoted by 1. The group of ideals is, in this case, an infinite cyclic group, generated by prime ideal $\pi\mathfrak{o}$.

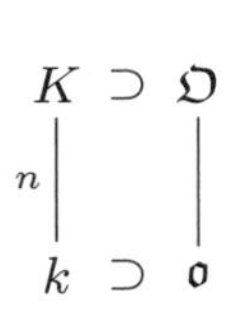

Let K be a field of degree n over k. The extension of the valuation onto K is unique; denote the valuation ring in K by $\mathfrak{O}$. $\mathfrak{o}$ is integrally closed in k, and $\mathfrak{O}$ is the integral closure of $\mathfrak{o}$ in K, as is seen by the place-theoretical characterization of integral elements. If the field extension is separable, then the different can be defined in the manner described before.

We here assume further, that the residue class field extension of $K|k$ is separable. There then exists an element α, which generates the field K, such that

$$\mathfrak{O} = \mathfrak{o} + \alpha\mathfrak{o} + \alpha^2\mathfrak{o} + \cdots + \alpha^{n-1}\mathfrak{o} = \mathfrak{o}[\alpha]$$

and hence

$$\mathfrak{O}' = \mathfrak{d}^{-1} = \frac{\mathfrak{o}[\alpha]}{f'(\alpha)} = \frac{1}{f'(\alpha)}\mathfrak{O}$$

or

$$\mathfrak{d} = f'(\alpha)\mathfrak{O}$$

where $\mathfrak{d} = \mathfrak{d}_{K|k}$ and $f(x) = \operatorname{Irr}(\alpha, k)$,

$f'(\alpha)$ can be written in the form $f'(\alpha) = \varepsilon \cdot \Pi^r$, with ε a unit of $\mathfrak{O}$, and Π a prime element of K. We can approximate this exponent r from above even without knowing an element α with the above property.

Let β be any integral element which generates K, and $f_1(x) = \operatorname{Irr}(\beta, k)$.

$$\Longrightarrow \mathfrak{o}[\beta] \subset \mathfrak{O} \Longrightarrow \mathfrak{O}' \subset \frac{\mathfrak{o}[\beta]}{f_1'(\beta)} \subset \frac{\mathfrak{O}}{f_1'(\beta)}$$

$$\Longrightarrow \mathfrak{d}^{-1} \subset \frac{1}{f_1'(\beta)}\mathfrak{O} \Longrightarrow \mathfrak{d} \supset f_1'(\beta)\mathfrak{O}$$

hence

$$f_1'(\beta)\mathfrak{O} \subset f'(\alpha)\mathfrak{O}.$$

$f'(\alpha)$ is therefore a factor of $f_1'(\beta)$, i.e. $f_1'(\beta) = cf'(\alpha)$, where $c \in \mathfrak{O}$.

This even continues to hold true if we replace $f_1(x)$ by any integral polynomial $g(x) \in k[x]$ for which $g(\beta) = 0$. For then, $g(x) = f_1(x)h(x)$, where $h(x)$ has integral coefficients. Thus

$$g'(\beta) = f_1'(\beta)h(\beta) = f'(\alpha) \cdot c \cdot h(\beta) \quad \text{where } c \cdot g(\beta) \in \mathfrak{O},$$

and we see that $g'(\beta)$ is divisible by $f'(\alpha)$.

We want to consider more closely the case of a separable residue class field extension:

1.) Let $K|k$ be unramified: $e = 1$, $f = n$. Then, $f'(\alpha) \not\equiv 0 \pmod{\pi}$, and $f'(\alpha)$ is a unit of $\mathfrak{O}$

$$\Longrightarrow \mathfrak{d} = \mathfrak{O} = 1.$$

2.) Let $K|k$ be fully ramified: $e = n$, $f = 1$. Then,

$$\mathfrak{O} = \mathfrak{o} + \Pi\mathfrak{o} + \cdots + \Pi^{e-1}$$

$$f(x) = \operatorname{Irr}(\Pi, k) = x^e + a_1 x^{e-1} + \cdots + a_n, \quad a_i \in \mathfrak{o} \ (i \in 1, 2, \ldots, n),$$

$$\mathfrak{d} = f'(\Pi)\mathfrak{O}.$$

The coefficients a_i of $f(x)$ are the elementary symmetric functions of Π and its conjugates, which all have values < 1. Hence

$$|a_i| < 1 \quad \text{or} \quad \pi \mid a_i \quad (i = 1, 2, \ldots, n).$$

For the coefficient a_n, more precisely: $\pi \mid a_n$, $\pi^2 \nmid a_n$, for, $a_n = \pm N\Pi$ (Norm with respect to k), and therefore:

$$\sqrt[e]{|\pi|} = |\Pi| = \sqrt[e]{|N\Pi|} = \sqrt[e]{|a_n|} \Longrightarrow |\pi| = |a_n|.$$

These are precisely the defining properties for the coefficients of an *Eisenstein Equation.*

We here introduce a proof of the *Eisenstein Theorem* using the valuation theory. Let k be a field with discrete valuation (not necessarily complete), π a prime element ($\operatorname{ord} \pi = 1$), and let

$$f(x) = x^n + a_1 x^{n-1} + \cdots + a_n, \quad a_i \in \mathfrak{o}, \ \pi \mid a_i \ (i = 1, 2 \ldots, n), \ \pi^2 \nmid a_n.$$

Then, it is maintained, $f(x)$ is irreducible.

Let Π be a root of $f(x) = 0$, and let the valuation of k be extended in some manner onto $K = k(\Pi)$. $|\Pi| < 1$; otherwise $f(\Pi) = 0$ would lead to a contradiction. In the equation

$$f(\Pi) = \Pi^n + a_1\Pi^{n-1} + \cdots + a_n = 0,$$

some two terms must have maximal values. Among the last n terms $|a_n|$ certainly is maximal, hence

$$|\Pi|^n = |a_n| = |\pi| \implies \pi = \Pi^n \varepsilon \quad \text{where } \varepsilon \text{ is a unit} \implies e \geq n;$$

therefore

$$e = n, \quad f = 1.$$

We have the result that the polynomial is irreducible, that $K|k$ is fully ramified and that the extension of the valuation onto K is unique, with Π prime in K.

In the case of a complete field, therefore, the fully ramified extensions are given by, and only by, Eisenstein equations.

We have

$$f'(\Pi) = e\Pi^{e-1} + (e-1)a_1\Pi^{e-2} + \cdots + a_{n-1} = \varepsilon\Pi^{\bar{e}} \quad (\varepsilon \text{ unit of } \mathfrak{O}),$$

$$\mathfrak{d} = \Pi^{\bar{e}}\mathfrak{O}.$$

If we take the absolute values of the first equation we see that the values of the non-zero individual terms of the middle expression are all different from each other. For the factors $e, (e-1)a_1, \ldots, a_{n-1}$, as long as they are not 0, divisible by powers of π, hence by powers of Π^e, and thus the orders of the various terms are even incongruent modulo e. Thus,

$$f'(\Pi) = |\Pi|^{\bar{e}} = \operatorname{Max}(|e\Pi^{e-1}|, |(e-1)a_1\Pi^{e-2}|, \ldots, |a_{n-1}|),$$

and then certainly

$$f'(\Pi) = |\Pi|^{\bar{e}} < 1,$$

as was to be expected, since the different is an integral ideal. In the case that e is a unit (and this will mostly be the case), $\bar{e} = e - 1$, and thus $\mathfrak{d} = \Pi^{e-1}\mathfrak{O}$.

$$\begin{array}{c} K \\ \Big|\, e \\ T \\ \Big|\, f \\ k \end{array}$$

3.) Separable Residue Class Field Extension in general: We have already proven: The residue class field $\bar{K}$ of K uniquely determines an unramified extension T of k ($\bar{T} = \bar{K}$) of degree f. T is contained in K, and K is necessarily of degree e over T, and fully ramified over T. Thus,

$$\mathfrak{d}_{K|k} = \mathfrak{d}_{K|T} \cdot \mathfrak{d}_{T|k} = \mathfrak{d}_{K|T},$$

since $\mathfrak{d}_{T|k} = 1$. Thus, the different is 1 if and only if the extension is unramified, i.e. $e = 1$.

CHAPTER 8

THE VALUATION-THEORETICAL APPROACH TO IDEAL THEORY

1. Valuation of Ideals

In the following we will lay the foundations for the development of Ideal Theory with the methods of Valuation Theory.

Let the field k be the quotient field of a ring $\mathfrak{o}$ (with unity); let the ring $\mathfrak{o}$ be contained in the valuation ring of a discrete valuation of k, i.e.

$$a \in \mathfrak{o} \implies |a| \leq 1 \quad \text{or equivalently,} \quad \operatorname{ord} a \geq 0.$$

A valuation is introduced into the set of ideals of $\mathfrak{o}$ by

$$\operatorname{ord} \mathfrak{a} = \operatorname{Min}_{a \in \mathfrak{a}}(\operatorname{ord} a) \quad \text{for every ideal } \mathfrak{a}.$$

This minimum always exists, for, by the definition of ideals, there exists an element $b \in k$, $b \neq 0$, with $b\mathfrak{a} \subset \mathfrak{o}$, hence $\operatorname{ord} ba \geq 0$ and $\operatorname{ord} a > -\operatorname{ord} b$ for all $a \in \mathfrak{a}$. Since these orders are integers, the minimum in the definition must exist.

THEOREM. *For any ideals $\mathfrak{a}$ and $\mathfrak{b}$ in $\mathfrak{o}$ we have:*

$$\operatorname{ord}(\mathfrak{a} + \mathfrak{b}) = \operatorname{Min}(\operatorname{ord} \mathfrak{a}, \operatorname{ord} \mathfrak{b}).$$

$$\operatorname{ord}(\mathfrak{a}\mathfrak{b}) = \operatorname{ord} \mathfrak{a} + \operatorname{ord} \mathfrak{b}, \quad \textit{and,}$$

for principal ideals $a\mathfrak{o}$ we have

$$\operatorname{ord}(a\mathfrak{o}) = \operatorname{ord} a.$$

PROOF. 1.) Because $\operatorname{ord}(a + b) = \operatorname{Min}(\operatorname{ord} a, \operatorname{ord} b)$ we have

$$\begin{aligned}\operatorname{ord}(\mathfrak{a} + \mathfrak{b}) &= \operatorname{Min}_{\substack{a \in \mathfrak{a} \\ b \in \mathfrak{b}}}(\operatorname{ord}(a + b), \operatorname{ord} a, \operatorname{ord} b) = \operatorname{Min}_{\substack{a \in \mathfrak{a} \\ b \in \mathfrak{b}}}(\operatorname{ord} a, \operatorname{ord} b) \\ &= \operatorname{Min}(\operatorname{ord} \mathfrak{a}, \operatorname{ord} \mathfrak{b}).\end{aligned}$$

2. Because $\operatorname{ord}\left(\sum_{ab} ab\right)$ we have:

$$\begin{aligned}\operatorname{ord}(\mathfrak{a}\mathfrak{b}) &= \operatorname{Min}_{\substack{a \in \mathfrak{a} \\ b \in \mathfrak{b}}}(\operatorname{ord} ab) = \operatorname{Min}_{\substack{a \in \mathfrak{a} \\ b \in \mathfrak{b}}}(\operatorname{ord} a + \operatorname{ord} b) \\ &= \operatorname{ord} \mathfrak{a} + \operatorname{ord} \mathfrak{b}.\end{aligned}$$

3.) Because $\operatorname{ord} x \geq 0$ for $x \in \mathfrak{o}$ and $\operatorname{ord} x = 0$ for $x = 1$ we have:

$$\operatorname{ord}(a\mathfrak{o}) = \operatorname{Min}_{x \in \mathfrak{o}}(\operatorname{ord} a + \operatorname{ord} x) = \operatorname{ord} a. \qquad \square$$

2. Divisors

Our goal is the answer to the following two questions: Does there exist a set of valuations, whose valuation rings contain $\mathfrak{o}$, such that the ideals of $\mathfrak{o}$ are fully characterized by their orders with respect to these valuations? Given a certain order, does there, necessarily exist an ideal with that order?

We introduce the following axioms for our field k:

AXIOM 1. The field k has a set $\mathfrak{M}$ of inequivalent, discrete valuations $\mathfrak{p}$ (notation: $|a|_\mathfrak{p}$ or $\operatorname{ord}_\mathfrak{p} a$) such that for every element $a \in k$:

$$|a|_\mathfrak{p} \leq 1 \quad \text{or} \quad \operatorname{ord}_\mathfrak{p} a \geq 0, \quad \text{respectively,}$$

for almost all $\mathfrak{p} \in \mathfrak{M}$.

This axiom can be shown to be stronger than it appears. For, letting $a \neq 0$, $a \in k$, application of the axiom to $\frac{1}{a}$ yields:

$$|a|_\mathfrak{p} = 1 \quad \text{or} \quad \operatorname{ord}_\mathfrak{p} a = 0 \quad \text{for almost all } \mathfrak{p} \in \mathfrak{M}.$$

New let $\mathfrak{o}$ be a ring contained in all the valuation rings corresponding to the $\mathfrak{p} \in \mathfrak{M}$, and let $\mathfrak{a}$ be an ideal of $\mathfrak{o}$ ($\mathfrak{a} \neq (0)$). There exists an element $b \neq 0$ with $b\mathfrak{a} \subset \mathfrak{o}$, hence $\operatorname{ord}_\mathfrak{p} \mathfrak{a} \geq -\operatorname{ord}_\mathfrak{p} b$ for all $\mathfrak{p} \in \mathfrak{M}$, and applying the axiom to b we see that $\operatorname{ord}_\mathfrak{p} \mathfrak{a} \geq 0$ for almost all $\mathfrak{p} \in \mathfrak{M}$. But the order of an ideal by a given valuation is less than that of any non-zero element; applying the axiom to some element we, see $\operatorname{ord}_\mathfrak{p} \mathfrak{a} \leq 0$ for almost all $\mathfrak{p}$, and hence

$$\operatorname{ord}_\mathfrak{p} \mathfrak{a} = 0 \quad \text{for almost all } \mathfrak{p} \in \mathfrak{M}.$$

DEFINITION. A *divisor* $\mathfrak{d}$ of K is a formal power product of valuations $\mathfrak{p} \in \mathfrak{M}$:

$$\mathfrak{d} = \prod_{\mathfrak{p} \in \mathfrak{M}} \mathfrak{p}^{\nu_\mathfrak{p}}, \qquad \nu_\mathfrak{p} \in \mathbb{Z},$$

for which $\nu_\mathfrak{p} = 0$ for almost all $\mathfrak{p}$.

By the convention of leaving away factors with exponent 0 we write:

$$\mathfrak{d} = \mathfrak{p}_1^{\nu_1} \mathfrak{p}_2^{\nu_2} \cdots \mathfrak{p}_r^{\nu_r};$$

then $\nu_\mathfrak{p} = 0$ for $\mathfrak{p} \neq \mathfrak{p}_1, \mathfrak{p}_2, \ldots, \mathfrak{p}_r$.

The *order of a divisor* $\mathfrak{d} = \prod_{\mathfrak{p} \in \mathfrak{M}} \mathfrak{p}^{\nu_\mathfrak{p}}$ at the place $\mathfrak{p}$ is defined as

$$\operatorname{ord}_\mathfrak{p} \mathfrak{d} = \nu_\mathfrak{p}.$$

A divisor $\mathfrak{d}$ is called *integral* if $\operatorname{ord}_\mathfrak{p} \mathfrak{d} \geq 0$ for all $\mathfrak{p}$.

For two divisors $\mathfrak{d}_1 = \prod_{\mathfrak{p} \in \mathfrak{M}} \mathfrak{p}^{\nu_\mathfrak{p}}$, $\mathfrak{d}_2 = \prod_{\mathfrak{p} \in \mathfrak{M}} \mathfrak{p}^{\mu_\mathfrak{p}}$ we define sum and product:

$$\mathfrak{d}_1 \mathfrak{d}_2 = \prod_{\mathfrak{p} \in \mathfrak{M}} \mathfrak{p}^{\nu_\mathfrak{p} + \mu_\mathfrak{p}}$$

$$\mathfrak{d}_1 + \mathfrak{d}_2 = \prod_{\mathfrak{p} \in \mathfrak{M}} \mathfrak{p}^{\operatorname{Min}(\nu_\mathfrak{p}, \mu_\mathfrak{p})}$$

and say that $\mathfrak{d}_1$ divides $\mathfrak{d}_2$ ($\mathfrak{d}_1 \mid \mathfrak{d}_2$) if $\nu_\mathfrak{p} \leq \mu_\mathfrak{p}$ for all $\mathfrak{p}$.

It is obvious that the divisors form a group by our definition of multiplication. It is a subgroup of the direct product $\prod_{\mathfrak{p} \in \mathfrak{M}} \mathbb{Z}_\mathfrak{p}$, where $\mathbb{Z}_\mathfrak{p}$ is isomorphic to the additive group of rational integers, in particular the subgroup containing those elements almost all of whose components are the unity element of the respective $\mathbb{Z}_\mathfrak{p}$. It is customary to refer to this subgroup as the "direct sum of the $\mathbb{Z}_\mathfrak{p}$". The group of divisors has thus, a quite trivial structure.

The set of ideals $\mathfrak{a}$ of $\mathfrak{o}$ ($\mathfrak{a} \neq (0)$) can be mapped into the group of divisors as follows:

$$\mathfrak{a} \to \prod_{\mathfrak{p} \in \mathfrak{M}} \mathfrak{p}^{\operatorname{ord}_\mathfrak{p} \mathfrak{a}}$$

and similarly the non-zero field elements $a \in k^*$:

$$a \to \prod_{\mathfrak{p}\in\mathfrak{M}} \mathfrak{p}^{\operatorname{ord}_\mathfrak{p} a}.$$

¿From what we have derived about $\operatorname{ord}\mathfrak{a}$ it follows that $\prod_{\mathfrak{p}\in\mathfrak{M}} \mathfrak{p}^{\operatorname{ord}_\mathfrak{p} \mathfrak{a}}$ is a divisor. Each element $a \in k$ is mapped onto the same divisor as the principal ideal $a\mathfrak{o}$. Sums and products of ideals are ordered onto the respective sums and products of divisors. Thus, the map defined is a homomorphism of the multiplicative semigroup of ideals into the multiplicative group of divisors, and of the additive semigroup of ideals into the additive semigroup of divisors.

In order to make this homomorphism a isomorphism onto the set of divisors, we must introduce another axiom, which, in effect, is a strengthening of the Approximation Theorem.

Axiom 2. Given a finite number of valuations $\mathfrak{p}_1, \mathfrak{p}_2, \dots, \mathfrak{p}_r \in \mathfrak{M}$, a real number $\varepsilon > 0$, and any r elements $a_1, a_2, \dots, a_r \in k$ there exists an element c such that:

$$|c - a_i|_{\mathfrak{p}_i} \le \varepsilon \quad (i = 1, 2, \dots, r)$$
$$|c|_\mathfrak{p} \le 1 \quad \text{for all } \mathfrak{p} \in \mathfrak{M},\ \mathfrak{p} \ne \mathfrak{p}_1, \mathfrak{p}_2, \dots, \mathfrak{p}_r.$$

These two axioms fully suffice to found ideal theory in valuation theory.

Let $\mathfrak{o}_\mathfrak{p}$ be the valuation ring of $\mathfrak{p}$: $\mathfrak{o}_\mathfrak{p} = \{a \mid a \in k,\ |a|_\mathfrak{p} \le 1\}$. We then define the ring $\mathfrak{o}$ as:

$$\mathfrak{o} = \bigcap_{\mathfrak{p}\in\mathfrak{M}} \mathfrak{o}_\mathfrak{p}.$$

That this ring $\mathfrak{o}$ is not too small, i.e., that its quotient field is, in fact, the field k, will be found to be a consequence of the axioms.

3. Diophantine Equations

We first consider the problem of the solvability of diophantine equations. This problem can be split into two parts: first, the local solvability at the place $\mathfrak{p}$ (i.e. in $\mathfrak{o}_\mathfrak{p}$), and second, the global solvability (i.e. in $\mathfrak{o}$),

We are given a system of linear forms:

$$(1) \qquad \left\{\begin{aligned} y_1 &= a_{11}x_1 + a_{12}x_2 + \cdots + a_{1n}x_n + b_1 \\ y_2 &= a_{21}x_1 + a_{22}x_2 + \cdots + a_{2n}x_n + b_2 \\ &\cdots\cdots\cdots\cdots\cdots\cdots \\ y_m &= a_{m1}x_1 + a_{m2}x_2 + \cdots + a_{mn}x_n + b_m \end{aligned}\right\} \quad a_{ij} \in k,\ b_i \in k.$$

Local problem. Sought, for some $\mathfrak{p} \in \mathfrak{M}$ are x_i $(i = 1, 2, \dots, n)$ in $\mathfrak{o}_\mathfrak{p}$ so that the y_j $(j = 1, \dots, m)$ all lie in $\mathfrak{o}_\mathfrak{p}$.

Global problem. Sought are elements x_i $(i = 1, 2, \dots, n)$, in $\mathfrak{o}$, so that the y_j $(j = 1, 2, \dots, m)$ all lie in $\mathfrak{o}$.

Fundamental Theorem. *The global problem is solvable if, and only if, the local problem is solvable for every place $\mathfrak{p} \in \mathfrak{M}$.*

Proof. I. One part of the theorem is trivial, for any global solution is, by the definition of $\mathfrak{o}$ a local solution for every $\mathfrak{p}$.

II. a) If $\mathfrak{p}$ is a valuation for which

$$|a_{ij}|_\mathfrak{p} \le 1, \quad |b_i|_\mathfrak{p} \le 1, \quad (i = 1, \dots, n;\ j = 1, \dots, m),$$

then all n-tuples $x_1, x_2, \ldots, x_n$ for which all $|x_i|_\mathfrak{p} \leq 1$ are local solutions at the place $\mathfrak{p}$.

b) If $x_i^\mathfrak{p}$ $(i = 1, 2, \ldots, n)$ is a local solution for the place $\mathfrak{p}$, then every $\bar{x}_i^\mathfrak{p}$ sufficiently close to $x_i^\mathfrak{p}$ is also a local solution for that place. The truth of this, as well as the size of the neighborhood, follows directly from (1).

c) Let there now be given for every $\mathfrak{p}$ a local solution $x_i^\mathfrak{p}$ $(i = 1, 2, \ldots, n)$. By axiom 1, the case a) holds true for almost all $\mathfrak{p}$, i.e. for all $\mathfrak{p}$ except the finite set $\mathfrak{p}_1, \mathfrak{p}_2, \ldots, \mathfrak{p}_r$ we have

$$|a_{ij}|_\mathfrak{p} \leq 1, \quad |b_i|_\mathfrak{p} \leq 1.$$

By axiom 2 we can then find an x_1 such that:

$$\begin{aligned} |x_1 - x_1^{\mathfrak{p}_i}|_{\mathfrak{p}_i} &\leq \varepsilon \quad \text{for } i = 1, 2, \ldots, r \\ |x_1|_\mathfrak{p} &\leq 1 \quad \text{for all other } \mathfrak{p}, \end{aligned}$$

where ε is chosen small enough to make b) applicable. Similarly we find elements $x_2, x_3, \ldots, x_n \in k$.

These $x_1, x_2, \ldots, x_n$ are now a local solution for all $\mathfrak{p}$:

$$x_i \in \mathfrak{o}_\mathfrak{p},\ y_j \in \mathfrak{o}_\mathfrak{p}\ (i = 1, \ldots, n;\ j = 1, \ldots, m) \text{ for all } \mathfrak{p},$$

and hence $x_i \in \mathfrak{o}$, $y_j \in \mathfrak{o}$ $(i = 1, \ldots, n;\ j = 1, \ldots, m)$. $x_1, \ldots, x_n$ is thus a global solution. □

REMARK. A slight refinement of this proof permits us to show that we can choose all the $x_i \neq 0$. For, if $x_i^{\mathfrak{p}_j} \neq 0$ for some $\mathfrak{p}_j$, then $x_i \neq 0$, if ε is chosen suitably small. If, however, $x_i^{\mathfrak{p}_j} = 0$ for all $j = 1, \ldots, r$, then replace $x_i^{\mathfrak{p}_1}$ by an element $\bar{x}_i^{\mathfrak{p}_1}$ nearby enough, so that $x_1^{\mathfrak{p}_1}, \ldots, \bar{x}_i^{\mathfrak{p}_1}, \ldots, x_n^{\mathfrak{p}_1}$ is still a local solution at $\mathfrak{p}_1$, and then the first case holds.

By a first degree diophantine equation we mean a system:

$$\text{(2)} \qquad \left.\begin{cases} a_{11}x_1 + a_{12}x_2 + \cdots + a_{1n}x_n = b_1 \\ a_{21}x_1 + a_{22}x_2 + \cdots + a_{2n}x_n = b_2 \\ \cdots\cdots\cdots\cdots\cdots\cdots\cdots\cdots \\ a_{m1}x_1 + a_{m2}x_2 + \cdots + a_{mn}x_n = b_m \end{cases}\right\} \quad a_{ij}, b_j \in k.$$

for which solutions $x_1, x_2, \ldots, x_n$ with $x_i \in \mathfrak{o}$ (global problem) or $x_i \in \mathfrak{o}_\mathfrak{p}$ (local problem) are sought. We can apply our fundamental theorem.

THEOREM. *The global solvability of* (2) *is equivalent to the local solvability for all* $\mathfrak{p}$.

One part of the theorem is again trivial. As for the other part: the local solvability of the system (2) means that the system certainly has a solution in k. Thus, we can apply the theory of linear equations: Either the system has one unique solution, which must then, since it coincides with all local solutions, be a global solution, or some of the x_i can be expressed as linear forms in the others, which can be freely chosen. Identifying the first class of these with the y_i, we are led to a system of type (1).

This result will be used by us only in the case of a single equation, i.e., for $m = 1$; we therefore reformulate the problem. Let

$$a_1x_1 + a_2x_2 + \cdots + a_nx_n = b, \quad a_i \in k,\ b \in k$$

be such an equation. We may assume at least one coefficient, say a_n, $\neq 0$:

$$x_n = -\frac{a_1}{a_n}x_1 - \frac{a_2}{a_n}x_2 - \cdots - \frac{a_{n-1}}{a_n}x_{n-1} - \frac{b}{a_n}.$$

First we investigate the conditions for local solvability of

$$y = a_1x_1 + a_2x_2 + \cdots + a_{n-1}x_{n-1} + b$$

in integral $x_1, \ldots, x_{n-1}, y$, at the place $\mathfrak{p}$.

¿From $b = y - a_1x_1 - \cdots - a_{n-1}x_{n-1}$ it follows that:

1.) The local solvability has as consequence:

$$|b|_{\mathfrak{p}} \leq \text{Max}(1, |a_1|_{\mathfrak{p}}, |a_2|_{\mathfrak{p}}, \ldots, |a_{n-1}|_{\mathfrak{p}}).$$

2.) Let, on the other hand, this inequality be satisfied:

Case I: The maximum is 1. Then $|a_i|_{\mathfrak{p}} \leq 1$ $(i = 1, \ldots, n-1)$ and $|b|_{\mathfrak{p}} \leq 1$. Every set $x_1, \ldots, x_n$ with $|x_i|_{\mathfrak{p}} \leq 1$ is a local solution.

Case II: Let, say $|a_1|$ be the maximum:

$$\Longrightarrow \quad |b|_{\mathfrak{p}} \leq |a_1|_{\mathfrak{p}} \quad \Longrightarrow \quad \left|\frac{b}{a_1}\right|_{\mathfrak{p}} \leq 1.$$

Setting $x_1 = -\frac{b}{a_1}$, and $x_2 = x_3 = \cdots = x_{n-1} = 0$, we have $y = 0$, and a local solution.

If we reformulate this inequality for the diophantine equation

$$a_1x_1 + a_2x_2 + \cdots + a_nx_n = b, \quad a_i \in k, \ b \in k \tag{3}$$

we arrive at

$$\left|\frac{b}{a_n}\right|_{\mathfrak{p}} \leq \text{Max}\left(1, \left|\frac{a_1}{a_n}\right|_{\mathfrak{p}}, \ldots, \left|\frac{a_{n-1}}{a_n}\right|_{\mathfrak{p}}\right)$$

and the

THEOREM. *Necessary and sufficient for the solvability of the diophantine equation* (3) *at the place* $\mathfrak{p}$ *is the condition*

$$|b|_{\mathfrak{p}} \leq \text{Max}_{1 \leq i \leq n} |a_i|_{\mathfrak{p}}.$$

Finally, it will be demonstrated this condition coincides with the familiar condition for solvability of diophantine equations in the ring of rational integers. In other notation our condition is:

$$\text{ord}_{\mathfrak{p}}\, b \geq \text{Min}_i(\text{ord}_{\mathfrak{p}}\, a_i).$$

Let $(b) = \prod_{\mathfrak{p}} \mathfrak{p}^{\text{ord}_{\mathfrak{p}} b}$ and $(a_i) = \prod_{\mathfrak{p}} \mathfrak{p}^{\text{ord}_{\mathfrak{p}} a_i}$ be the divisors onto which the elements b and a_i are mapped, so that the condition becomes:

$$(b) \text{ divisible by } (a_1) + (a_2) + \cdots + (a_n),$$

The definition of the sum of divisors allows us to construe it as the greatest common divisor of $(a_1), (a_2), \ldots, (a_n)$. If we identify the divisors with the corresponding principal ideals or with the elements themselves, we are lead to the familiar conditions for solvability of diophantine equations.

4. Derivation of Multiplicative Ideal Theory from Valuation Theory

With the help of this theorem about diophantine equations, ideal theory can be shown to follow the two axioms.

1) First it must be shown that k is, in fact, the quotient field of the ring $\mathfrak{o} = \bigcap_{\mathfrak{p}} \mathfrak{o}_{\mathfrak{p}}$. Let $a \in k^*$, we can find an element $x \in k$, with

$$\left|x - \frac{1}{a}\right| < \varepsilon \quad \text{if } |a|_{\mathfrak{p}} > 1,$$
$$|x|_{\mathfrak{p}} \leq 1 \quad \text{otherwise,}$$

for some small ε. Since $|a|_{\mathfrak{p}} > 1$ implies $\left|\frac{1}{a}\right|_{\mathfrak{p}} < 1$, we have, for sufficiently small ε, that $|a|_{\mathfrak{p}} > 1$ implies $|x|_{\mathfrak{p}} < 1$, and $x \neq 0$. Thus, $x \in \mathfrak{o}$, and

$$|ax - 1|_{\mathfrak{p}} < \varepsilon \cdot |a|_{\mathfrak{p}} \quad \text{if } |a|_{\mathfrak{p}} > 1,$$
$$|ax|_{\mathfrak{p}} \leq 1 \qquad \text{otherwise,}$$

so that for sufficiently small ε also $ax \in \mathfrak{o}$. Hence,

$$a = \frac{ax}{x}, \quad ax \in \mathfrak{o},\ x \in \mathfrak{o};$$

every element of k is a quotient of elements of $\mathfrak{o}$.

(Another proof: The diophantine equation $y - ax = 0$ is solvable, for $0 = |0|_{\mathfrak{p}} \leq \operatorname{Max}(1, |a|_{\mathfrak{p}})$ for every $\mathfrak{p}$. As was proven, there even exists a solution with $x \neq 0$, and then, $a = \frac{y}{x}$, $y \in \mathfrak{o}$, $x \in \mathfrak{o}$.)

We want to characterize the ideals by their orders $\nu_{\mathfrak{p}}$ for $\mathfrak{p} \in \mathfrak{M}$.

Let $\mathfrak{a}$ be an ideal of $\mathfrak{o}$ and $\nu_{\mathfrak{p}} = \operatorname{ord}_{\mathfrak{p}} \mathfrak{a}$. Let a be any non-zero element in $\mathfrak{a}$. Then,

$$\operatorname{ord}_{\mathfrak{p}} a \geq \nu_{\mathfrak{p}} \quad \text{for all } \mathfrak{p}.$$

Let $\mathfrak{p}_1, \mathfrak{p}_2, \ldots, \mathfrak{p}_r$ be a subset of $\mathfrak{M}$, so that for all other $\mathfrak{p}$, $\operatorname{ord}_{\mathfrak{p}} a = \nu_{\mathfrak{p}}$. Such a finite set certainly exists, for both $\operatorname{ord}_{\mathfrak{p}} \mathfrak{a}$ and $\operatorname{ord}_{\mathfrak{p}} a$ are 0 for almost all $\mathfrak{p}$. To each $i = 1, 2, \ldots, r$, we choose an $a_i \in \mathfrak{a}$, such that $\operatorname{ord}_{\mathfrak{p}_i} a_i = \nu_{\mathfrak{p}_i}$, and consider the diophantine equation

$$ax + a_1 x_1 + \cdots + a_r x_r = b, \quad b \in k.$$

By our choice of a and a_i, and the condition for the solvability of such equations, it follows that this equation is solvable if and only if

$$\operatorname{ord}_{\mathfrak{p}} b \geq \nu_{\mathfrak{p}} \quad \text{for all } \mathfrak{p}.$$

Every b satisfying this inequality can be represented in the form

$$b = ax + a_1 x_1 + \cdots + a_r x_r; \quad x, x_i \in \mathfrak{o}$$

and since the elements a and a_i lie in $\mathfrak{a}$, so does b.

Thus, the ideal $\mathfrak{a}$ consists precisely of all $b \in k$ with $\operatorname{ord}_{\mathfrak{p}} b \geq \nu_{\mathfrak{p}}$ for all $\mathfrak{p}$, and is therefore uniquely determined by the values of the order function, $\nu_{\mathfrak{p}}$. This, however, means that the map

$$\mathfrak{a} \to \prod_{\mathfrak{p}} \mathfrak{p}^{\nu_{\mathfrak{p}}}, \quad \nu_{\mathfrak{p}} = \operatorname{ord}_{\mathfrak{p}} \mathfrak{a}$$

of the set of ideals into the set of divisors is one-to-one; $\mathfrak{a}$ is uniquely determined by its divisor.

3.) Finally, it will be shown that the map of the ideals into the divisors is in fact a map onto the divisors, i.e., that to any set of values for the order function there exists a corresponding ideal.

Let there be given for every $\mathfrak{p}$ a value $\nu_{\mathfrak{p}}$, such that $\nu_{\mathfrak{p}} = 0$ for almost all $\mathfrak{p}$. Axiom 2 can now be strengthened:

LEMMA. *To any given rational integers $\nu_{\mathfrak{p}_1}, \nu_{\mathfrak{p}_2}, \dots, \nu_{\mathfrak{p}_r}$, there exists an element $a \in k$ such that*

$$\operatorname{ord}_{\mathfrak{p}_i} a = \nu_{\mathfrak{p}_i} \quad (i = 1, 2, \dots, r)$$

and

$$\operatorname{ord}_{\mathfrak{p}} a \geq 0 \quad \textit{for all other } \mathfrak{p}.$$

PROOF. To every $i = 1, 2, \dots, r$, we choose an $a_i \in k$ with $\operatorname{ord}_{\mathfrak{p}_i} a_i = \nu_{\mathfrak{p}_i}$, and then determine an $a \in k$ with the properties:

$$\begin{cases} |a - a_i|_{\mathfrak{p}_i} < |a_i|_{\mathfrak{p}_i} & (i = 1, 2, \dots, r) \\ |a|_{\mathfrak{p}} \leq 1 & \text{for all other } \mathfrak{p}. \end{cases}$$

Then, since $|a|_{\mathfrak{p}_i} = |a_i + (a - a_i)|_{\mathfrak{p}_i} = |a_i|_{\mathfrak{p}_i}$, the lemma is proven by rewriting in terms of order. □

We now choose $\mathfrak{p}_1, \mathfrak{p}_2, \dots, \mathfrak{p}_r$, so that $\nu_{\mathfrak{p}} = 0$ for all other $\mathfrak{p} \neq \mathfrak{p}_1, \mathfrak{p}_2, \dots, \mathfrak{p}_r$. By axiom 2 we can find an $a \in k^*$ such that

$$\begin{cases} \operatorname{ord}_{\mathfrak{p}_i} a \geq \nu_{\mathfrak{p}} & (i = 1, 2, \dots, r) \\ \operatorname{ord}_{\mathfrak{p}} a \geq 0 & \text{otherwise}, \end{cases}$$

and thus $\operatorname{ord}_{\mathfrak{p}} a \geq \nu_{\mathfrak{p}}$ for all $\mathfrak{p}$. (Remark: Assuming we had already proven the existence of an ideal with these orders, we could have chosen for a any element of $\mathfrak{a}$.) Then, choose $\mathfrak{q}_1, \mathfrak{q}_2, \dots, \mathfrak{q}_s \in \mathfrak{M}$ so that

$$\operatorname{ord}_{\mathfrak{p}} a = 0 \quad \mathfrak{p} \neq \mathfrak{p}_1, \mathfrak{p}_2, \dots, \mathfrak{p}_r, \mathfrak{q}_1, \mathfrak{q}_2, \dots, \mathfrak{q}_s.$$

By the lemma there then exists a $b \in k$ with

$$\begin{cases} \operatorname{ord}_{\mathfrak{p}} b = \nu_{\mathfrak{p}} & \mathfrak{p} = \mathfrak{p}_1, \mathfrak{p}_2, \dots, \mathfrak{p}_r, \mathfrak{q}_1, \mathfrak{q}_2, \dots, \mathfrak{q}_s, \\ \operatorname{ord}_{\mathfrak{p}} b \geq 0 & \text{otherwise}. \end{cases}$$

For the sum of the divisors (a) and (b) we have:

$$\operatorname{ord}_{\mathfrak{p}}((a) + (b)) = \operatorname{Min}(\operatorname{ord}_{\mathfrak{p}} a, \operatorname{ord}_{\mathfrak{p}} b) = \nu_{\mathfrak{p}}$$

for all $\mathfrak{p}$. But the divisor $(a) + (b)$ is the image of the ideal $\mathfrak{a} = a\mathfrak{o} + b\mathfrak{o}$, which therefore has the given order numbers.

In addition to our principal result we have thus also proven that any ideal can be generated by two elements

$$\mathfrak{a} = a\mathfrak{o} + b\mathfrak{o} = (a, b),$$

where one can even be chosen freely in $\mathfrak{a}$.

We have now demonstrated that the set of divisors are fully isomorphic with the set of ideals, so that we can replace the notion of ideals by divisors. Because the divisors form a multiplicative group, so do the ideals.

The integral ideals $\mathfrak{a} \subset \mathfrak{o} = \bigcap_{\mathfrak{p}} \mathfrak{o}_{\mathfrak{p}}$ are precisely those for which $\operatorname{ord}_{\mathfrak{p}} \mathfrak{a} \geq 0$ for all $\mathfrak{p}$, thus precisely those corresponding to integral divisors. We must still see which are prime ideals.

For two ideals $\mathfrak{a}, \mathfrak{b}$ we say that $\mathfrak{b}$ divides $\mathfrak{a}$ ($\mathfrak{b} \mid \mathfrak{a}$), if the ideal $\mathfrak{b}^{-1}\mathfrak{a}$ is integral. This is equivalent to:

$$\mathfrak{c} = \mathfrak{b}^{-1}\mathfrak{a} \subset \mathfrak{o} \iff \mathfrak{a} \subset \mathfrak{b}.$$

The statement $\mathfrak{b} \mid \mathfrak{a}$ means that $\mathfrak{a}$ is representable as $\mathfrak{a} = \mathfrak{b}\mathfrak{c}$ with integral $\mathfrak{c}$. The corresponding statement is true for divisors.

Now let $\mathfrak{a}, \mathfrak{b}, \mathfrak{c}$, be integral ideals with $\mathfrak{b} \neq \mathfrak{o}$, $\mathfrak{c} \neq \mathfrak{o}$, and

$$\mathfrak{a} = \mathfrak{b}\mathfrak{c}, \quad \text{and thus} \quad \mathfrak{b} \mid \mathfrak{a}, \; \mathfrak{c} \mid \mathfrak{a}.$$

We then have

$$\mathfrak{a} \subset \mathfrak{b}, \quad \mathfrak{a} \neq \mathfrak{b}$$

for, if $\mathfrak{a} = \mathfrak{b}$ then $\mathfrak{a} = \mathfrak{a}\mathfrak{c}$ and $\mathfrak{o} = \mathfrak{c}$. Similarly

$$\mathfrak{a} \subset \mathfrak{c}, \quad \mathfrak{a} \neq \mathfrak{c}.$$

Thus we can find elements $b \in \mathfrak{b}$ and $c \in \mathfrak{c}$ such that $b, c \notin \mathfrak{a}$. Then,

$$bc \in \mathfrak{b}\mathfrak{c} = \mathfrak{a}, \quad \text{but} \quad b, c \notin \mathfrak{a}.$$

Therefore, $\mathfrak{a}$ is not a prime ideal.

¿From this it can be seen that the only possible prime ideals other than (0) and $\mathfrak{o}$ are those whose divisors have only one factor, $\mathfrak{p}$. But these are, in fact, prime ideals. For, let $\bar{\mathfrak{p}}$ be such an ideal and $\mathfrak{p}$ the corresponding divisor. $\bar{\mathfrak{p}}$ is the set of all $a \in k$ with

$$\operatorname{ord}_{\mathfrak{p}} a \geq 1, \quad \operatorname{ord}_{\mathfrak{q}} a \geq 0 \quad \text{for } \mathfrak{q} \neq \mathfrak{p}.$$

Then, if $ab \in \bar{\mathfrak{p}}$, with $a \in \mathfrak{o}$, $b \in \mathfrak{o}$, we have

$$\operatorname{ord}_{\mathfrak{p}}(ab) \geq 1, \quad \text{and} \quad \operatorname{ord}_{\mathfrak{q}}(ab) \geq 0 \quad \text{for } \mathfrak{q} \neq \mathfrak{p}.$$

This, however, implies that either

$$\operatorname{ord}_{\mathfrak{p}} a \geq 1, \quad \text{and} \quad \operatorname{ord}_{\mathfrak{q}} a \geq 0 \quad \text{for } \mathfrak{q} \neq \mathfrak{p},$$

or that

$$\operatorname{ord}_{\mathfrak{p}} b \geq 1, \quad \text{and} \quad \operatorname{ord}_{\mathfrak{q}} n \geq 0 \quad \text{for } \mathfrak{q} \neq \mathfrak{p},$$

i.e., either $a \in \bar{\mathfrak{p}}$ or $b \in \bar{\mathfrak{p}}$. $\bar{\mathfrak{p}}$ is thus a prime ideal, which we will again designate by $\mathfrak{p}$.

Except for the zero ideal and the unity ideal, the ideals corresponding to the valuations $\mathfrak{p}$ are the only prime ideals of $\mathfrak{o}$. The isomorphism of ideals and divisors then shows that every ideal is uniquely decomposable into the product of prime ideals.

Finally, it will be demonstrated that non-trivial places φ of k which are finite in $\mathfrak{o}$ are precisely those places corresponding to the valuations $\mathfrak{p} \in \mathfrak{M}$.

It is obvious that every $\mathfrak{p} \in \mathfrak{M}$ yields such a place. On the other hand, let φ be a place of k finite in $\mathfrak{o}$. The set

$$\{a \mid a \in \mathfrak{o}, \; \varphi(a) = 0\}$$

is a prime ideal of $\mathfrak{o}$, the kernel of a homomorphism being an ideal, and the primeness following from $\varphi(ab) = \varphi(a)\varphi(b)$, φ being a homomorphism into a field. This prime ideal must be such a $\mathfrak{p}$. It cannot be $\mathfrak{o}$ because $\varphi(1) = 1 \neq 0$, and, as the place is not trivial, cannot be (0). The homomorphism

$$\mathfrak{o} \to \mathfrak{o}/\mathfrak{p}$$

could, in general, be extendable in many ways to a place; we will however show, that in the case at hand the place φ is uniquely determined.

To this end we investigate the valuation ring of φ. Let $a \in k$, $a \neq 0$. We distinguish two cases:

Case 1. $\operatorname{ord}_{\mathfrak{p}} a = m \geq 0$.

We can then find an $x \in k$ such that:

$$|x-1|_{\mathfrak{p}} \leq \varepsilon$$
$$|x|_{\mathfrak{q}} < \varepsilon \quad \text{if } \mathfrak{q} \neq \mathfrak{p} \text{ and } |a|_{\mathfrak{q}} > 1$$
$$|x|_{\mathfrak{q}} \leq 1 \quad \text{otherwise,}$$

with a sufficiently small ε. Then, both x and ax are in $\mathfrak{o}$, and since $\varphi|\mathfrak{o}$ is the canonical homomorphism

$$\mathfrak{o} \to \mathfrak{o}/\mathfrak{p}$$

we have $\varphi(x-1) = 0 \implies \varphi(x) = 1$.

Since $\varphi(ax)$ is finite, so is $\varphi(a) = \varphi\left(\frac{ax}{x}\right) = \varphi(ax)$. The statement $\varphi(a) = 0$ is equivalent to any of the following:

$$\varphi(ax) = 0 \iff ax \in \mathfrak{p} \iff m = \operatorname{ord}_{\mathfrak{p}} a = \operatorname{ord}_{\mathfrak{p}} ax > 0.$$

Case 2. $\operatorname{ord}_{\mathfrak{p}} a < 0$

$$\operatorname{ord}_{\mathfrak{p}} \frac{1}{a} > 0 \implies \varphi\left(\frac{1}{a}\right) = 0 \implies \varphi(a) = \infty.$$

Thus, we have uniquely determined the valuation ring of φ:

$$\{a \mid a \in k,\ \operatorname{ord}_{\mathfrak{p}} a \geq 0\};$$

it coincides with the valuation ring $\mathfrak{o}_{\mathfrak{p}}$ of the valuation $\mathfrak{p}$.

Hence, the only places of k finite in $\mathfrak{o}$ are those corresponding to the valuations $\mathfrak{p}$. This, and the definition of integrity in terms of places, show that the ring $\mathfrak{o} = \bigcap_{\mathfrak{p}} \mathfrak{o}_{\mathfrak{p}}$ is integrally closed.

As incidental result of these considerations, we see that $\mathfrak{o}/\mathfrak{p}$ is the full residue class field of the place φ, for, since $\varphi(a) = \varphi(ax)$, where $ax \in \mathfrak{o}$, all possible images of the place are exhausted in $\mathfrak{o}/\mathfrak{p}$.

5. The Necessity of the Axioms

The most important result of the Axioms 1 and 2 was, that the ideals of $\mathfrak{o}$ form a group. We want to show that this property alone implies the validity of the axioms.

Let $\mathfrak{o}$ be a ring with unity, with quotient field k, and whose ideals form a group.

In any integral domain $\mathfrak{o}$ with quotient field k, we have the

THEOREM. *q An ideal $\mathfrak{a}$ of $\mathfrak{o}$, to which there exists an (inverse) ideal $\mathfrak{b}$ with $\mathfrak{a}\mathfrak{b} = \mathfrak{o}$, is finitely generated.*

PROOF. Since $1 \in \mathfrak{o}$, there must exist a representation

$$1 = \sum_{i=1}^{r} a_i b_i, \quad a_i \in \mathfrak{a},\ b_i \in \mathfrak{b}.$$

The elements $a_1, a_2, \ldots, a_r$ in this representation generate $\mathfrak{a}$. For,

$$\bar{\mathfrak{a}} = a_1\mathfrak{o} + a_2\mathfrak{o} + \cdots + a_n\mathfrak{o} \in \mathfrak{a},$$

and thus $\bar{\mathfrak{a}}\mathfrak{b} \subset \mathfrak{a}\mathfrak{b} = \mathfrak{o}$. But $1 \in \bar{\mathfrak{a}}\mathfrak{b} \implies \mathfrak{o} \subset \bar{\mathfrak{a}}\mathfrak{b}$ and thus

$$\bar{\mathfrak{a}}\mathfrak{b} = \mathfrak{o} = \mathfrak{a}\mathfrak{b} \implies \bar{\mathfrak{a}} = \mathfrak{a},$$

by multiplication by $\mathfrak{b}^{-1}$. □

In our ring $\mathfrak{o}$ every ideal is therefore finitely generated. This is, however, equivalent with the maximal property for ideals: Every set of integral ideals contains maximal ideals.

We have shown the equivalence of the statements:

$$\mathfrak{a} \subset \mathfrak{b} \iff \mathfrak{c} = \mathfrak{a}\mathfrak{b}^{-1} \subset \mathfrak{o} \iff \mathfrak{a} = \mathfrak{b}\mathfrak{c},\ \mathfrak{c} \subset \mathfrak{o} \iff \mathfrak{b} \mid \mathfrak{a}.$$

If $\mathfrak{a}$ is an integral ideal, and $\mathfrak{a} \neq \mathfrak{o}$, then there exists a maximal ideal $\mathfrak{p} \supset \mathfrak{a}$: i.e. $\mathfrak{a} = \mathfrak{p}\mathfrak{c}$, $\mathfrak{c} \subset \mathfrak{o}$, $\mathfrak{p}$ maximal. We have $\mathfrak{a} \subset \mathfrak{c}$, and, since $\mathfrak{p} \neq \mathfrak{o}$, also $\mathfrak{c} \neq \mathfrak{a}$. It can be shown that all integral ideals of $\mathfrak{o}$ can be decomposed into maximal prime ideals, or more precisely: For any integral ideal $\mathfrak{a}$ of $\mathfrak{o}$ one of the three statements must be true:

$$\mathfrak{a} = \mathfrak{o}$$

or, $\mathfrak{a}$ is a maximal prime ideal,

or, $\mathfrak{a}$ is a product of maximal prime ideals.

Proof. Consider the set A of ideals for which this is not so, which, if not empty, must, contain a maximal ideal, $\mathfrak{a}$. Then $\mathfrak{a} \neq \mathfrak{o}$, and $\mathfrak{a}$ is itself not a maximal prime ideal of $\mathfrak{o}$. Thus, there exists a maximal prime ideal $\mathfrak{p}$ in $\mathfrak{o}$ with $\mathfrak{a} \subset \mathfrak{p}$; we have

$$\mathfrak{a} = \mathfrak{p}\mathfrak{c}, \quad \mathfrak{a} \subset \mathfrak{c}, \quad \mathfrak{c} \neq \mathfrak{a}, \quad \mathfrak{c} \neq \mathfrak{o}.$$

Because $\mathfrak{a}$ is maximal in A, $\mathfrak{a} \notin A$, and the statement to be proved holds, in any case, for $\mathfrak{c}$. But then it obviously also holds for $\mathfrak{a}$, contradicting the choice of $\mathfrak{a}$. A is empty. □

The following considerations yield the uniqueness of the decomposition into prime ideals:

(1) Let $\mathfrak{a}, \mathfrak{b}$ be integral ideals, and $\mathfrak{p}$ a maximal prime ideal satisfying $\mathfrak{p} \mid \mathfrak{a}\mathfrak{b}$ i.e. $\mathfrak{a}\mathfrak{b} \subset \mathfrak{p}$. Let $\mathfrak{p} \nmid \mathfrak{a}$, i.e. $\mathfrak{a} \not\subset \mathfrak{p}$. There then exists an $a \in \mathfrak{a}$, $a \notin \mathfrak{p}$ with

$$a\mathfrak{b} \subset \mathfrak{p}_1, a \notin \mathfrak{p} \implies b \subset \mathfrak{p} \text{ i.e. } \mathfrak{p} \mid \mathfrak{b}$$

since $\mathfrak{p}$ is prime.

(This can also be shown in another way: $\mathfrak{a} + \mathfrak{p} = \mathfrak{o}$ because $\mathfrak{p}$ is maximal; therefore $\mathfrak{a}\mathfrak{b} + \mathfrak{p}\mathfrak{b} = \mathfrak{b}$ which implies $\mathfrak{p} \mid \mathfrak{b}$.)

(2) Let $\mathfrak{p}_1\mathfrak{p}_2 \dots \mathfrak{p}_r = \mathfrak{q}_1\mathfrak{q}_2 \dots \mathfrak{q}_s$, with maximal prime ideals $\mathfrak{p}_i, \mathfrak{q}_i$. Since $\mathfrak{p}_1 \mid \mathfrak{q}_1\mathfrak{q}_2 \dots \mathfrak{q}_s$, $\mathfrak{p}_1$ must, by (1) divide one factor, say $\mathfrak{q}_1$. The, because of the maximal property $\mathfrak{p}_1 = \mathfrak{q}_1$, and then also

$$\mathfrak{p}_2\mathfrak{p}_3 \dots \mathfrak{p}_r = \mathfrak{q}_2\mathfrak{q}_3 \dots \mathfrak{q}_s.$$

Induction on the number r yields $r = s$ and $\mathfrak{p}_i = \mathfrak{q}_i$ $(i = 1, 2, \dots, r)$ if the $\mathfrak{q}_i$ are accordingly rearranged.

Thus, every integral ideal $\mathfrak{a}$ of $\mathfrak{o}$ is uniquely representable as the product of powers of prime ideals:

$$\mathfrak{a} = \prod_{\mathfrak{p}} \mathfrak{p}^{\nu_\mathfrak{p}}, \quad \nu_\mathfrak{p} \geq 0, \text{ almost all } \nu_\mathfrak{p} = 0.$$

This also holds for fractional ideals. For, if $\mathfrak{a}$ is a fractional ideal, then there exists a $c \in \mathfrak{o}$ so that $c\mathfrak{a} \subset \mathfrak{o}$. Using the maximal prime ideals $\mathfrak{p}$, the following decompositions follow:

$$c\mathfrak{a} = (c\mathfrak{o})\mathfrak{a} = \prod_{\mathfrak{p}} \mathfrak{p}^{\mu_{\mathfrak{p}}}, \quad \mu_{\mathfrak{p}} \geq 0, \text{ almost all } \mu_{\mathfrak{p}} = 0,$$

$$c\mathfrak{o} = \prod_{\mathfrak{p}} \mathfrak{p}^{\rho_{\mathfrak{p}}}, \quad \rho_{\mathfrak{p}} \geq 0, \text{ almost all } \rho_{\mathfrak{p}} = 0,$$

$$\Longrightarrow \mathfrak{a}(c\mathfrak{o})^{-1}c\mathfrak{a} = \prod_{\mathfrak{p}} \mathfrak{p}^{\nu_{\mathfrak{p}}}, \quad \nu_{\mathfrak{p}} = \mu_{\mathfrak{p}} - \rho_{\mathfrak{p}}, \text{ almost all } \nu_{\mathfrak{p}} = 0.$$

This last representation is also unique, for, the fractional ideal $\mathfrak{a}$ can be written in the form

$$\mathfrak{a} = \mathfrak{b}\mathfrak{c}^{-1} = \frac{\mathfrak{b}}{\mathfrak{c}},$$

where $\mathfrak{a}$ and $\mathfrak{b}$ are integral ideals without common factors in their decompositions. Then,

$$\frac{\mathfrak{b}}{\mathfrak{c}} = \frac{\mathfrak{b}_1}{\mathfrak{c}_1} \Longrightarrow \mathfrak{b}\mathfrak{c}_1 = \mathfrak{b}_1\mathfrak{c} \Longrightarrow \mathfrak{b} \mid \mathfrak{b}_1\mathfrak{c} \Longrightarrow \mathfrak{b} \mid \mathfrak{b}_1 \Longrightarrow \mathfrak{b}_1 \subset \mathfrak{b}$$

and similarly $\mathfrak{b} \subset \mathfrak{b}_1$, and hence

$$\mathfrak{b} = \mathfrak{b}_1 \Longrightarrow \mathfrak{c} = \mathfrak{c}_1,$$

from which the uniqueness of the decomposition follows.

To every maximal prime ideal we introduce a valuation of k.

Let $a \neq 0$ be an element of k, and let

$$a\mathfrak{o} = \prod_{\mathfrak{p}} \mathfrak{p}^{\nu_{\mathfrak{p}}}.$$

We define:

$$\operatorname{ord}_{\mathfrak{p}} a = \nu_{\mathfrak{p}}, \quad \operatorname{ord}_{\mathfrak{p}} 0 = +\infty.$$

The definition is meaningful, because of the uniqueness of the decomposition, and will be shown to yield a discrete valuation of k for every $\mathfrak{p}$. The first valuation axiom is obviously satisfied, the second follows from

$$ab\mathfrak{o} = a\mathfrak{o} \cdot b\mathfrak{o} \Longrightarrow \operatorname{ord}_{\mathfrak{p}} ab = \operatorname{ord}_{\mathfrak{p}} a + \operatorname{ord}_{\mathfrak{p}} b$$

for $a, b \neq 0$. (If either $a = 0$, or $b = 0$ it is trivially satisfied.) In case $a = 0$ or $a = -1$, we immediately have the third valuation axiom:

$$\operatorname{ord}_{\mathfrak{p}} a \geq 0 \Longrightarrow \operatorname{ord}_{\mathfrak{p}}(1 + a) \geq 0.$$

But $a \neq 0$, $a \neq -1$, $\operatorname{ord}_{\mathfrak{p}} a \geq 0 \Longrightarrow a\mathfrak{o} = \frac{\mathfrak{b}}{\mathfrak{c}}$, $\mathfrak{b}, \mathfrak{c}$ integral, $\mathfrak{p} \nmid \mathfrak{c}$.

$$\Longrightarrow (1 + a)\mathfrak{o} \subset \mathfrak{o} + a\mathfrak{o} = \frac{\mathfrak{c}}{\mathfrak{c}} + \frac{\mathfrak{b}}{\mathfrak{c}} = \frac{\mathfrak{c} + \mathfrak{b}}{\mathfrak{c}},$$

since the distributive law for addition and multiplication holds for the ideals of a ring. The last equation implies

$$(1 + a)\mathfrak{o} = \frac{\mathfrak{c} + \mathfrak{b}}{\mathfrak{c}}\mathfrak{d} \quad \text{with integral } \mathfrak{d},\ \mathfrak{p} \nmid \mathfrak{c},$$

$$\Longrightarrow \operatorname{ord}_{\mathfrak{p}}(1 + a) \geq 0.$$

The thus defined order $\operatorname{ord}_{\mathfrak{p}}$ is indeed, adapted, as a discrete valuation, to the field k, i.e. it assumes all integers (and $+\infty$) as values:

$$\mathfrak{p} \mid \mathfrak{p}^2 \Longrightarrow \mathfrak{p}^2 \subset \mathfrak{p}, \quad \mathfrak{p}^2 \neq \mathfrak{p},$$

and thus there exists an element $\pi \in \mathfrak{p}$ with $\pi \notin \mathfrak{p}^2$, i.e. $\mathfrak{p} \mid \pi\mathfrak{o}$, $\mathfrak{p}^2 \nmid \pi\mathfrak{o}$, and then $\operatorname{ord}_\mathfrak{p} \pi = 1$, which proves our statement.

Finally, we demonstrate that different prime ideals, $\mathfrak{p}$ and $\mathfrak{q}$, yield different valuations. The ideals are maximal, hence $\mathfrak{p} + \mathfrak{q} = \mathfrak{o}$, and there exist

$$\pi \in \mathfrak{p}, \ \kappa \in \mathfrak{q} \quad \text{with} \quad \pi + \kappa = 1.$$

Since $1 \notin \mathfrak{p}$, we have $\kappa \notin \mathfrak{p}$, and therefore

$$\operatorname{ord}_\mathfrak{p} \kappa \leq 0, \quad \operatorname{ord}_\mathfrak{q} \kappa > 0$$

so that the valuations must differ.

We now show that the set of valuations thus arrived at satisfy axioms 1 and 2:

For the first axiom it is trivial, for, in the factorization of the ideal $a\mathfrak{o}$ only finitely many prime ideals have non-zero exponents; thus for non-zero $a \in k$:

$$\operatorname{ord}_\mathfrak{p} a = 0 \quad \text{for almost all } \mathfrak{p}.$$

An element a of k belongs to $\mathfrak{o}$ if and only if $a\mathfrak{o} \subset \mathfrak{o}$, i.e. $\operatorname{ord}_\mathfrak{p} a \geq 0$ for all $\mathfrak{p}$. Let $\mathfrak{o}_\mathfrak{p}$ be the valuation ring of the valuation corresponding to $\mathfrak{p}$, then

$$\mathfrak{o} = \bigcap_\mathfrak{p} \mathfrak{o}_\mathfrak{p}.$$

If $\mathfrak{p}_1, \mathfrak{p}_2, \ldots, \mathfrak{p}_s$ are any finite number of different prime ideals, then for any $N \geq 1$ we have

$$\mathfrak{p}_1^N + \mathfrak{p}_2^N \mathfrak{p}_3^N \ldots \mathfrak{p}_s^N = \mathfrak{o},$$

for the left side of the equation is the greatest common divisor of its two summands, and this must, indeed, be $\mathfrak{o}$. There thus exists a representation

$$\alpha + x = 1 \quad \text{with } \alpha \in \mathfrak{o},\ x \in \mathfrak{o},\ \mathfrak{p}_1^N \mid \alpha\mathfrak{o},\ \mathfrak{p}_i^N \mid x\mathfrak{o}\ (i = 2, 3, \ldots, s).$$

By choosing a sufficiently large N we have:

$$\begin{cases} |x - 1|_{\mathfrak{p}_1} \text{ small} & \\ |x|_{\mathfrak{p}_i} \text{ small} & \text{for } i = 2, 3, \ldots, s \\ |x|_\mathfrak{p} \leq 1 & \text{for all other } \mathfrak{p}. \end{cases}$$

Thus, given an element $a \in k$, and r different prime ideals $\mathfrak{p}_1, \mathfrak{p}_2, \ldots, \mathfrak{p}_r$, we can find an element $x \in k$ which satisfies

$$\begin{cases} |x - 1|_{\mathfrak{p}_1} \leq \varepsilon' & \\ |x|_{\mathfrak{p}_i} \leq \varepsilon' & (i = 2, \ldots, r) \\ |x|_\mathfrak{p} \leq \varepsilon' & \mathfrak{p} \neq \mathfrak{p}_i, \text{ but with } |a|_\mathfrak{p} > 1 \\ |x|_\mathfrak{p} \leq 1 & \text{otherwise.} \end{cases}$$

For the element $y = ax \in k$, we then have:

$$\begin{cases} |y - a|_{\mathfrak{p}_1} \leq \varepsilon' |a|_{\mathfrak{p}_1} & \\ |y|_{\mathfrak{p}_i} \leq \varepsilon' |a|_{\mathfrak{p}_i} & (i = 2, 3, \ldots, r) \\ |y|_\mathfrak{p} \leq \varepsilon' |a|_\mathfrak{p} & \mathfrak{p} \neq \mathfrak{p}_i \text{ but with } |a|_\mathfrak{p} > 1 \\ |y|_\mathfrak{p} \leq 1 |a|_\mathfrak{p} \leq 1 & \text{otherwise.} \end{cases}$$

Now, let r different prime ideals of $\mathfrak{o}$, $\mathfrak{p}_1, \dots, \mathfrak{p}_r$, r elements of k, $a_1, \dots, a_r$, and a real number $\varepsilon > 0$ be given. By the above we can find $y_1, y_2, \dots, y_r \in k$ satisfying:

$$\begin{cases} |y_i - a_i|_{\mathfrak{p}_i} \leq \varepsilon & \\ |y_i|_{\mathfrak{p}_j} \leq \varepsilon & i \neq j,\ j = 1, 2, \dots, r \\ |y_i|_{\mathfrak{p}} \leq 1 & \text{otherwise}, \end{cases}$$

for every $i = 1, 2, \dots, r$. The element $z = y_1 + y_2 + \cdots + y_r \in k$ then satisfies

$$\begin{cases} |z - a_i|_{\mathfrak{p}_i} \leq \varepsilon & (i = 1, 2, \dots, r) \\ |z|_{\mathfrak{p}} \leq 1 & \text{otherwise}. \end{cases}$$

The set $\mathfrak{M}$ of valuations $|\ |_{\mathfrak{p}}$ thus satisfies the Axiom 2.

Thus, we have shown that the axioms 1 and 2 are equivalent to the statement that the ideals of the ring $\mathfrak{o}$ form a group.

6. Transition to a Finite Extension

THEOREM. *If the axioms 1 and 2 are satisfied in a field k, and K is a field finite over k then the axioms are also satisfied in K.*

PROOF. Ler $\mathfrak{o}$ be the valuation ring corresponding to the valuation $\mathfrak{p} \in \mathfrak{M}$ of k, and let

$$\mathfrak{o} = \bigcap \mathfrak{o}_{\mathfrak{p}}.$$

The ideals of $\mathfrak{o}$ form a group.

The set of valuations $\overline{\mathfrak{M}}$ of K is defined as the set of all possible extensions $\mathfrak{P}$ of the valuations $\mathfrak{p} \in \mathfrak{M}$ of k. The valuations $\mathfrak{P} \in \overline{\mathfrak{M}}$ are then also discrete. If $\mathfrak{P}$ is an extension of $\mathfrak{p}$ we will write

$$\mathfrak{P} \mid \mathfrak{p} \quad (\mathfrak{P} \text{ divides } \mathfrak{p}).$$

1.) As we saw earlier, to every $\mathfrak{p}$ there are only finitely many $\mathfrak{P}$ with $\mathfrak{P} \mid \mathfrak{p}$.

2.) Let $\alpha \in K$, which satisfies

$$\alpha^n + a_1\alpha^{n-1} + \cdots + a_n, \quad a_i \in k.$$

The axioms hold in k, hence $|a_i|_{\mathfrak{p}} \leq 1$ for at most finite many $\mathfrak{p} \in \mathfrak{M}$. But then by 1.), we have $\text{Max}_i\, |a_i|_{\mathfrak{P}} \leq 1$ for almost all $\mathfrak{P} \in \overline{\mathfrak{M}}$, and thus

$$|\alpha|_{\mathfrak{P}} \leq 1 \quad \text{for almost all } \mathfrak{P} \in \overline{\mathfrak{M}}.$$

Axiom 1 is satisfied in K, where $\overline{\mathfrak{M}}$ is taken as the set of valuations of K extending those in $\mathfrak{M}$.

3.) Let $\mathfrak{P}_1, \mathfrak{P}_2, \dots, \mathfrak{P}_r$ be different valuations of the set $\overline{\mathfrak{M}}$, and let the elements $\alpha_1, \alpha_2, \dots, \alpha_r$, and a real number $\varepsilon > 0$ be given. Let $\omega_1, \omega_2, \dots, \omega_n$ be a basis of $K|k$, i.e. $K = \omega_1 k + \omega_2 k + \cdots + \omega_n k$.

Choose valuations $\mathfrak{P}_{r+1}, \mathfrak{P}_{r+2}, \dots, \mathfrak{P}_s$ of $\overline{\mathfrak{M}}$ so that

$$|\omega_i|_{\mathfrak{P}} \leq 1 \quad (i = 1, 2, \dots, n) \text{ for } \mathfrak{P} \neq \mathfrak{P}_1, \dots, \mathfrak{P}_r, \mathfrak{P}_{r+1}, \dots, \mathfrak{P}_s.$$

Let the system of valuations $\mathfrak{p}_1, \mathfrak{p}_2, \dots, \mathfrak{p}_q$ of k be such that each of the valuations $\mathfrak{P}_i$ $(i = 1, \dots, s)$ is an extension of one of the $\mathfrak{p}_j$ $(j = 1, 2, \dots, q)$. Then, let $\mathfrak{P}_1, \dots, \mathfrak{P}_t \in \overline{\mathfrak{M}}$ be all possible extensions of the valuations $\mathfrak{p}_j$ $(j = 1, \dots, q)$, which then certainly contains the valuations $\mathfrak{P}_1, \dots, \mathfrak{P}_s$. We can set $\alpha_{r+1} = \alpha_{r+2} = \cdots = \alpha_t = 0$, and assume $\varepsilon < 1$.

By the approximation theorem for valuations of rank 1, there exists a $\beta \in K$ with

$$|\beta - \alpha_i|_{\mathfrak{P}_i} \leq \varepsilon \quad (i = 1, 2, \ldots, t).$$

Let

$$\beta = x_1\omega_1 + x_2\omega_2 + \cdots + x_n\omega_n, \quad x_i \in k$$

and set

$$\gamma = y_1\omega_1 + y_2\omega_2 + \cdots + y_n\omega_n,$$

where the y_i are chosen in k by Axiom 2, so that

$$|y_j - x_j|_{\mathfrak{p}_i} \leq \varepsilon' \quad (i = 1, 2, \ldots, q)$$
$$|y_j|_{\mathfrak{p}} \leq 1 \quad \text{otherwise},$$

with some given $\varepsilon' > 0$. Then we have, with sufficiently small ε',

$$|\gamma - \beta|_{\mathfrak{P}_i} \leq \varepsilon' \operatorname{Max}_j |\omega_j|_{\mathfrak{P}_i} \leq \varepsilon (i = 1, 2, \ldots, t),$$

and hence

$$|\gamma - \alpha_i|_{\mathfrak{P}_i} = |(\gamma - \beta) + (\beta - \alpha_i)|_{\mathfrak{P}_i} \leq \varepsilon \quad (i = 1, 2, \ldots, t),$$
$$|\gamma|_{\mathfrak{P}} \leq 1 \quad \text{otherwise}.$$

Since $\alpha_i = 0$ for $i \geq r+1$, and $\varepsilon < 1$, this implies

$$|\gamma - \alpha_i|_{\mathfrak{P}_i} \leq \varepsilon \quad (i = 1, 2, \ldots, r)$$
$$|\gamma|_{\mathfrak{P}} \leq 1 \quad \text{otherwise},$$

and Axiom 2 holds in K with the valuation set $\overline{\mathfrak{M}}$. □

The ring corresponding to the set $\overline{\mathfrak{M}}$,

$$\mathfrak{O} = \bigcap_{\mathfrak{P}} \mathfrak{O}_{\mathfrak{P}}$$

is also fully determined as the integral closure of $\mathfrak{o}$ in K:

An element $\alpha \in K$ is integral with respect to $\mathfrak{o}$ if for every place (valuation) finite on $\mathfrak{o}$, the image of α is finite ($|\alpha| \leq 1$). These places are precisely the places corresponding to the extensions of the valuations $\mathfrak{p} \in \mathfrak{M}$, i.e. to the $\mathfrak{P} \in \overline{\mathfrak{M}}$. α is thus integral with respect to $\mathfrak{o}$ if $|\alpha|_{\mathfrak{P}} \leq 1$ for all $\mathfrak{P} \in \overline{\mathfrak{M}}$, and thus the integral closure of $\mathfrak{o}$ in K coincides with $\mathfrak{O}$.

Thus we see that if our multiplicative ideal theory holds in a ring $\mathfrak{o}$, then it holds in the integral closure of $\mathfrak{o}$ in a finite extension field K.

In most applications, though, ideal theory is superfluous; divisor theory suffices. To use divisors one need only assume Axiom 1.

EXAMPLES. 1.) The above theory holds in any principal ideal ring.

2.) The ideal theory holds In the field of rational numbers using the ring of the usual integers; it therefore holds in every algebraic number field, with the ring of algebraic integers of the field. The set $\mathfrak{M}$ of valuations consists of all possible valuations leaving out the archimedean valuation, or, in the case of algebraic number fields, the extensions of these valuations, i.e. all possible non-archimedean valuations.

3.) Let k be the field of rational functions over the field $\mathbb{C}$ of complex numbers. All possible valuations of k trivial on $\mathbb{C}$ are discrete, they are given by the irreducible, thus linear, polynomials $x - \alpha$, and by $\frac{1}{x}$; they correspond one-to-one with the points of the Riemann number sphere.

Taking the set of all these valuations, Axiom 1 is satisfied, for a function has only a finite number of poles, but Axiom 2 could not hold. For, the corresponding ring $\mathfrak{o} = \bigcap_{\mathfrak{p}} \mathfrak{o}_{\mathfrak{p}}$ would be much too small, consisting of all the constants — the only functions with no poles whatsoever. If, however, one leaves out just one of these valuations, say the one corresponding to the infinite place, then both axioms hold, with, in our case, the ring $\mathfrak{o}$ being the ring of polynomials.

CHAPTER 9

THE DIVISOR GROUP

1. Divisor Classes

We want to study the group of divisors of k more closely.

Let k be a field, $\mathfrak{M} = \{\mathfrak{p}\}$ a set of discrete valuations of k that satisfy Axiom 1. The divisor group corresponding to $\mathfrak{M}$ will be denoted by $\mathfrak{D}$. There then exists a map of k^* into $\mathfrak{D}$ given by:

$$a \to \prod \mathfrak{p}^{\mathrm{ord}_{\mathfrak{p}} a}$$

The kernel of the map is the set

$$\{a \mid a \in k, \ \mathrm{ord}_{\mathfrak{p}} a = 0 \text{ for all } \mathfrak{p} \in \mathfrak{M}\};$$

the elements of this multiplicative group are called the *units* of this theory. We can, in general, gain no further information on these.

The images of elements $a \in k^*$ in $\mathfrak{D}$ are called principal divisors of $\mathfrak{D}$: they form a subgroup $\mathfrak{Y}$ of $\mathfrak{D}$. The elements of the factor group $\mathfrak{D}/\mathfrak{Y}$ (the cokernel of the map) are called divisor classes.

In case $\mathfrak{Y} = \mathfrak{D}$, there exists only one divisor class, all divisors are principal divisors, and theorem of unique factorization into prime elements holds in k^*. In the event that $\mathfrak{D} \neq \mathfrak{Y}$, this theorem no longer holds; but then, if the second axiom holds, we have unique decomposition of ideals.

The units and the cokernel are the most important notions for describing the field k. If k is an algebraic number field, and $\mathfrak{M}$ the set of all discrete valuations, then the group of divisor classes is finite, and the group of units finitely generated. If k is the field of rational functions over the constant field Ω, and $\mathfrak{M}$ the set of valuations of k trivial on Ω, then the units are precisely the non-zero constants.

For the case of algebraic number fields we will, in the last chapter, consider more closely the group of units and the group of divisor classes.

2. Extension of a Valuation with Base Field not Complete

We again consider the transition to a finite extension field K of k.

It was shown, in the case of a complete field k, how to arrive at the extension of the valuation on a field K of degree n over k. We want to do the same for a non-complete field. We will always assume that the valuation is non-trivial, the trivial case posing no further difficulties.

Consider the field k, with valuation, and the separable extension field K of degree n over k. Similar results to those that follow hold for inseparable extensions; the proofs, however, are not as simple.) We assume that some extension of the valuation onto K is given. We can then go over to the perfect completions, $\tilde{K}$ and $\tilde{k}$.

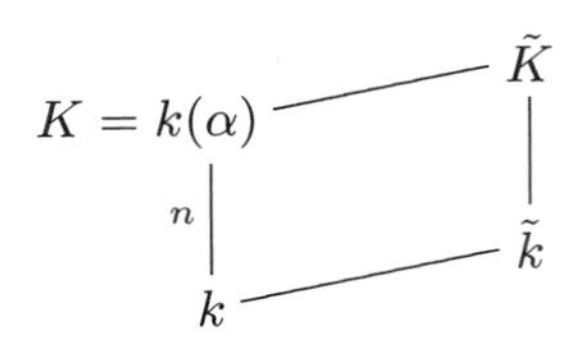

Let $K = k(\alpha)$, then $\tilde{K} = \tilde{k}\tilde{K} = \tilde{k}(\alpha)$. The polynomial $f(x) = \operatorname{Irr}(\alpha, k)$ is separable. $f(x)$ is, however, not necessarily irreducible in $\tilde{k}$. Let

$$f(x) = P_1(x)P_2(x)\dots P_r(x)$$

be the decomposition to irreducible factors in $\tilde{k}$. Then,

$$\operatorname{Irr}(\alpha, \tilde{k}) \mid f(x), \quad \text{and, say,} \quad P_i(x) = \operatorname{Irr}(\alpha, \tilde{k}).$$

Thus, to every valuation of K there corresponds uniquely some factor $P_i(x)$, and the valuation is uniquely determined on K by that factor, for, the factor uniquely determines the perfect completion $\tilde{K} = \tilde{k}(\alpha)$.

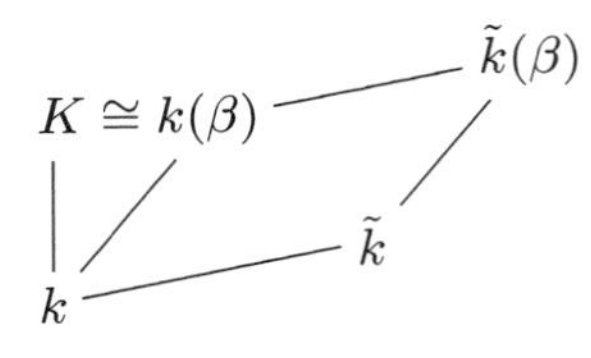

On the other hand, let $P_i(x)$ be some factor, and consider the field $\tilde{k}(\beta)$, where $P_i(\beta) = 0$.

The valuation of $\tilde{k}$ can be uniquely extended onto $\tilde{k}(\beta)$, and then

$$k \subset \tilde{k}(\beta) \implies k(\beta) \subset \tilde{k}(\beta).$$

Since $P_i(x) \mid f(x)$, we have $f(\beta) = 0$, and then, because $f(x)$ is irreducible in k

$$K = k(\alpha) \cong k(\beta),$$

where the isomorphism σ leaves k fixed and maps α onto β. $\tilde{k}(\beta)$ then induces a valuation on $k(\beta)$, which in turn induces a valuation on K by:

$$|\theta| = |\sigma^{-1}\theta| \quad \text{for } \theta \in K.$$

Each of the factors $P_i(x)$, therefore, yield an extension of the valuation, and these valuations are necessarily different from each other, as, due to the separability of K, the factors P_i must all be different. (One sees that in the inseparable case, there may be less than r extensions.)

EXAMPLE. Let $\mathbb{Q}$ be the field of rational numbers, with the usual absolute valuation, and $K = \mathbb{Q}(\alpha)$ with $\alpha^3 = 2$. In $\tilde{\mathbb{Q}} = \mathbb{R}$ we have the decomposition

$$x^3 - 2 = (x - \sqrt[3]{2})(x^2 + \sqrt[3]{2}\,x + \sqrt[3]{4}).$$

There thus correspond exactly two extensions of the valuation on $\mathbb{Q}(\alpha)$:

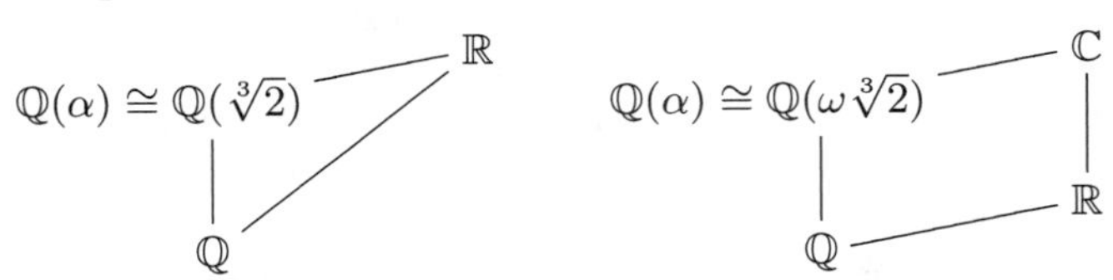

ω any primitive 3rd root of unity, and these are the only possible extensions of the valuation. They are given by the isomorphism between $\mathbb{Q}(\alpha)$ and a real, and complex, respectively, subfield of the field $\mathbb{C}$.

Let $\mathfrak{P}_i \mid \mathfrak{p}$, i.e. $\mathfrak{P}_i$ be an extension of the valuation $\mathfrak{p}$, and $\mathfrak{P}_i$ correspond to the factor $P_i(x)$. Then, using the *local degree*

$$n_{\mathfrak{P}_i} = [\tilde{K} : \tilde{k}] = \text{degree } P_i(x),$$

we have the following formula for the *global degree* $[K : k]$:

$$n = \sum_{\mathfrak{P}\mid\mathfrak{p}} n_{\mathfrak{P}}.$$

Thus, if the valuation $\mathfrak{p}$ is discrete, it follows that

$$n = \sum_{\mathfrak{P}|\mathfrak{p}} e_{\mathfrak{P}} f_{\mathfrak{P}}.$$

Consideration of only the last and the second highest coefficients of the polynomials $f(x)$ and $P_i(x)$ yields: The *global norm* $N = N_{K|k}$ and the *local norm* $N_{\mathfrak{P}} = N_{\tilde{k}(\beta)|\tilde{k}}$ are connected by the formula

$$N\alpha = \prod_{\mathfrak{P}|\mathfrak{p}} N_{\mathfrak{P}}\alpha.$$

The *global trace* S is, similarly, the sum of *local traces* $S_{\mathfrak{P}}$:

$$S\alpha = \sum_{\mathfrak{P}|\mathfrak{p}} S_{\mathfrak{P}}\alpha.$$

These formulas obviously holds for generating elements α. To show that they always hold, we let α be a generating element of the field K over k, and β any element. The field k necessarily has infinitely many elements, finite fields having only trivial valuations.

1.) Consider all elements

$$\gamma_c = \beta + c\alpha, \quad c \in k,$$

and the corresponding field, $k(\gamma_c)$. Since there can be only finitely many intermediate fields between k and $k(\alpha)$, there exists a pair $c, d \in k$, $c \neq d$, with

$$\begin{gathered} k(\gamma_c) = k(\gamma_d) = E \\ \beta + c\alpha \in E, \quad \beta + d\alpha \in E, \\ \implies (c-d)\alpha \in E \implies \alpha \in E \\ \implies E = k(\alpha) = k(\gamma_c). \end{gathered}$$

The trace formula is correct for α and for $\gamma_c = \beta + c\alpha$, and therefore, because of the linearity of the trace function, also for β.

2.) The norm formula is trivial for $\beta = 0$; let $\beta \neq 0$, and consider all elements

$$\begin{gathered} \delta_c = \beta(\alpha + c), \quad c \in k. \\ \implies k(\delta_c) = k(\delta_d) = E' \end{gathered}$$

for some pair $c, d \in k$, $c \neq d$ as in 1.)

$$\begin{gathered} \beta(\alpha + c) \in E', \quad \beta(\alpha + d) \in E' \\ \implies (c-d)\beta \in E' \implies \beta \in E' = k(\alpha) = k(\delta_c). \end{gathered}$$

The formula holding for $\alpha + c$ and $\beta(\alpha + c)$, it holds for β.

REMARK. The method of this proof can also be used to easily prove the

THEOREM OF STEINITZ. *An algebraic extension field K over k can be generated by a single element if, and only if, there exist only a finite number of intermediate fields between k and K.*

That the "only if" is true, is not difficult. If, on the other hand, there are only finitely many intermediate fields, then K can be generated by a finite number of elements. If, now, $K = k(\alpha, \beta)$, consider all elements

$$\gamma_c = \alpha + \beta c, \quad c \in k,$$

and the method of 1.) yields $K = k(\gamma_c)$ for some $c \in k$. If K is generated by more than two elements, the same procedure call be repeated.

3. The Norm of Divisors

We now assume that Axiom 1 holds in the field k with some set $\mathfrak{M}$ of valuations. Then Axiom 1 also holds in K with the set $\overline{\mathfrak{M}}$ of all extensions of the valuations of $\mathfrak{M}$.

The corresponding divisor groups $\mathfrak{D}_k$ and $\mathfrak{D}_K$ are free abelian groups, generated by the $\mathfrak{p} \in \mathfrak{M}$ and the $\mathfrak{P} \in \overline{\mathfrak{M}}$, respectively. To each $a \in k^*$ $(\alpha \in K^*)$ there corresponds a divisor in $\mathfrak{D}_k$ $(\mathfrak{D}_K)$. The following diagram describes the situation:

$$\begin{array}{ccc} K^* & \xrightarrow{\phi} & \mathfrak{D}_K \\ {\scriptstyle i}\uparrow & & \uparrow{\scriptstyle i'}\,? \\ k^* & \xrightarrow{\varphi} & \mathfrak{D}_k \end{array}$$

i is an injection, φ, ϕ are maps into.

Our goal: to map $\mathfrak{D}_k$ into $\mathfrak{D}_K$ by a map i' so that the diagram is commutative, i.e.

$$\phi i = i'\varphi.$$

Let $a \in k^*$, then $\varphi(a) = \prod_{\mathfrak{p}} \mathfrak{p}^{\operatorname{ord}_{\mathfrak{p}} a}$ and

$$\phi i(a) = \phi(a) = \prod_{\mathfrak{P}} \mathfrak{P}^{\operatorname{ord}_{\mathfrak{P}} a} = \prod_{\mathfrak{p}} \left(\prod_{\mathfrak{P}|\mathfrak{p}} \mathfrak{P}^{\operatorname{ord}_{\mathfrak{P}} a} \right)$$
$$= \prod_{\mathfrak{p}} \left(\prod_{\mathfrak{P}|\mathfrak{p}} \mathfrak{P}^{e_{\mathfrak{P}}} \right)^{\operatorname{ord}_{\mathfrak{p}} a}.$$

This suggests that we define

$$i'(\mathfrak{p}) = \prod_{\mathfrak{P}|\mathfrak{p}} \mathfrak{P}^{e_{\mathfrak{P}}}$$

for the generating elements of the free group $\mathfrak{D}_k$, thus getting a homomorphism

$$i' : \mathfrak{D}_k \to \mathfrak{D}_K.$$

The diagram is then commutative and furthermore, i' is an isomorphism of $\mathfrak{D}_k$ into $\mathfrak{D}_K$.

If we make the further assumption that Axiom 2 is satisfied in k (and thereby also in K), then we can further find an isomorphism (which we will also denote as i) of the group of ideals $\mathfrak{J}_k$ of k into that of K, $\mathfrak{J}_K$, so that the following diagram is commutative:

$$\begin{array}{ccc} \mathfrak{J}_K & \xrightarrow{\phi} & \mathfrak{D}_K \\ {\scriptstyle i}\uparrow & & \uparrow{\scriptstyle i'} \\ \mathfrak{J}_k & \xrightarrow{\varphi} & \mathfrak{D}_k \end{array}$$

(The maps denoted as φ and ϕ coincide with those maps of non-zero field elements in the last diagram, if we associate elements with principal ideals.)

The map i is defined by

$$i(\mathfrak{a}) = \mathfrak{a}\mathfrak{O} \quad \text{for all } \mathfrak{a} \in \mathfrak{I}_k$$

$$\Longrightarrow \varphi(a) = \prod_{\mathfrak{p}} \mathfrak{p}^{\operatorname{ord}_{\mathfrak{p}} \mathfrak{a}}$$

$$\phi i(a) = \prod_{\mathfrak{p}} \left(\prod_{\mathfrak{P}|\mathfrak{p}} \mathfrak{P}^{\operatorname{ord}_{\mathfrak{P}}(i(\mathfrak{a}))} \right).$$

But, $\operatorname{ord}_{\mathfrak{P}}(i(\mathfrak{a})) = \operatorname{Min}_{\alpha \in i(\mathfrak{a})}(\operatorname{ord}_{\mathfrak{P}} \alpha) = \operatorname{Min}_{a \in \mathfrak{a}}(\operatorname{ord}_{\mathfrak{P}} a) = e_{\mathfrak{P}} \operatorname{ord}_{\mathfrak{P}} \mathfrak{a}$, so that the diagram is commutative.

The maps φ and ϕ are isomorphisms of $\mathfrak{I}_k$ onto $\mathfrak{D}_k$ and $\mathfrak{I}_K$ onto $\mathfrak{D}_K$, respectively. Thus, i' is uniquely determined by the commutative property, i.e. the i' given above is the only possibility (by this choice of i).

Remark. In the previous diagram the i' can also be shown to be fully determined by commutativity if it is further stipulated that the image $i'(\mathfrak{p})$ of $\mathfrak{p} \in \mathfrak{M}$ should only be a product of those $\mathfrak{P} \in \mathfrak{M}$ which extend $\mathfrak{p}$ onto K. The uniqueness is then proven by the Approximation Theorem.

We now want to construct a homomorphism N' of the divisor group $\mathfrak{D}_K$ into the divisor group $\mathfrak{D}_k$, corresponding to the norm map of K^* into k^*. We will need only the assumption of Axiom 1.

$$\begin{array}{ccc} K^* & \xrightarrow{\phi} & \mathfrak{D}_K \\ {\scriptstyle N}\downarrow & & \downarrow{\scriptstyle N'} \\ k^* & \xrightarrow{\varphi} & \mathfrak{D}_k \end{array}$$

To meaningfully define the norm of a divisor we will require the commutativity of this diagram. We have for $\alpha \in K^*$:

$$\varphi N(\alpha) = \prod_{\mathfrak{p}} \mathfrak{p}^{\operatorname{ord}_{\mathfrak{p}} N\alpha} = \prod_{\mathfrak{p}} \mathfrak{p}^{\operatorname{ord}_{\mathfrak{p}}(\prod_{\mathfrak{P}|\mathfrak{p}} N_{\mathfrak{P}}\alpha)} = \prod_{\mathfrak{p}} \mathfrak{p}^{\sum_{\mathfrak{P}|\mathfrak{p}} \operatorname{ord}_{\mathfrak{p}}(N_{\mathfrak{P}}\alpha)}$$

$$= \prod_{\mathfrak{p}} \prod_{\mathfrak{P}|\mathfrak{p}} \mathfrak{p}^{\operatorname{ord}_{\mathfrak{p}} N_{\mathfrak{P}}\alpha}.$$

But, we have

$$|\alpha|_{\mathfrak{P}} = |N_{\mathfrak{P}}\alpha|_{\mathfrak{p}}^{\frac{1}{e_{\mathfrak{P}} f_{\mathfrak{P}}}} \Longrightarrow \operatorname{ord}_{\mathfrak{P}} \alpha = e_{\mathfrak{P}} \cdot \frac{1}{e_{\mathfrak{P}} f_{\mathfrak{P}}} \operatorname{ord}_{\mathfrak{p}}(N_{\mathfrak{P}}\alpha)$$

and hence

$$\varphi N(\alpha) = \prod_{\mathfrak{p}} \prod_{\mathfrak{P}|\mathfrak{p}} \mathfrak{p}^{f_{\mathfrak{P}} \operatorname{ord}_{\mathfrak{P}} \alpha}.$$

On the other hand,

$$N'\phi(\alpha) = \prod_{\mathfrak{p}} \prod_{\mathfrak{P}|\mathfrak{p}} N'\mathfrak{P}^{\operatorname{ord}_{\mathfrak{P}} \alpha}.$$

We therefore define:

$$N'\mathfrak{P} = \mathfrak{p}^{f_{\mathfrak{P}}}$$

for the basis elements $\mathfrak{P}$ of $\mathfrak{D}_K$, which yields the homomorphism

$$N' \colon \mathfrak{D}_K \to \mathfrak{D}_k.$$

This homomorphism is called the norm of the divisors. We will replace the notation N' by N.

It is customary to identify the image of the divisors of $\mathfrak{D}_k$ by the isomorphism i' with their inverse images, writing, for example

$$\mathfrak{p} = \prod_{\mathfrak{P}|\mathfrak{p}} \mathfrak{P}^{e_{\mathfrak{P}}}$$

and saying that this is the decomposition of the prime divisor $\mathfrak{p}$ of the base field into prime divisors of the extension field.

CHAPTER 10

DIFFERENT AND DISCRIMINANT

1. The Ramification Divisor of an Extension

Let Axioms 1 and 2 be satisfied by a field k with a set $\mathfrak{M}$ of valuations. Then the axioms also hold in a finite separable field K over k, with set $\overline{\mathfrak{M}}$ of extensions of the valuations $\mathfrak{p} \in \mathfrak{M}$.

Let $\mathfrak{o}$ and $\mathfrak{O}$ be the corresponding rings, the intersection of the valuation rings of all the $\mathfrak{p} \in \mathfrak{M}$ and the $\mathfrak{P} \in \overline{\mathfrak{M}}$ respectively.

$$\begin{array}{ccc} K & \text{———} & \mathfrak{O} \\ | & & | \\ k & \text{———} & \mathfrak{o} \end{array}$$

For this case we have defined the different $\mathfrak{d}$ of K over k by

$$\mathfrak{d}^{-1} = \mathfrak{O}',$$

where $\mathfrak{O}'$ is the complementary set to $\mathfrak{O}$. $\mathfrak{d}$ is the *global different* of K over k.

If $\mathfrak{P} \mid \mathfrak{p}$, then the perfect completion $\tilde{K}_{\mathfrak{P}}$ of K with respect to $\mathfrak{P}$ contains the perfect completion $\tilde{k}_{\mathfrak{p}}$ of k with respect to $\mathfrak{p}$ as subfield. Let the corresponding valuation rings be $\mathfrak{O}_{\mathfrak{P}}$ and $\mathfrak{o}_{\mathfrak{p}}$. The equation

$$\begin{array}{cccc} \mathfrak{P} & K & \text{———} & \mathfrak{O} \\ | & | & & | \\ \mathfrak{p} & k & \text{———} & \mathfrak{o} \end{array}$$

$$\mathfrak{d}_{\mathfrak{P}}^{-1} = \mathfrak{O}'_{\mathfrak{P}}$$

then defines the *local different* $\mathfrak{d}_{\mathfrak{P}}$ at the place $\mathfrak{P}$. We want to derive a relationship between the local differents and the global different.

1.) $\mathfrak{d}^{-1} \subset \mathfrak{d}_{\mathfrak{P}}^{-1}$.

For, if $\alpha \in \mathfrak{d}^{-1}$ and β is some element in $\mathfrak{O}_{\mathfrak{P}}$, we need only show

$$S_{\mathfrak{P}}(\alpha\beta) \in \mathfrak{o}_{\mathfrak{p}}.$$

Decompose $\mathfrak{p}$ in K:

$$\mathfrak{p} = \mathfrak{P}_1^{e_1}\mathfrak{P}_2^{e_2}\dots\mathfrak{P}_r^{e_r}, \quad \mathfrak{P} = \mathfrak{P}_1.$$

For any given $\varepsilon > 0$ we can then find a $\beta' \in K$, such that

$$\begin{cases} |\beta' - \beta|_{\mathfrak{P}_1} < \varepsilon & \\ |\beta'|_{\mathfrak{P}_i} < \varepsilon & (i = 2, \dots, r) \\ |\beta'|_{\mathfrak{Q}} \leq 1 & \text{otherwise.} \end{cases}$$

For, since K is dense in $\tilde{K}_{\mathfrak{P}}$ one can substitute some $\beta_0 \in K$, sufficiently near β, for β in these conditions, and then Axiom 2 in K guarantees the existence of β'. If, further, ε is chosen sufficiently small, $\beta' \in \mathfrak{O}$, and in

$$S_{\mathfrak{P}_1}(\alpha\beta) = S_{\mathfrak{P}_1}(\alpha(\beta - \beta')) + S(\alpha\beta') + \sum_{i=2}^{r} S_{\mathfrak{P}_i}(\alpha\beta')$$

each of the summands is integral, and $S_{\mathfrak{P}_1}(\alpha\beta) \in \mathfrak{o}_{\mathfrak{p}}$. Since β was chosen arbitrarily in $\mathfrak{O}_{\mathfrak{P}}$ we have $\alpha \in \mathfrak{d}_{\mathfrak{P}}^{-1}$, and, since α is arbitrary in $\mathfrak{d}^{-1}$,

$$\mathfrak{d}^{-1} \subset \mathfrak{d}_{\mathfrak{P}}^{-1}.$$

2.) If $\alpha \neq 0$, and $\alpha \in \mathfrak{d}_{\mathfrak{P}}^{-1}$, there exists an $\alpha' \in \mathfrak{d}^{-1}$ with $\operatorname{ord}_{\mathfrak{P}} \alpha = \operatorname{ord}_{\mathfrak{P}} \alpha'$. To prove this, we choose, for a given $\varepsilon > 0$ an element $\alpha' \in K$, such that

$$|\alpha' - \alpha|_{\mathfrak{P}} < \varepsilon$$
$$\alpha'|_{\mathfrak{Q}} \leq 1 \quad \text{for } \mathfrak{Q} \neq |\mathfrak{P}.$$

For sufficiently small ε we then have: $\operatorname{ord}_{\mathfrak{P}} \alpha = \operatorname{ord}_{\mathfrak{P}} \alpha'$, $\alpha' \in \mathfrak{d}_{\mathfrak{P}}^{-1}$. We must now show that $\alpha' \in \mathfrak{d}^{-1}$.

If $\beta \in \mathfrak{O} \subset \mathfrak{O}_{\mathfrak{P}}$, then

$$\begin{aligned} S(\alpha'\beta) &= S_{\mathfrak{P}_1}(\alpha'\beta) + \sum_{i=2}^{r} S_{\mathfrak{P}_i}(\alpha'\beta) \\ &= S_{\mathfrak{P}_1}(\alpha\beta) + S_{\mathfrak{P}_1}((\alpha' - \alpha)\beta) + \sum_{i=2}^{r} S_{\mathfrak{P}_i}(\alpha'\beta), \end{aligned}$$

and since, by sufficiently small ε, every summand on the right side is integral at the place $\mathfrak{p}$, we have $S(\alpha'\beta) \in \mathfrak{o}_{\mathfrak{p}}$: $S(\alpha'\beta)$ is integral at $\mathfrak{p}$.

If, now, $\mathfrak{q} = \mathfrak{Q}_1^{e_1'}\mathfrak{Q}_2^{e_2'} \ldots \mathfrak{Q}_s^{e_s'}$ is the decomposition of a place $\mathfrak{q} \neq \mathfrak{p}$ in K, then, by our choice of α', $\alpha'\beta \in \mathfrak{O}_{\mathfrak{Q}_i}$ $(i = 1, \ldots, s)$, and therefore

$$S(\alpha'\beta) = \sum_{i=1}^{s} S_{\mathfrak{Q}_i}(\alpha'\beta) \in \mathfrak{o}_{\mathfrak{q}},$$

which shows that $S(\alpha'\beta)$ is also integral at every other place of k,

$$S(\alpha'\beta) \in \mathfrak{o} \implies \alpha' \in \mathfrak{d}^{-1}$$

for β was arbitrarily chosen in $\mathfrak{O}$.

The different $\mathfrak{d}^{-1}$ is uniquely determined by all its orders. By 2.),

$$\operatorname{ord}_{\mathfrak{P}} \mathfrak{d}^{-1} = \operatorname{Min}_{\alpha \in \mathfrak{d}^{-1}} \operatorname{ord}_{\mathfrak{P}} \alpha = \operatorname{ord}_{\mathfrak{P}} \mathfrak{d}_{\mathfrak{P}}^{-1}.$$

Taking inverse ideals, we have

$$\mathfrak{d} = \prod_{\mathfrak{P}} \mathfrak{d}_{\mathfrak{P}}.$$

This is to be interpreted as follows: Each of the $\mathfrak{d}_{\mathfrak{P}}$ is to be represented in powers of the prime ideal $\tilde{\mathfrak{P}}$ of $\mathfrak{O}_{\mathfrak{P}}$, and then the corresponding prime ideals $\mathfrak{P}$ of $\mathfrak{O}$ substituted for the $\tilde{\mathfrak{P}}$. This yields the product representation of the different $\mathfrak{d}$.

Thus, it is seen, that since the product representation of $\mathfrak{d}$ can contain only a finite number of prime ideals with non-zero exponents, $\mathfrak{d}_{\mathfrak{P}} = \mathfrak{O}_{\mathfrak{P}} = 1$ must hold for almost all $\mathfrak{P}$. But, with our previous results, this implies that there are only a finite number of places $\mathfrak{P}$ of K ramified over k $(e_{\mathfrak{P}} > 1)$, and these are precisely those prime ideals that divide the different. The different, considered as divisor, is therefore also termed the *ramification divisor* of $K|k$.

We can introduce the notion of *ramification divisor* for the more general case in which only Axiom 1 holds in the field k with finite separable extension field K. We can then define

$$\mathfrak{d} = \prod_{\mathfrak{P}} \mathfrak{d}_{\mathfrak{P}},$$

for the local differents, $\mathfrak{d}_{\mathfrak{P}}$, which are defined here.

We need only show that $\mathfrak{d}$ is a divisor. Let $K = k(\alpha)$, then,

$$|\alpha|_{\mathfrak{P}} \leq 1, \quad \text{for almost all } \mathfrak{P},$$

thus $\alpha \in \mathfrak{O}_{\mathfrak{P}}$ for almost all $\mathfrak{P}$. Let $f = \operatorname{Irr}(\alpha, k)$, and $\alpha \in \mathfrak{O}_{\mathfrak{P}}$, then

$$\mathfrak{d}_{\mathfrak{P}} \mid f'(\alpha),$$

but, since, for almost all places $\mathfrak{P}$ we have $\operatorname{ord}_{\mathfrak{P}} f'(\alpha) = 0$, it follows that

$$\mathfrak{d}_{\mathfrak{P}} = 1 \quad \text{for almost all } \mathfrak{P}.$$

$\mathfrak{d}$ is thus a divisor. By its definition, it has the same meaning for ramifications as the different in the previous case. Now, however, $\mathfrak{d}$ can no longer be determined directly; the local differents are needed first.

2. The Discriminant of a Module

In most applications the different is impractical, for, in general the ring $\mathfrak{O}$ is unknown. It is therefore that we introduce the coarser, but more easily calculatable *discriminant*, which, although it allows more general definition, we define only for the case of a module with n generators.

Let K be finite and separable of degree n over k, in which Axioms 1 and 2 hold. Let $\alpha_1, \alpha_2, \ldots, \alpha_n$ be some field basis of $K|k$, and

$$\mathfrak{A} = \alpha_1 \mathfrak{o} + \alpha_2 \mathfrak{o} + \cdots + \alpha_n \mathfrak{o}$$

be an $\mathfrak{o}$-module with the minimal basis of $\alpha_1, \ldots, \alpha_n$, where $\mathfrak{o}$ is the subring of k determined by the set of valuations $\mathfrak{M}$. If the n conjugates of α_i are denoted $\alpha_i = \alpha_i^{(1)}, \alpha_i^{(2)}, \ldots, \alpha_i^{(n)}$ $(i = 1, 2, \ldots, n)$, the discriminant $D(\mathfrak{A})$ of the module $\mathfrak{A}$ is defined as:

$$D(\mathfrak{A}) = (\det(\alpha_i^{(j)}))^2 = \begin{vmatrix} \alpha_1^{(1)} & \alpha_1^{(2)} & \ldots & \alpha_1^{(n)} \\ \alpha_2^{(1)} & \alpha_2^{(2)} & \ldots & \alpha_2^{(n)} \\ \cdots & \cdots & \cdots & \cdots \\ \alpha_n^{(1)} & \alpha_n^{(2)} & \ldots & \alpha_n^{(n)} \end{vmatrix}^2$$

$D(\mathfrak{A})$ is an element of the base field k, for it is invariant under all the relative isomorphisms of $K|k$. If $\mathfrak{A} \subset \mathfrak{O}$, then $D(\mathfrak{A}) \in \mathfrak{o}$, and thus $D(\mathfrak{A})$ is integral.

This definition still depends, to a certain extent, upon the basis chosen for the module $\mathfrak{A}$. We investigate this dependence:

1.) Let $\mathfrak{L}$ be another module with n generators:

$$\mathfrak{L} = \beta_1 \mathfrak{o} + \beta_2 \mathfrak{o} + \cdots + b_n \mathfrak{o}.$$

Since both the α_i and the β_i are bases of the field K over k, one can write

$$\alpha_i = \sum_{\nu=1}^{n} a_{i\nu} \beta_\nu, \quad a_i \in k \ (i = 1, \ldots, n; \ \nu = 1, \ldots, n);$$

$$\alpha_i^{(j)} = \sum_{\nu=1}^{n} a_{i\nu} \beta_\nu^{(j)}$$

and thus

$$D(\mathfrak{A}) = (\det(a_{ij}))^2 D(\mathfrak{L}), \quad \det(a_{ij}) \in k.$$

The quotient $\frac{D(\mathfrak{A})}{D(\mathfrak{L})}$ is a perfect square in the base field. If, further, $\mathfrak{A} \subset \mathfrak{L}$, then the $a_{ij} \in \mathfrak{o}$ and we have: $D(\mathfrak{L}) \mid D(\mathfrak{A})$, and $\frac{D(\mathfrak{A})}{D(\mathfrak{L})}$ is the square of an integral element in k.

2.) If, further, $\mathfrak{A} = \mathfrak{L}$, then also $D(\mathfrak{A}) \mid D(\mathfrak{L})$, and the quotient $\frac{D(\mathfrak{A})}{D(\mathfrak{L})}$ is the square of a unit in the base field — itself a unit. It follows that $D(\mathfrak{A})$ is uniquely defined up to the square of a unit as factor.

3.) Let $\mathfrak{A} \subset \mathfrak{L}$ and $\frac{D(\mathfrak{A})}{D(\mathfrak{L})}$ be a unit. The matrix (a_{ij}) is unimodular (its determinant is a unit), and therefore we also have $\mathfrak{L} \subset \mathfrak{A}$, i.e. $\mathfrak{A} = \mathfrak{L}$.

The notion of the discriminant can be helpful in determining the ring $\mathfrak{O}$ of integers with respect to $\mathfrak{o}$ in K. For, if one knows that $\mathfrak{O}$ is a module with n generators, and one has found an $\mathfrak{o}$-module with n generators, $\mathfrak{A}$, which consists only of integral elements of the field, and so that its discriminant is not divisible by a perfect square in the base field, then necessarily $\mathfrak{O} = \mathfrak{A}$.

In some cases the discriminant can be easily computed. Let $K = k(\alpha)$, and consider the case where $\mathfrak{L}$ is the module with n generators:

$$\mathfrak{L} = \mathfrak{o} + \alpha\mathfrak{o} + \alpha^2\mathfrak{o} + \cdots + \alpha^{n-1}\mathfrak{o}.$$

Then

$$D(\mathfrak{L}) = \begin{vmatrix} 1 & 1 & \dots & 1 \\ \alpha & \alpha^{(2)} & \dots & \alpha^{(n)} \\ \alpha^2 & (\alpha^{(2)})^2 & \dots & (\alpha^{(n)})^2 \\ \dots & \dots & \dots & \dots \\ \alpha^{n-1} & (\alpha^{(2)})^{n-1} & \dots & (\alpha^{(n)})^{n-1} \end{vmatrix}^2,$$

which is the square of the Vandermond determinant of α and it conjugates, and thus equal to the discriminant of the polynomial $f(x) = \operatorname{Irr}(\alpha, k)$. One need, therefore, only compute the discriminant of this polynomial.

For the discriminant of a polynomial of degrees 2 and 3 there are at hand explicit formulas (cf. Perron, Algebra). Thus, the general third degree polynomial

$$f(x) = x^3 + a_1x^2 + a_2x + a_3$$

has the discriminant

$$D(f) = a_1^2a_2^2 - 4a_1^3a_3 + 18a_1a_2a_3 - 4a_2^3 - 27a_3^2.$$

The case $n = 4$ can be computed from the formula for $n = 3$, if one first calculates the cubic Lagrangian resolvent (Perron, Algebra) of $f(x)$, which has the same discriminant as $f(x)$.

A formula for the discriminant can also be found for trinomials of the form

$$f(x) = x^n + ax + b.$$

For, if α_i $(i = 1, \dots, n)$ are the roots of the equation, we must essentially calculate the product $\prod_{i=1}^n f'(\alpha_i)$, which can, for this polynomial, be brought to a form where only the coefficients a and b occur.

EXAMPLE. For polynomial $f(x) = x^3 + px + q$ has the discriminant

$$D = -(4p^3 + 27q^2),$$

so that, for example, $x^3 - x - 1$ has the discriminant $D = -23$ and $x^3 + x + 1$ has the discriminant $D = -31$.

3. Discriminant of an Ideal, Field Discriminant

As the notion of discriminant was introduced only for $\mathfrak{o}$-modules with n generators, which ideals, in general, are not, we must make some limiting assumption about k in order to use this notion for the ideals of k.

ASSUMPTION. The ring $\mathfrak{o}$ is a principal ideal ring.

For the case of rational numbers or rational functions this assumption holds true (ring of integers, ring of polynomials).

These $\mathfrak{o}$-modules are additive abelian groups, with the elements of the ring $\mathfrak{o}$ as operators. In the case where $\mathfrak{o}$ is a principal ideal ring, all the theorems on usual abelian groups hold true (cf. van der Waerden, Algebra; Bourbaki, Algebra, Chapter 7; Kurosh, Group Theory), e.g. theorems on free abelian groups.

$\mathfrak{o}a$ is an infinite (additive) cyclic group, if $\alpha a \neq 0$ whenever $\alpha \in \mathfrak{o}$, $\alpha \neq 0$ (torsion-free group). The direct sum of such infinite cyclic groups:

$$\mathfrak{o}a_1 + \mathfrak{o}a_2 + \cdots + \mathfrak{o}a_r,$$

is a free module; r is called the rank of this module. The following theorem is now applicable: Subgroups of free groups are free. (Only such subgroups are considered, as permit the elements of $\mathfrak{o}$ as operators.) (cf., e.g., Eilenberg–Steenrod, Foundations of Algebraic Topology.)[1]

The ring $\mathfrak{O}$ of integers of K with respect to $\mathfrak{o}$ is, under the above assumption, an $\mathfrak{o}$-module with n generators. For, $\mathfrak{O}$ is an additive abelian group with $\mathfrak{o}$ as operator domain. We saw earlier that for a generating element α:

$$\mathfrak{o}[\alpha] \subset \mathfrak{O} \subset \frac{\mathfrak{o}[\alpha]}{f'(\alpha)}.$$

$\mathfrak{O}$ is, as subgroup of the rank n free module $\frac{\mathfrak{o}[\alpha]}{f'(\alpha)}$, a free module of rank $\leq n$; on the other hand, it contains $\mathfrak{o}[\alpha]$, also a free module of rank n, so that the rank of $\mathfrak{O}$ is exactly n.

Every ideal $\mathfrak{A}$ of $\mathfrak{O}$ is also an $\mathfrak{o}$-module with n generators, since

$$\beta\mathfrak{O} \subset \mathfrak{A} \subset \frac{1}{c}\,\mathfrak{O},$$

where β is any element of $\mathfrak{A}$ and c a common denominator of the elements of $\mathfrak{A}$.

EXAMPLE. Let $\mathbb{Q}$ be the field of rational numbers, $\mathbb{Z}$ the ring of rational integers and

$$K = \mathbb{Q}(\alpha),$$

where α is a root of the polynomial: 1) $x^3 - x - 1$; 2) $x^3 + x + 1$.

We want to determine the ring $\mathfrak{O}$ of integers. From the equations we have

$$\alpha \in \mathfrak{O}, \quad \mathbb{Z} + \alpha\mathbb{Z} + \alpha^2\mathbb{Z} \subset \mathfrak{O}.$$

The discriminant $D(\mathfrak{O})$ is integral, and a divisor of

$$D(\mathbb{Z} + \alpha\mathbb{Z} + \alpha^2\mathbb{Z}) = \begin{cases} -23 & \text{for 1.)} \\ -31 & \text{for 2.),} \end{cases}$$

[1] For a more recent reference see "Algebra" by Serge Lang or "Basic Algebra" by Nathan Jacobson. *Editor's note.*

and the quotient is a square in $\mathbb{Q}$. But obviously then, the quotient must, in both cases, be 1, so that we have for both cases:

$$\begin{aligned} D(\mathfrak{O}) &= D(\mathbb{Z} + \alpha\mathbb{Z} + \alpha^2\mathbb{Z}) \\ \Longrightarrow \mathfrak{O} &= \mathbb{Z} + \alpha\mathbb{Z} + \alpha^2\mathbb{Z}. \end{aligned}$$

In general, the determination of $\mathbb{Q}$ is not quite as simple as this. In such cases, the following *theorem of Stickelberger* can be useful:

If K is a separable extension of degree n over the field $k = \mathbb{Q}$ of rational numbers, with the ring $\mathbb{Z}$ of rational integers, and $\mathfrak{A}$ is a $\mathbb{Z}$-module with n generators in the ring of integers of K (with respect to $\mathbb{Z}$):

$$\mathfrak{A} = \alpha_1\mathbb{Z} + \alpha_2\mathbb{Z} + \cdots + \alpha_n\mathbb{Z} \subset \mathfrak{O},$$

then

$$\begin{aligned} D(\mathfrak{A}) &= \begin{vmatrix} \alpha_1 & \alpha_1^{(2)} & \dots & \alpha_1^{(n)} \\ \alpha_2 & \alpha_2^{(2)} & \dots & \alpha_2^{(n)} \\ \dots & \dots & \dots & \dots \\ \alpha_n & \alpha_n^{(2)} & \dots & \alpha_n^{(n)} \end{vmatrix}^2 = (P - N)^2 \\ &= (P + N)^2 - 4PN, \end{aligned}$$

where P is the sum of all terms occuring in the determinant with positive sign, and N the sum of terms with the negative sign. From the Galois Theory we know, then, that both $P + N$ and PN are numbers in $\mathbb{Q}$, in fact integers, and hence

$$D(\mathfrak{A}) \equiv 0 \text{ or } 1 \pmod 4.$$

The possible importance of this theorem for the determination of $\mathfrak{O}$ can be seen in the following example. Say we have a module $\mathfrak{A} \subset \mathfrak{O}$ with $D(\mathfrak{A}) \equiv 44$. Then, $D(\mathfrak{A}) = D(\mathfrak{O})$, and $\mathfrak{A} = \mathfrak{O}$, for the only positive square by which 44 is divisible is 4, but $11 \not\equiv 0, 1 \pmod 4$.

$D(\mathfrak{O})$ is termed the discriminant D of $K|k$.

The field, K, is, in general, not uniquely determined by its discriminant; e.g., the polynomial

$$x^3 + 10x - 1$$

generates a cubic field over $\mathbb{Q}$ with the discriminant $D = -4027$ (4027 is a prime number!). There exist, however, three other cubic fields with this discriminant, and that they are not identical is seen in their arithmetic, in that the rational primes have different decompositions in each.

4. Connection between Different and Discriminant

If $\mathfrak{A}$ is an ideal of the extension field K, then both $\mathfrak{A}$ and $\mathfrak{O}$ are $\mathfrak{o}$-modules with n generators and we have

$$D(\mathfrak{A}) = x^2 D(\mathfrak{O}),$$

where x is an element of the base field k, and a certain determinant. We want another characterization of the element x.

1.) Let $\alpha \in K^*$, and $\mathfrak{A} = \alpha\mathfrak{O}$. If $\omega_1, \dots, \omega_n$ is a minimal basis of $\mathfrak{O}$, then $\alpha\omega_1, \dots, \alpha\omega_n$, is a minimal basis of $\mathfrak{A}$. Then,

$$D(\alpha\mathfrak{O}) = (N\alpha)^2 \cdot D(\mathfrak{O}).$$

We can interpret $N\alpha$ as the norm of the divisor α.

2.) The conjecture can be made that, as in 1.) for any ideal

$$D(\mathfrak{A}) = (N\mathfrak{A})^2 D(\mathfrak{O}),$$

in the sense of either a divisor or ideal equation.

To prove this, let $\mathfrak{A}$ be some ideal (divisor). Let the finite set of places $\mathfrak{P}_1, \mathfrak{P}_2, \ldots, \mathfrak{P}_r$ of K contain all those that enter into $\mathfrak{A}$ with a non-zero exponent. By axiom 2 we can then find an element $\alpha \in K^*$ with

$$\begin{cases} \operatorname{ord}_{\mathfrak{P}_i} \alpha = \operatorname{ord}_{\mathfrak{P}_i} \mathfrak{A} & (i = 1, 2, \ldots, r) \\ \operatorname{ord}_{\mathfrak{P}} \alpha \geq \operatorname{ord}_{\mathfrak{P}} \mathfrak{A} = 0 & \text{otherwise.} \end{cases}$$

Applying this to the divisor $\mathfrak{A}^{-1}$, this implies the existence of a $\beta^{-1} \in K^*$ with

$$\begin{cases} \operatorname{ord}_{\mathfrak{P}_i} \beta^{-1} = \operatorname{ord}_{\mathfrak{P}_i} \mathfrak{A}^{-1} & (i = 1, 2, \ldots, r) \\ \operatorname{ord}_{\mathfrak{P}} \beta^{-1} \geq \operatorname{ord}_{\mathfrak{P}} \mathfrak{A}^{-1} & \text{otherwise,} \end{cases}$$

or,

$$\begin{cases} \operatorname{ord}_{\mathfrak{P}_i} \beta = \operatorname{ord}_{\mathfrak{P}_i} \mathfrak{A} & (i = 1, 2, \ldots, r) \\ \operatorname{ord}_{\mathfrak{P}} \beta \leq \operatorname{ord}_{\mathfrak{P}} \mathfrak{A} & \text{otherwise,} \end{cases}$$

$$\Longrightarrow \mathfrak{A} \mid \alpha, \ \beta \mid \mathfrak{A} \quad \text{and} \quad N\mathfrak{A} \mid N\alpha, \ N\beta \mid N\mathfrak{A},$$

and for the corresponding ideals,

$$\alpha\mathfrak{O} \subset \mathfrak{A} \subset \beta\mathfrak{O},$$

$$D(\mathfrak{A}) \mid D(\alpha\mathfrak{O}), \quad D(\beta\mathfrak{O}) \mid D(\mathfrak{A}).$$

But

$$\left.\begin{cases} D(\alpha\mathfrak{O}) = (N\alpha)^2 D(\mathfrak{O}) \\ D(\beta\mathfrak{O}) = (N\beta)^2 D(\mathfrak{O}) \\ D(\mathfrak{A}) = x^2 D(\mathfrak{O}) \end{cases}\right\} \Longrightarrow x \mid N\alpha, \ N\beta \mid x.$$

If $\mathfrak{p}$ is a place of k, such that all places $\mathfrak{P}$ of K with $\mathfrak{P} \mid \mathfrak{p}$ are among the $\mathfrak{P}_1, \mathfrak{P}_2, \ldots, \mathfrak{P}_r$, then the exponent of $\mathfrak{p}$ in the decomposition of $N\alpha$ is the same as in that of $N\beta$, and therefore also the same as in $N\mathfrak{A}$ and x. If we choose the $\mathfrak{P}_1, \mathfrak{P}_2, \ldots, \mathfrak{P}_r$ from the start in such a manner as to include all extensions $\mathfrak{P}$ of valuations $\mathfrak{p}$ for which either $\operatorname{ord}_{\mathfrak{p}} x \neq 0$ or $\operatorname{ord}_{\mathfrak{p}}(N\mathfrak{A}) \neq 0$, we have

$$x = N\mathfrak{A}.$$

The connection between the different $\mathfrak{d}$ and the discriminant $D = D(\mathfrak{O})$ of $K|k$ is now easily ascertained. We have

$$\mathfrak{d}^{-1} = \mathfrak{O}' = \omega_1'\mathfrak{o} + \omega_2'\mathfrak{o} + \cdots + \omega_n'\mathfrak{o},$$

where $\omega_1', \ldots, \omega_n'$, the complementary basis to $\omega_1, \ldots, \omega_n$, is defined by

$$S(\omega_i \omega_j') = \delta_{ij} \quad (i, j = 1, 2, \ldots, n).$$

We thus have the matrix equation

$$\begin{pmatrix} \omega_1 & \omega_1^{(2)} & \cdots & \omega_1^{(n)} \\ \omega_2 & \omega_2^{(2)} & \cdots & \omega_2^{(n)} \\ \cdots & \cdots & \cdots & \cdots \\ \omega_n & \omega_n^{(2)} & \cdots & \omega_n^{(n)} \end{pmatrix} \begin{pmatrix} \omega_1' & \omega_2' & \cdots & \omega_n' \\ \omega_1'^{(2)} & \omega_2'^{(2)} & \cdots & \omega_n'^{(2)} \\ \cdots & \cdots & \cdots & \cdots \\ \omega_1'^{(n)} & \omega_2'^{(n)} & \cdots & \omega_n'^{(n)} \end{pmatrix} = E$$

($E = n$-row unit matrix). By taking the determinant and squaring, we see that

$$D(\mathfrak{O}) \cdot D(\mathfrak{O}') = 1$$
$$\Longrightarrow d(\mathfrak{O})(N\mathfrak{d}^{-1})^2 D(\mathfrak{O}) = 1 \Longrightarrow (N\mathfrak{d})^2 = (D(\mathfrak{O}))^2$$
$$N\mathfrak{d} = D(\mathfrak{O}) = D.$$

The discriminant is the norm of the different. D is divisible by exactly those $\mathfrak{p}$ to which there exists an extension $\mathfrak{P}$ (i.e. $\mathfrak{P} \mid \mathfrak{p}$) by which $\mathfrak{d}$ is divisible. In terms of ramification, the discriminant is divisible by $\mathfrak{p}$ if there exists an extension $\mathfrak{P} \mid \mathfrak{p}$ with ramification $e_{\mathfrak{P}} > 1$.

Thus, the discriminant gives less information than the different, the latter giving all $\mathfrak{P}$ ramified over k, the former only the corresponding $\mathfrak{p}$. If, for example, $\mathfrak{p} = \mathfrak{P}^2\mathfrak{P}_1$ the discriminant is divisible by $\mathfrak{p}$, but we still do not know whether $\mathfrak{P}$ or $\mathfrak{P}_1$ is ramified over k.

5. Examples. Cubic and Quadratic Number Fields

1.) The two cubic number fields with discriminants -23 and -31 have been studied. The cubic number field with next lowest discriminant is generated by a root α of the polynomial

$$f(x) = x^3 - 2x + 2; \quad D(f) = -44.$$

But this is, indeed, also the discriminant of the generated field, for $f(x)$ is an Eisenstein equation for the prime number 2, so that the field is fully ramified and the discriminant divisible by 2. But then $D(f)$ contains no further square > 1, so that

$$D = D(\mathfrak{O}) = -44 \quad \text{and} \quad \mathfrak{O} = \mathbb{Z} + \alpha\mathbb{Z} + \alpha^2\mathbb{Z}.$$

(Remark: The Stickelberger theorem is of no use here, since $-11 \equiv 1 \pmod 4$.)

2.) We will consider an example of a cubic field dating back to Dedekind, for which no minimal basis of the form $1, \alpha, \alpha^2$ exists. Consider the field generated by the root α of the polynomial

$$f(x) = x^3 + x^2 - 2x + 8.$$

The irreducibility is easily established, since $f(x)$ can have no rational roots, as is seen without difficulty. One computes:

$$D(f) = -4 \cdot 503$$

(503 is prime!). As before, the Stickelberger theorem is of no use here. Another method, though, shows us that the discriminant

$$D = D(\mathfrak{O}) = -503.$$

$$f(\alpha) = \alpha^3 + \alpha^2 - 2\alpha + 8 = 0 \tag{1}$$

informing us that $\alpha \mid 8$. Multiplying (1) by 8 and dividing by α^3 yields

$$8 + \frac{8}{\alpha} - \frac{16}{\alpha^2} + \frac{64}{\alpha^3} = 0$$

so that the element $\beta = \frac{4}{\alpha}$ satisfies the equation with integral coefficients

$$\beta^3 - \beta^2 + 2\beta + 8 = 0; \tag{2}$$

thus β is integral: $\beta \in \mathfrak{O}$. We divide (1) and (2) by α and β respectively

$$\alpha^2 + \alpha - 2 + 2\beta = 0, \quad 2\alpha + 2 - \beta + \beta^2 = 0,$$

and thus

$$\alpha^2 = -\alpha + 2 - 2\beta, \quad \beta^2 = -2\alpha - 2 + \beta, \quad \alpha\beta = 4. \tag{3}$$

Thus,

$$\mathbb{Z} + \alpha\mathbb{Z} + \alpha^2\mathbb{Z} \subset \mathbb{Z} + \alpha\mathbb{Z} + \beta\mathbb{Z};$$

but since $\beta = \frac{-(\alpha^2+\alpha)}{2} + 1 \notin \mathbb{Z} + \alpha\mathbb{Z} + \alpha^2\mathbb{Z}$, we see

$$\mathbb{Z} + \alpha\mathbb{Z} + \alpha^2\mathbb{Z} \neq \mathbb{Z} + \alpha\mathbb{Z} + \beta\mathbb{Z}.$$

¿From (3) it is apparent that $\mathbb{Z} + \alpha\mathbb{Z} + \beta\mathbb{Z}$ is a ring; the discriminant of this $\mathbb{Z}$-module, contained in $\mathfrak{O}$ must be less than that of $\mathbb{Z} + \alpha\mathbb{Z} + \alpha^2\mathbb{Z}$, thus it must be -503. But then, we necessarily have

$$D = D(\mathfrak{O}) = -503,$$

and

$$\mathfrak{O} = \mathbb{Z} + \alpha\mathbb{Z} + \beta\mathbb{Z}. \tag{4}$$

The ring $\mathfrak{O}$ of integers and its structure is fully given by (4) together with (1), (2), and (3). That, in this case, there can be no number γ so that $\mathfrak{O} = \mathbb{Z} + \gamma\mathbb{Z} + \gamma^2\mathbb{Z}$, will be demonstrated later.

3.) Quadratic Number Field: Let m be a rational integer without square factors and

$$K = \mathbb{Q}(\sqrt{m})$$

(a) Let either $m \equiv 2 \pmod 4$ or $m \equiv 3 \pmod 4$. Then

$$D(\mathbb{Z} + \mathbb{Z}\sqrt{m}) = \begin{vmatrix} 1 & \sqrt{m} \\ 1 & -\sqrt{m} \end{vmatrix}^2 = 4m.$$

It then follows from the theorem of Stickelberger that

$$D = D(\mathfrak{O}) = 4m, \quad \mathfrak{O} = \mathbb{Z} + \mathbb{Z}\sqrt{m}.$$

(b) Let $m \equiv 1 \pmod 4$.

The element $\theta = \frac{1+\sqrt{m}}{2}$ is integral, satisfying

$$\theta^2 - \theta + \frac{1-m}{4} = 0,$$

an equation with integral coefficients, thus $\theta \in \mathfrak{O}$. Hence:

$$D(\mathbb{Z} + \theta\mathbb{Z}) = \begin{vmatrix} 1 & \frac{1+\sqrt{m}}{2} \\ 1 & \frac{1-\sqrt{m}}{2} \end{vmatrix}^2 = m,$$

and thus

$$D = D(\mathfrak{O}) = m, \quad \mathfrak{O} = \mathbb{Z} + \mathbb{Z}\frac{1+\sqrt{m}}{2}.$$

CHAPTER 11

FACTORIZATION OF PRIME IDEALS IN FIELD EXTENSIONS

1. Factorization: Ring of Integers Fully Known

If we have, by some method, determined the ring $\mathfrak{O}$ of integers with respect to $\mathfrak{o}$ of the extension field K, the problem arises to find the prime ideals $\mathfrak{P}$ of K, by which the given prime ideal $\mathfrak{p}$ of the base field k is divisible. The best way of finding them is to first seek the corresponding places

$$\mathfrak{O} \to \mathfrak{O}/\mathfrak{P}$$

of K, i.e. the homomorphisms of $\mathfrak{O}$ which extend the given homomorphism

$$\mathfrak{o} \to \mathfrak{o}/\mathfrak{p}.$$

The prime ideals $\mathfrak{P}$ are then the kernels of these homomorphisms.

We are given $\mathfrak{O}$. Let $\omega_1, \omega_2, \ldots, \omega_m$, be a system of generators of $\mathfrak{O}$ over $\mathfrak{o}$ (not necessarily a minimal basis: $m \geq n$).

$$\mathfrak{O} = \omega_1\mathfrak{o} + \omega_2\mathfrak{o} + \cdots + \omega_m\mathfrak{o}$$

is then completely determined by $\mathfrak{o}$ and the multiplication table

$$\omega_i\omega_j = \sum_{\nu=1}^{m} c_{ij}^{(\nu)}\omega_\nu, \quad c_{ij}^{(\nu)} \in \mathfrak{o} \ (\nu, i, j = 1, 2, \ldots, m)$$

Let $\mathfrak{p}$ be a prime ideal of $\mathfrak{o}$ and

$$\mathfrak{o} \to \mathfrak{o}/\mathfrak{p} = F$$

the corresponding homomorphism onto the field F, by which an element $a \in \mathfrak{o}$ goes into $\bar{a} \in F$. To determine an extension

$$\mathfrak{O} \to \mathfrak{O}/\mathfrak{P}$$

of this homomorphism it suffices to select the images $\bar{\omega}_i$ of the elements ω_i ($i = 1, 2, \ldots, m$) in such a manner that

$$\bar{\omega}_1 F + \bar{\omega}_2 F + \cdots + \bar{\omega}_m F$$

is a field, and that

$$\sum_{\nu=1}^{m} c_{ij}^{(\nu)}\bar{\omega}_\nu = \bar{\omega}_i\bar{\omega}_j \quad (i, j = 1, 2, \ldots, m).$$

For then, $\overline{\omega_i\omega_j} = \bar{\omega}_i\bar{\omega}_j$, and the map is a homomorphism.

EXAMPLE. We again consider the field $K = \mathbb{Q}(\alpha)$, generated by a root α of

$$f(x) = x^3 + x^2 - 2x + 8,$$

With the discriminant $D = -503$. The ring $\mathfrak{O}$ of K is generated by $1, \alpha, \beta$, with

$$\alpha^2 = -\alpha + 2 - 2\beta, \quad \beta^2 = -2\alpha - 2 + \beta, \quad \alpha\beta = 4. \tag{1}$$

Let $\mathfrak{p}$ the prime ideal generated by the prime number 2 in $\mathbb{Q}$, i.e. $\mathfrak{p} = 2\mathbb{Z}$. The homomorphism $\mathbb{Z} \to \mathbb{Z}/\mathfrak{p}$ maps $\mathbb{Z}$ onto the field $F = \{0, 1\}$ of characteristic 2. (Even numbers mapped onto 0, odd onto 1.) By (3), the images $\bar{\alpha}$ and $\bar{\beta}$ under an extension of the homomorphism must satisfy:

$$\bar{\alpha}^2 = \bar{\alpha}, \quad \bar{\beta}^2 = \bar{\beta}, \quad \bar{\alpha}\bar{\beta} = 0.$$

Thus $\bar{\alpha}$ and $\bar{\beta}$ can only be 0 or 1, with one of them 0, and $\mathfrak{O}/\mathfrak{P} = F$, i.e. the residue class degree $f = 1$ for any possible extension. We have the following possibilities for an extension:

1.) $\alpha \to 0$, $\beta \to 0$ (corresponding prime ideal $\mathfrak{P}_2$),
2.) $\alpha \to 1$, $\beta \to 0$ (corresponding prime ideal $\mathfrak{P}_2'$),
3.) $\alpha \to 0$, $\beta \to 1$ (corresponding prime ideal $\mathfrak{P}_2''$).

Whether or not a given element belongs to one of the prime ideals $\mathfrak{P}_2$, $\mathfrak{P}_2'$, or $\mathfrak{P}_2''$ can be seen directly. Recalling that $\sum e_i f_i = \sum e_i \leq 3$, we have the factorization of the prime number 2 (the prime ideal $2\mathbb{Z}$) in K:[1]

$$2 = 2\mathbb{Z} = \mathfrak{P}_2\mathfrak{P}_2'\mathfrak{P}_2''.$$

To factor α and β we utilize the norm map.

If α generates the field K and $f = \operatorname{Irr}(\alpha, k)$, then $N\alpha = (-1)^n f(0)$, and, for any $a \in k$ we have

$$N(\alpha - a) = (-1)^n f(a).$$

In our example, therefore,

$$N(\alpha) = N(\beta) = -8 = -2^3 \quad N\mathfrak{P}_2 = N\mathfrak{P}_2' = N\mathfrak{P}_2'' = 2$$

α and β are thus only divisible by prime ideals that divide 2 (α and β being integral), hence by $\mathfrak{P}_2, \mathfrak{P}_2', \mathfrak{P}_2''$. From the definitions of these ideals we have

$$\mathfrak{P}_2' \nmid \alpha, \quad \mathfrak{P}_2 \mid \alpha, \quad \mathfrak{P}_2'' \mid \alpha, \quad \mathfrak{P}_2'' \nmid \beta, \quad \mathfrak{P}_2 \mid \beta, \quad \mathfrak{P}_2' \mid \beta.$$

But because $\alpha\beta = 4$

$$\mathfrak{P}_2^2 \nmid \alpha, \quad \mathfrak{P}_2^2 \nmid \beta,$$

since $4 = 2 \cdot 2 = \mathfrak{P}_2^2 \mathfrak{P}_2'^2 \mathfrak{P}_2''^2$. Only one possibility remains for the norm to come out right:

$$\alpha = \mathfrak{P}_2\mathfrak{P}_2''^2, \quad \beta = \mathfrak{P}_2\mathfrak{P}_2'^2.$$

For any odd rational integer, a, $\alpha - a$ is divisible by $\mathfrak{P}_2'$ (since $\alpha - a$ goes onto $1 - 1 = 0$ by the corresponding homomorphism), but not divisible by $\mathfrak{P}_2$ and $\mathfrak{P}_2''$; similarly $\beta - a$ is divisible by $\mathfrak{P}_2''$, but not by $\mathfrak{P}_2$ and $\mathfrak{P}_2'$. For integers, a, divisible by only the first power of 2, $\alpha - a$ is divisible by only the first power of $\mathfrak{P}_2''$ and not

[1] Here, and elsewhere, Artin identifies an element α with the principal ideal (divisor) determined by it. *Editor's note.*

by $\mathfrak{P}_2'$, and $\beta - a$ is divisible by only the first power of $\mathfrak{P}_2'$ and not by $\mathfrak{P}_2''$. In this manner, and by using the norms, one easily sees that

$$\alpha - 1 = \mathfrak{P}_2'^3, \qquad \alpha - 2 = \mathfrak{P}_2^3\mathfrak{P}_2'',$$
$$\alpha + 3 = \mathfrak{P}_2'^2, \qquad \alpha + 2 = \mathfrak{P}_2^2\mathfrak{P}_2''.$$

Prime ideals, other than the three we have, cannot enter here, for the integers and the norms are powers of two. It follows that

$$\mathfrak{P}_2 = \frac{\alpha-2}{\alpha+2}, \quad \mathfrak{P}_2' = \frac{\alpha-1}{\alpha+3}, \quad \mathfrak{P}_2'' = \frac{2}{\mathfrak{P}_2\mathfrak{P}_2'} = \frac{2(\alpha+2)(\alpha+3)}{(\alpha-2)(\alpha-1)},$$

i.e. $\mathfrak{P}_2, \mathfrak{P}_2', \mathfrak{P}_2''$ are principal ideals.

¿From our calculations we can also see that this field has no minimal basis of the form $1, \gamma, \gamma^2$. For, this would imply

$$\mathfrak{O} = \mathbb{Z} + \gamma\mathbb{Z} + \gamma^2\mathbb{Z},$$

and then the prime number 2 could not be decomposed into three prime ideals of the first degree. Since the residue class field of each of these prime ideals consists only of the elements 0 and 1, there would be only two possibilities

$$\gamma \to 0 \quad \text{or} \quad \gamma \to 1$$

for the respective homomorphisms. Since the homomorphism would be completely determined by the image of γ, 2 could be divisible by only two prime ideals of the first degree.

2. Factorization: Ring of Integers not Fully Known

We want to investigate what can be said about the prime ideals of K, even if the ring $\mathfrak{O}$ of integers is not fully known. Let α be an integral generator of the separable extension K of degree n over k:

$$K = k(\alpha).$$

We consider the subring

$$\mathfrak{O}_0 = \mathfrak{o} + \mathfrak{o}\alpha + \mathfrak{o}\alpha^2 + \cdots + \mathfrak{o}\alpha^{n-1},$$

and will show that this subring suffices to decompose the prime ideals of k into prime ideals of K, if, we leave out a finite number of prime ideals (in the case just considered these would be the divisors of 2; $\mathfrak{P}_2, \mathfrak{P}_2', \mathfrak{P}_2''$).

Let θ be an element of $\mathfrak{O}$, θ can be represented in the form

$$\theta = x_0 + x_1\alpha + x_2\alpha^2 + \cdots + x_{n-1}\alpha^{n-1}$$

with $x_i \in k$ $(i = 0, 1, \ldots, n-1)$. If we similarly represent the conjugates of θ, the resulting system of n equations in the n unknowns x_i has the determinant

$$D^{\frac{1}{2}}(\mathfrak{O}_0) = D^{\frac{1}{2}}(1, \alpha, \alpha^2, \ldots, \alpha^{n-1}) \neq 0.$$

By Kramer's rule each of the x_i $(i = 0, \ldots, n-1)$ can be written as the quotient of two determinants with elements in $\mathfrak{O}$, symbolically

$$x_i = \frac{D^{\frac{1}{2}}(1, \alpha, \ldots \alpha^{i-1}, \theta, \alpha^{i+1}, \ldots, \alpha^{n-1})}{D^{\frac{1}{2}}(1, \alpha, \alpha^2, \ldots, \alpha^{n-1})} \quad (i = 0, 1, \ldots, n-1).$$

If $\omega_1, \omega_2, \ldots, \omega_n$ is a minimal basis of $K|k$, then

$$x_i = \frac{\det(b_{ij})D^{\frac{1}{2}}(\omega_1, \ldots, \omega_n)}{\det(a_{ij})D^{\frac{1}{2}}(\omega_1, \ldots, \omega_n)} = \frac{\det(b_{ij})}{\det(a_{ij})}$$

with certain $a_{ij}, b_{ij} \in \mathfrak{O}$ $(i, j = 0, 1, \ldots, n-1)$. This, the denominator of the elements x_i divides

$$\det(a_{ij}) = \frac{D^{\frac{1}{2}}(1, \alpha, \ldots, \alpha^{n-1})}{D^{\frac{1}{2}}(\omega_1, \omega_2, \ldots, \omega_n)} = \frac{D^{\frac{1}{2}}(1, \alpha, \ldots, \alpha^{n-1})}{D^{\frac{1}{2}}(\mathfrak{O})}$$

(in our last example, $\det(a_{ij}) = \frac{\sqrt{-4 \cdot 503}}{\sqrt{-503}} = 2$, thus the denominator of the elements x_i can, at most, be 2.)

If $\mathfrak{p}$ is a prime ideal of k not dividing

$$\frac{D^{\frac{1}{2}}(1, \alpha, \alpha^2, \ldots, \alpha^{n-1})}{D^{\frac{1}{2}}(\mathfrak{O})},$$

then any denominators appearing in the representation of integers by the basis $1, \alpha, \alpha^2, \ldots, \alpha^{n-1}$, are relatively prime to $\mathfrak{p}$. We can find all possible divisors $\mathfrak{P}$ of $\mathfrak{p}$ in K.

The homomorphism corresponding to any such $\mathfrak{P}$, an extension of

$$\mathfrak{o} \to \mathfrak{o}/\mathfrak{p} = F,$$

maps $\mathfrak{O}$ onto some field, and thereby induces a homomorphism of $\mathfrak{O}_0$. On the other hand, any homomorphism of $\mathfrak{O}_0$ onto a field, extending the homomorphism $\mathfrak{o} \to \mathfrak{o}/\mathfrak{p}$, can be extended to $\mathfrak{O}$ in a unique manner. For, such a homomorphism of $\mathfrak{O}_0$ maps elements prime to $\mathfrak{p}$ onto non-zero elements. But, by our choice of $\mathfrak{p}$, all elements θ of $\mathfrak{O}$ have the form

$$\theta = \frac{\theta_0}{c}, \quad \text{with } \theta_0 \in \mathfrak{O}_0,\ c \in \mathfrak{o},\ c \notin \mathfrak{p}.$$

Then, the homomorphism φ could at most be extended onto $\mathfrak{O}$ by setting

$$\varphi(\theta) = \frac{\varphi(\theta_0)}{\varphi(c)},$$

which is indeed a homomorphism of $\mathfrak{O}$. The image field of $\mathfrak{O}$ is, furthermore, the image field of $\mathfrak{O}_0$.

In therefore suffices to find the extensions of the homomorphism

$$\mathfrak{o} \to \mathfrak{o}/\mathfrak{p} = F$$

on $\mathfrak{O}_0$. Since $\mathfrak{O}_0 = \mathfrak{o}[\alpha]$, the image field of $\mathfrak{O}_0$ under such a map extension, φ, is

$$F(\bar{\alpha}) \quad \text{where } \bar{\alpha} = \varphi(\alpha).$$

If $f = \operatorname{Irr}(\alpha, k) \in \mathfrak{o}[x]$, and $\bar{f}$ is the image polynomial in $F[x]$, then $\bar{\alpha}$ satisfies the equation $\bar{f}(\bar{\alpha}) = 0$. Now, $\bar{f}$ need no longer be irreducible in F, and decomposes into irreducible factors, say

$$\bar{f}(\bar{x}) = \bar{P}_1^{e_1}(x)\bar{P}_2^{e_2}(x)\ldots\bar{P}_r^{e_r}(x).$$

But this is equivalent to a decomposition of $f(x)$ into irreducible factors modulo $\mathfrak{p}$ in the form

$$f(x) = P_1^{e_1}(x)P_2^{e_2}(x)\ldots P_r^{e_r}(x) \pmod{\mathfrak{p}},$$

where the $P_i(x)$ are determined only up to their images, and thus up to addition by any element of $\mathfrak{p}[x]$.

Since $\bar{f}(\bar{\alpha}) = 0$, we have $\bar{P}_i(\bar{\alpha}) = 0$ for exactly one i. Thus each $\mathfrak{P}$ which divides $\mathfrak{p}$ determines a unique $\bar{\alpha}$ and one $P_i(x)$. If, on the other hand, $P_i(x)$ is given, and $\bar{\alpha}$ is a root of $\bar{P}_i(x)$, then the extension field $F(\bar{\alpha})$ is uniquely determined, and the map

$$\mathfrak{O}_0 = \mathfrak{o}[\alpha] \to F(\bar{\alpha})$$

which maps $\mathfrak{o}$ onto F and α onto $\bar{\alpha}$ is a well-defined homomorphism φ of $\mathfrak{O}_0$. The various roots of $\bar{P}_i(x)$ yield, up to isomorphism, the same homomorphism.

The unique extension of this homomorphism to $\mathfrak{O}$ yields a uniquely determined prime ideal $\mathfrak{P}_i$ (the kernel of the map), with $\mathfrak{P}_i \mid \mathfrak{p}$. Thus, the prime ideals $\mathfrak{P}$ with $\mathfrak{P} \mid \mathfrak{p}$ stand in one-to-one correspondence with the factors $P_i(x)$ in the decomposition of $f(x)$ modulo $\mathfrak{p}$.

The map of $\mathfrak{o}$ into $\mathfrak{o}/\mathfrak{p} = F$, and therefore the map of $\mathfrak{O}$ into $F(\bar{\alpha})$, is actually a map onto F, $F(\bar{\alpha})$, respectively, so that the residue class degree of $\mathfrak{P}_i$

$$\begin{aligned} f_i &= [F(\bar{\alpha}) : F] = \text{degree } \bar{P}_i(x) \\ \Longrightarrow n &= [K : k] = \text{degree } f(x) = \text{degree } \bar{f}(x) \\ &= \sum_{i=1}^{r} e_i \cdot \text{degree } \bar{P}_i(x) = \sum_{i=1}^{r} e_i f_i, \end{aligned}$$

where the e_i still denote only the multiplicity of the factors $\bar{P}_i(x)$ in $\bar{f}(x)$.

These multiplicities e_i are the exact ramification indices of the prime ideals $\mathfrak{P}_i$. For, if these ramifications were e_i', then we would have

$$\mathfrak{p} = \mathfrak{P}_1^{e_1'} \mathfrak{P}_2^{e_2'} \ldots \mathfrak{P}_r^{e_r'} \quad \text{and} \quad \sum_{i=1}^{r} e_i' f_i = n.$$

We will now show that $e_i' = e_i$ for $i = 1, 2, \ldots, r$. Let θ be an element of $\mathfrak{O}_0 = \mathfrak{o}[\alpha]$, i.e. of the form

$$\theta = G(\alpha), \quad G(x) \in \mathfrak{o}[x].$$

θ lies in $\mathfrak{P}_i$ if, by the corresponding homomorphism, $\bar{G}(\bar{\alpha}) = 0$. But then also,

$$\begin{aligned} \bar{G}(x) &= \bar{P}_i(x)\bar{g}(x), \quad G(x) \equiv P_i(x)g(x) \pmod{\mathfrak{p}} \\ G(x) &= P_i(x)g(x) + h(x), \quad h(x) \in \mathfrak{p}[x]. \end{aligned}$$

Since, now, $G(\alpha) = P_i(\alpha)g(\alpha) + h(\alpha)$, $g(\alpha) \in \mathfrak{O}_0$, $h(\alpha) \in \mathfrak{p}\mathfrak{O}_0$, the kernel, $\mathfrak{P}_i^{(0)}$ of

$$\mathfrak{O}_0 \to \bar{F}(\bar{\alpha})$$

is generated by $\mathfrak{p}$ and $P_i(\alpha)$.

$\mathfrak{P}_i$ contains an element with the order 1; it can be put in the form $\frac{\theta}{c}$. $\theta_0 \in \mathfrak{O}_0$, $c \in \mathfrak{o}$, c prime to $\mathfrak{p}$. By multiplying by the denominator we see that $\mathfrak{P}_i^{(0)}$ contains an element of order 1. Thus, either $P_i(\alpha)$ or some element of $\mathfrak{p}$ has order 1 at the place $\mathfrak{P}_i$. We can even assume that

$$\operatorname{ord}_{\mathfrak{P}_i} P_i(\alpha) = 1,$$

for $P_i(\alpha)$ was only determined up to addition of any element of $\mathfrak{p}[\alpha]$.

For $j \neq i$ we have $\bar{P}_j(\bar{\alpha}) \neq 0$, i.e. $\mathfrak{P}_i \nmid P_j(\alpha)$. Now, we know

$$P_1^{e_1}(\alpha)P_2^{e_2}(\alpha) \ldots P_r^{e_r}(\alpha) \equiv f(\alpha) \equiv 0 \pmod{\mathfrak{p}},$$

as well as the divisor representation,

$$P_1^{e_1}(\alpha)P_2^{e_2}(\alpha)\dots P_r^{e_r}(\alpha) = \mathfrak{P}_1^{e_1}\mathfrak{P}_2^{e_2}\dots\mathfrak{P}_r^{e_r}\mathfrak{D},$$

where $\mathfrak{D}$ is an integral divisor prime to $\mathfrak{p}$. Therefore

$$\mathfrak{P}_1^{e_1'}\mathfrak{P}_2^{e_2'}\dots\mathfrak{P}_r^{e_r'} = \mathfrak{p} \mid \mathfrak{P}_1^{e_1}\mathfrak{P}_2^{e_2}\dots\mathfrak{P}_r^{e_r}\mathfrak{D}$$

$$\Longrightarrow e_i' \leq e_i \ (i=1,2,\dots,r), \quad \text{and since } \sum e_i' f_i = n = \sum e_i f_i,$$

we have

$$e_i' = e_i \quad (i=1,2,\dots,r),$$

The exponents e_i are the ramification indices of the prime ideals $\mathfrak{P}_i$.

Thus, the irreducible equation for α yields the decomposition of all prime ideals of k into prime ideals of $K = k(\alpha)$, except those $\mathfrak{p}$ for which

$$\mathfrak{p} \mid \frac{D(1,\alpha,\alpha^2,\dots,\alpha^{n-1})}{D(\mathfrak{D})}.$$

As an example we again consider the field generated by a root of

$$f(x) = x^3 + x^2 - 2x + 8, \quad D = -503.$$

The element $\alpha + 1$ has norm 10, and is thus divisible by a divisor of 5 and a divisor of 2: in fact

$$\alpha + 1 = \mathfrak{P}_2'\mathfrak{P}_5, \quad \mathfrak{P}_5 \mid 5.$$

$\mathfrak{P}_5$ is uniquely determined by the fact that the corresponding homomorphism maps α onto -1. $\mathfrak{P}_5$ has norm 5, so that the corresponding residue class degree $f_5 = 1$. Similarly,

$$\alpha - 3 = \mathfrak{P}_2'\mathfrak{P}_{19}, \quad \mathfrak{P}_{19} \mid 19.$$

This, with our earlier decompositions:

$$\alpha - 1 = \mathfrak{P}_2'^3, \quad \alpha + 3 = \mathfrak{P}_2'^2, \quad \alpha - 2 = \mathfrak{P}_2^3\mathfrak{P}_2'', \quad \alpha + 2 = \mathfrak{P}_2^2\mathfrak{P}_2'', \quad \alpha = \mathfrak{P}_2\mathfrak{P}_2'',$$

implies that by the homomorphism corresponding to a prime divisor of 7, none of the elements $\alpha-3$, $\alpha-2$, $\alpha-1$, α, $\alpha+1$, $\alpha+2$, $\alpha+3$ and therefore none of the elements $\alpha + a$, $a \in \mathbb{Z}$, are mapped onto 0. This implies that $f(x) \pmod 7$ can have no linear factors; $f(x) \pmod 7$ has no rational roots. Thus, as a third degree polynomial, $f(x)$ is irreducible modulo 7, and the prime number 7 cannot be decomposed in K. These considerations are more than sufficient to show that 3 also remains prime in K. Several further decompositions show that the prime numbers 11 and 13 remain prime in K.

For the prime number 5, we found a first degree divisor $\mathfrak{P}_5$, but no others. Thus, $f(x)$ decomposes, modulo 5, into a linear and a quadratic factor:

$$5 = \mathfrak{P}_5\mathfrak{P}_5', \quad f_5 = 1, \quad f_5' = 2,$$

$$\mathfrak{P}_5 = \frac{\alpha+1}{\mathfrak{P}_2'} = \frac{(\alpha+1)(\alpha+3)}{\alpha-1}, \quad \mathfrak{P}_5' = \frac{5}{\mathfrak{P}_5}.$$

We have demonstrated that the prime divisors of all prime numbers < 11 correspond to principal ideals. We will later show that this suffices to prove that all prime ideals, and thus all ideals, of K are principal ideals, and that hence K has the class number 1.

The only ramified prime ideals of K divide 503. Since 503 is the norm of the different and a prime number, the different must be a prime ideal $\mathfrak{P}$. The ramification can only have a value 1, 2, or 3, thus $e \not\equiv 0 \pmod{503}$, hence the different is divisible by exactly $\mathfrak{P}^{e-1}$, and thus $e-1=1$ or $e=2$, and the polynomial $f(x)$ has a double root γ modulo 503;

$$\begin{cases} \gamma^3+\gamma^2-2\gamma+8 \equiv 0 \pmod{503} \\ \quad 3\gamma^2+2\gamma-2 \equiv 0 \pmod{503}. \end{cases}$$

The solution,

$$\gamma \equiv 149 \pmod{503}$$

yields the decomposition, $f(x) \equiv (x-149)^2(x+299) \pmod{503}$, and

$$503 = \mathfrak{P}^2\mathfrak{P}'.$$

CHAPTER 12

THE FINITENESS OF THE CLASS NUMBER

All problems discussed thus far could be referred to the corresponding local problems. Certain questions arise, however, which are truly global in nature, and cannot be treated by local considerations. In the remaining chapters we will consider the two most important topics of this sort, for algebraic number fields: the theorem of the finiteness of the class number (the order of the divisor group), and the Dirichlet theorem on units.

1. Divisor Norm in Algebraic Number Fields

We want to establish the connection between the norm introduced earlier for ideals and divisors, and the usual special definition of norm for algebraic number fields.

Let k be a field in which Axioms 1 and 2 hold, and $\mathfrak{o}$ be the corresponding ring. If $\mathfrak{a}$ is an integral ideal of $\mathfrak{o}$, we define

$$\mathfrak{N}_k(\mathfrak{a}) = \text{Number of elements in the residue class ring } \mathfrak{o}/\mathfrak{a}.$$

This number (it can be infinity) is a multiplicative ideal function:

$$\mathfrak{N}_k(\mathfrak{a}\mathfrak{b}) = \mathfrak{N}_k(\mathfrak{a})\mathfrak{N}_k(\mathfrak{b})$$

far any two integral ideals $\mathfrak{a}$ and $\mathfrak{b}$ of $\mathfrak{o}$.

To show this we choose to the given ideals $\mathfrak{a}$ and $\mathfrak{b}$ an $\alpha \in \mathfrak{a}$ with

$$\operatorname{ord}_{\mathfrak{p}} \alpha = \operatorname{ord}_{\mathfrak{p}} \mathfrak{a} \quad \text{for all } \mathfrak{p} \text{ with } \mathfrak{p} \mid \mathfrak{a}\mathfrak{b}.$$

LEMMA. *The congruence $\alpha x \equiv a \pmod{\mathfrak{a}\mathfrak{b}}$ is solvable with $x \in \mathfrak{o}$ for every $a \in \mathfrak{a}$.*

PROOF. We have $a\mathfrak{o} + \mathfrak{a}\mathfrak{b} = \mathfrak{a}$, for both sides of this equation have the same order at every place $\mathfrak{p}$. This proves the lemma. □

Let ξ run through a system of representatives of the residue classes modulo $\mathfrak{a}$ and η run through such a system modulo $\mathfrak{b}$. The statement

$$\mathfrak{N}_k(\mathfrak{a}\mathfrak{b}) = \mathfrak{N}_k(\mathfrak{a})\mathfrak{N}_k(\mathfrak{b})$$

is then equivalent with the statement: $\xi + \alpha\eta$ runs through a system of class representatives modulo $\mathfrak{a}\mathfrak{b}$. We prove the latter statement:

1.) Let $\theta \in \mathfrak{o}$ and ξ the representative with $\theta \equiv \xi \pmod{\mathfrak{a}}$, i.e. $\theta - \xi \in \mathfrak{a}$. Let $x \in \mathfrak{o}$ be a solution of the congruence

$$\alpha x \equiv \theta - \xi \pmod{\mathfrak{a}\mathfrak{b}},$$

and η the representative such that $x \equiv \eta \pmod{\mathfrak{b}}$.

$$\Longrightarrow \eta - x \in \mathfrak{b},\ \alpha \in \mathfrak{a} \Longrightarrow \alpha\eta - \alpha x \in \mathfrak{ab}$$
$$\alpha\eta \equiv \alpha x \pmod{\mathfrak{ab}} \Longrightarrow \alpha\eta \equiv \theta - \xi \pmod{\mathfrak{ab}}$$
$$\Longrightarrow \theta \equiv \xi + \alpha\eta \pmod{\mathfrak{ab}}.$$

2.) Let ξ, ξ' and η, η' be representatives of two residue classes modulo $\mathfrak{a}$ and modulo $\mathfrak{b}$, respectively, and let

$$\xi + \alpha\eta \equiv \xi' + \alpha\eta' \pmod{\mathfrak{ab}}.$$
$$\Longrightarrow \xi \equiv \xi' \pmod{\mathfrak{a}}, \quad \text{since } \alpha\eta \in \mathfrak{a}, \alpha\eta' \in \mathfrak{a}$$
$$\Longrightarrow \xi = \xi' \Longrightarrow \alpha\eta \equiv \alpha\eta' \pmod{\mathfrak{ab}}$$
$$\Longrightarrow \mathfrak{ab} \mid \alpha(\eta - \eta')$$
$$\Longrightarrow \operatorname{ord}_{\mathfrak{p}}(\eta - \eta') \geq \operatorname{ord}_{\mathfrak{p}} \mathfrak{b}, \quad \text{by our choice of } \alpha$$
$$\Longrightarrow \mathfrak{b} \mid \eta - \eta', \quad \text{as the above congruence is trivial for } \mathfrak{p} \nmid \mathfrak{ab}$$
$$\Longrightarrow \eta \equiv \eta' \pmod{\mathfrak{b}} \Longrightarrow \eta = \eta'.$$

The second, and therefore also the first, of the two equivalent statements follow immediately from 1.) and 2.).

$$\begin{array}{ccc} K & \text{———} & \mathfrak{O} \\ | & & | \\ k & \text{———} & \mathfrak{o} \end{array}$$

Let K be an extension finite over k, and $\mathfrak{O}$ the ring of elements integral with respect to $\mathfrak{o}$. Assume the function $\mathfrak{N}_k$ in the base field k is known. Then, in the field K we have

$$\mathfrak{N}_K(\mathfrak{A}) = \mathfrak{N}_k(N_{K|k}\mathfrak{A})$$

for every integral ideal $\mathfrak{A}$ of $\mathfrak{O}$.

PROOF. Since both sides of the equation are multiplicative ideal functions, it will suffice to prove it for prime ideals $\mathfrak{P}$ of $\mathfrak{O}$. Let, therefore, $\mathfrak{P}$ be a prime ideal of $\mathfrak{O}$, $\mathfrak{P} \mid \mathfrak{p}$, and let f be the residue class degree of $\mathfrak{P}$ with respect to k. Then $\mathfrak{N}_K(\mathfrak{A})$ is the number of elements of $\mathfrak{O}/\mathfrak{P}$, and since $\mathfrak{O}/\mathfrak{P}$ is an extension field of degree f over $\mathfrak{o}/\mathfrak{p}$,

$$\mathfrak{N}_K(\mathfrak{P}) = (\mathfrak{N}_k(\mathfrak{p}))^f = \mathfrak{N}_k(\mathfrak{p}^f) = \mathfrak{N}_k(N_{K|k}\mathfrak{P}). \qquad \square$$

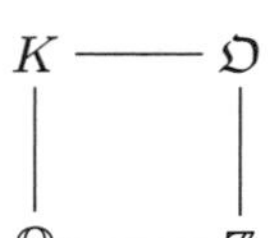

We consider the special case of an algebraic number field K. For an integral ideal $\mathfrak{A}$ of $\mathfrak{O}$ we have

$$\mathfrak{N}_K(\mathfrak{A}) = \mathfrak{N}_{\mathbb{Q}}(N_{K|\mathbb{Q}}\mathfrak{A}).$$

We will continue to use the tacitly assumed convention of identifying the numbers of the field with the corresponding principal ideals and principal divisors. The units, which could appear as factors, are lost thereby; in the rational number field these are only ± 1. As is customary, we identify divisors in $\mathbb{Q}$ with corresponding positive rational numbers.

Under this convention, $N_{K|\mathbb{Q}}$ becomes a positive rational number. If $a \neq 0$, $a \in \mathbb{Z}$, then there are a residue classes modulo a in $\mathbb{Z}$; $\mathfrak{N}_{\mathbb{Q}}(a) = a$, so that we can write:

$$\mathfrak{N}_K(\mathfrak{A}) = N_{K|\mathbb{Q}}\mathfrak{A}.$$

This was the original definition of the norm. It is disadvantageous, for, it is defined only for integral ideals, and then only in the case of algebraic number fields, and even then the base field is restricted to be the field $\mathbb{Q}$ of rational numbers. In what follows we will let N signify the norm taken with respect to $\mathbb{Q}$, considered as a positive rational number.

2. The Minkowski Bound

¿From now on k will always be an algebraic number field. It is easy to present an elementary proof that the number of ideal classes is finite. This would give us no method, though, to calculate or approximate this number in a given field, and we prefer to show directly that in every class of ideals there is an integral ideal, whose norm lies below a fixed, determined, bound. The best such bound was found by Minkowski, who proved the

THEOREM. *In every class of ideals of k there exists an integral ideal $\mathfrak{a}$ with*

$$N\mathfrak{a} \leq \left(\frac{4}{\pi}\right)^{r_2} \frac{n!}{n^n} \sqrt{|D|},$$

where N is the norm with respect to the field of rational numbers $\mathbb{Q}$, D the discriminant of k with respect to $\mathbb{Q}$, and $2r_2$ the number of non-real number fields conjugate to k over $\mathbb{Q}$.

The aim of the following investigations is the proof of this theorem.

For the case $n = 2$ this approximation is of no use in calculations; the literature for this case, however, is plentiful, and there exists a fixed algorithm for the determination of the class number. For $n \geq 3$ the bound of the theorem is quite good, and provides valuable help in calculation of the class number.

3. Theorems on Lattices in Vector Spaces

We first need several theorems about lattices. Let V be an n-dimensional vector space over the field $\mathbb{R}$ of real numbers. The r vectors

$$A_1, A_2, \ldots, A_r$$

are independent (over $\mathbb{R}$). We call the subset

$$\Lambda = \mathbb{Z}A_1 + \mathbb{Z}A_2 + \cdots + \mathbb{Z}A_r \quad (\mathbb{Z} \text{ rational integers})$$

of V an r-dimensional lattice. Λ is a free abelian group, generated by the vectors A_i, which are not only independent over $\mathbb{Z}$ but also over $\mathbb{R}$.

THEOREM. *An additive subgroup of V is a lattice if and only if every bounded domain in V contains only finite many elements of Λ.*

PROOF. That the "only if" part of the theorem is true, is trivial. To demonstrate the converse, we assume the condition is satisfied, and that $A_1, A_2, \ldots, A_r$ is some maximal independent system of vectors of Λ over $\mathbb{R}$. Induction will show that Λ is an r-dimensional lattice.

a) This is trivial for $r = 0$. For $r = 1$ let $X \in \Lambda$, then

$$X = x_1 A_1, \quad x_1 \in R.$$

By the hypothesis we can find a B_1 in Λ with

$$B_1 = b_1 A_1, \quad b_1 > 0,\ b_1 \text{ minimal},$$

$$\Longrightarrow \Lambda = \mathbb{Z}B_1$$

b) Let $r > 1$, W be the subspace of V spanned by $A_1, A_2, \ldots, A_{r-1}$ and

$$\Lambda' = \Lambda \cap W.$$

Then the condition of the theorem certainly also holds for Λ' in W. If we assume the theorem to be correct if the maximal number of independent vectors is $\le r-1$, then Λ' is an $r-1$ dimensional lattice, say

$$\Lambda' = \mathbb{Z}B_1 + \mathbb{Z}B_2 + \cdots + \mathbb{Z}B_{r-1}$$

the vectors B_i $(i = 1, 2, \ldots, r-1)$ being linearly independent over R, and spanning W.

Obviously then, any vector $X \in \Lambda$ can be written in the form

$$X = x_1B_1 + x_2B_2 + \cdots + x_{r-1}B_{r-1} + yA_r, \quad x_u, y \in \mathbb{R}.$$

The map

$$\varphi\colon X \to yA_r$$

is a homomorphism of Λ into V: denote the image group Λ''. If $yA_r \in \Lambda''$ and

$$x_1B_1 + x_2B_2 + \cdots + x_{r-1}B_{r-1} + yA_r \in \varphi^{-1}(yA_r)$$

then so is

$$(x_1 - [x_1])B_1 + \cdots + (x_{r-1} - [x_{r-1}])B_{r-1} + yA_r \in \varphi^{-1}(yA_r).$$

Every element $yA_r \in \Lambda''$ in some bounded domain (say $a \le y \le b$) is then the inverse image of some element

$$x = z_1B_1 + \cdots + z_{r-1}B_{r-1} + yA_r$$

of a bounded domain of V $(0 \le z_i \le 1,\ a \le y \le b)$. By hypothesis, therefore, there can be only a finite number of vectors $yA_r \in \Lambda''$ in the bounded image domain, and by inductive hypothesis Λ'' is a one-dimensional lattice:

$$\Lambda'' = \mathbb{Z}\cdot\varphi(B_r), \quad B_r = x_1B_1 + \cdots + x_{r-1}B_{r-1} + yA_r.$$
$$X \in \Lambda,\ \varphi(X) = x_r\varphi(B_r),\ x_r \in \mathbb{Z} \implies \varphi(X - x_rB_r) = 0$$
$$\implies \Lambda' \ni X - x_rB_r = z_1B_1 + \cdots + z_{r-1}B_{r-1},\ z_i \in \mathbb{Z},$$

and since conversely any vector of this form obviously belongs to Λ, we have

$$\Lambda = \mathbb{Z}B_1 + \mathbb{Z}B_2 + \cdots + \mathbb{Z}B_r. \qquad \square$$

The maximal dimension for a lattice in a number field is n, and we will, for the time being, limit ourselves to lattices of this dimension.

Let

$$\Lambda = \mathbb{Z}A_1 + \mathbb{Z}A_2 + \cdots + \mathbb{Z}A_n$$

be an n-dimensional lattice, and choose some lattice point

$$X_0 = x_1A_1 + x_2A_2 + \cdots + x_nA_n.$$

To this point we attach the *lattice loop* M,

$$M = \{Y \mid Y = y_1A_1 + \cdots + y_nA_n,\ y_i \in \mathbb{R},\ x_i \le y_i < x_i + 1\},$$

a fundamental domain with respect to the translations of the vectors of Λ. V is then the disjoint union of all the loops

$$V = \bigcup M.$$

Let S be a measurable (in the usual sense, with respect to some fixed coordinate system) subset of V.

THEOREM (Minkowski). *If the volume $\mu(S)$ is greater than the volume of a lattice loop, then S contains at least two distinct points, X_1 and X_2, congruent with respect to Λ ($X_1 - X_2 \in \Lambda$).*

PROOF. Assume the contrary. Then, the set $S \cap M$ is measurable, for all loops M, and

$$\mu(S) = \sum_M \mu(S \cap M)$$

If M_0 is some fixed loop, then each of these sets $S \cap M$ can be brought onto a subset $(S \cap M)_0$ of M_0 by translation with some element of Λ. The measure is invariant under such a translation, so that

$$\mu(S) = \sum_M \mu((S \cap M)_0).$$

But, since there are no congruent points in S, the sets $(S \cap M)_0$ are disjoint, and

$$\mu(S) = \mu\left(\bigcup_M (S \cap M)_0\right) \leq \mu(M_0),$$

contrary to the hypothesis about S. □

Let the set T contain, with X and Y, also $\frac{X-Y}{2}$. If we set $S = \frac{1}{2}T$, then T contains the difference of any two points in S, and therefore some lattice point other than 0, if it satisfies the corresponding condition:

$$\mu(T) > 2^n \mu(M_0).$$

These conditions for T are equivalent to the following: T is *convex* (i.e., with two points T also contains the connecting line), T is *point symmetric* (with X, T contains $-X$). An open, bounded, convex point set is called a convex domain. Thus, a convex domain with point symmetry (with respect to 0), with $\mu(T) > 2^n \mu(M_0)$ contains at least one lattice point other than 0.

If T is *compact* (closed and bounded), convex and point symmetric, then there is a lattice point other than 0 in T if

$$\mu(T) = 2^n \mu(M_0),$$

To show this (the Minkowski lattice-point theorem), we replace T by $(1+\varepsilon)T$, with some real $\varepsilon > 0$. This set also satisfies the hypothesis, and

$$\mu((1+\varepsilon)T) = (1+\varepsilon)^n 2^n \mu(M_0) > 2^n \mu(m_0).$$

Thus, for any $\varepsilon > 0$, the set $(1+\varepsilon)T$ contains a non-zero lattice point. But a bounded domain can contain only a finite number of lattice points, thus one non-zero lattice point must lie in all the sets $(1+\varepsilon)T$, and therefore in the closed set T. We see the necessity of T being compact.

That this theorem cannot be improved is seen by the example of a cube with width 2 and the usual cartesian unit lattice.

4. The Volume of Certain Convex Domains

In applications of the Minkowski lattice theorem we will need the volumes of certain convex domains, which will be calculated here.

Let V be the cartesian product of r_1 fields, each isomorphic to $\mathbb{R}$ (field of real numbers), and r_2 fields, each isomorphic to $\mathbb{C}$ (complex numbers):

$$V = \mathbb{R}^{r_1}\mathbb{C}^{r_2}.$$

V is thus a space spanned by r_1 real axes and r_2 complex planes. Considering each complex coordinate to be two real coordinates, V has dimension

$$n = r_1 + 2r_2$$

over $\mathbb{R}$. A point of V can be given by the $(r_1 + r_2)$-tuple

$$(x_1, x_2, \dots, x_{r_1}, x_{r_1+1}, \dots, x_{r_1+r_2}), \quad x_i \in \begin{cases} \mathbb{R} & 0 < i \leq r_1 \\ \mathbb{C} & r_1 < i \leq r_1 + r_2. \end{cases}$$

We, however, prefer to designate the point x by the coordinates:

$$(x) = (x_1, x_2, \dots, x_{r_1+r_2}, x_{r_1+r_2+1}, \dots, x_n),$$

where the additional (superfluous) r_2 coordinates are the complex conjugates of the complex coordinates:

$$x_{r_1+r_2+i} = \bar{x}_{r_1+i} \quad (i = 1, 2, \dots, r_2).$$

Thus, for $n = 2$ our space can be either the real or the complex plane, while for $n = 3$ it must be either the usual 3-dimensional space or a space spanned by a single real axis and a complex plane.

We define norm and trace for elements $x \in V$:

$$Nx = x_1 x_2 \dots x_n, \qquad Sx = x_1 + x_2 + \cdots + x_n.$$

We will later have to consider the existence of lattice points in that part of the space defined by $|Nx| \leq t$, for real, positive t ($|\ |$ here, and in what follows, signifies the usual absolute value.) The Minkowski lattice point theorem is, since this is generally not a convex domain, not immediately applicable. We will therefore inscribe a convex domain as large as possible into this part of the space, and, carry out our considerations for this domain.

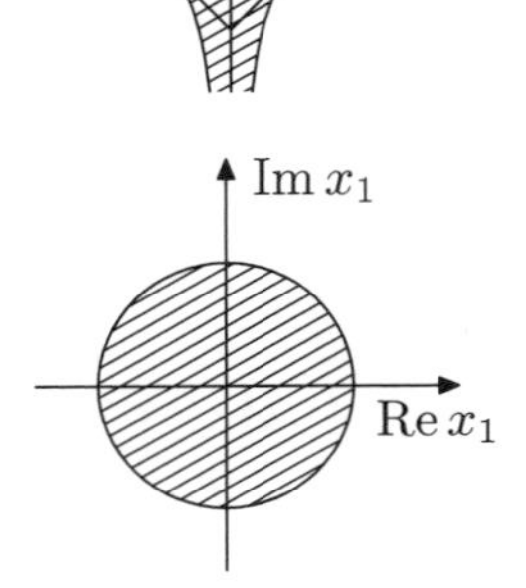

EXAMPLES. $n = 2$: *The real plane*:

$$|Nx| = |x_1| \cdot |x_2| \leq t.$$

This yields a non-convex area, bounded by hyperbolae, into which a square is then inscribed.

The complex plane:

$$|Nx| = |x_1|^2 \leq t.$$

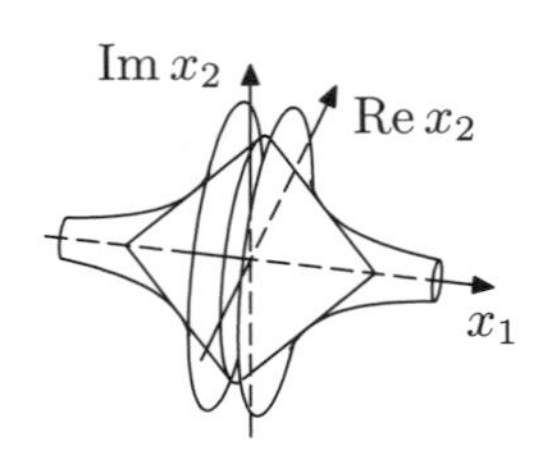

This yields the interior and boundary of a circle, already a convex domain.

$n = 3$: 3-*dimensional space*:

$$|Nx| = |x_1| \cdot |x_2| \cdot |x_3| \leq t.$$

This case is analogous to two dimensional real spaces. The figure to be inscribed here is a cube.

Complex plane and real axis:

$$|Nx| = |x_1| \cdot |x_2|^2 \leq t.$$

In this case we get a figure with rotational symmetry. Two cones with coinciding bases can be inscribed.

In general, we define the convex domain I by the inequality

$$\sum_{\nu=1}^{n} |x_\nu| \leq t,$$

and proceed to the calculation of the volume, μ, of this domain.

Let dv be the volume element in V. Then

$$\mu = \int_I dv.$$

Since the domain is symmetric with respect to the r_1 real axes:

$$\mu = 2^{r_1} \int_{I'} dv, \quad I' = \left\{ x \mid x_1, \ldots, x_{r_1} \geq 0, \sum_{\nu=1}^{n} |x_\nu| \leq t \right\}.$$

Since our domain is rotation symmetric with respect to the complex coordinates, we replace these with polar coordinates:

$$x_\nu = \frac{1}{2}\rho_\nu \cos\varphi_\nu + \frac{1}{2} i\rho_\nu \sin\varphi_v \quad (\nu = r_1+1, \ldots, r_1+r_2)$$
$$x_{r_1+r_2+i} = \bar{x}_{r_1+i} \quad (i = 1, 2, \ldots, r_2)$$

The functional determinant with respect to the transformation is

$$\begin{vmatrix} \frac{1}{2}\cos\varphi_\nu & \frac{1}{2}\sin\varphi_\nu \\ -\frac{1}{2}\rho_\nu \sin\varphi_\nu & \frac{1}{2}\rho_v \cos\varphi_\nu \end{vmatrix} = \frac{1}{4}\rho_\nu.$$

If the integration over φ $(0 \leq \varphi \leq 2\pi)$ is carried out, the result is

$$\frac{2^{r_1}(2\pi)^{r_2}}{2^{2r_2}} \int_{I''} \rho_{r_1+1} \cdots \rho_{r_1+r_2}\, dx_1 \ldots dx_{r_1}\, d\rho_{r_1+1} \ldots d\rho_{r_1+r_2}$$

$$I'' = \left\{ x \mid x_1, \ldots x_{r_1}, \rho_{r_1+1}, \ldots, \rho_{r_1+r_2} \geq 0, \sum_{\nu=1}^{r_1} x_\nu + \sum_{r_1+1}^{r_1+r_2} \rho_\nu \leq t \right\}$$

This integral, which now has only real variables, is of the form

$$C(\alpha_1, \ldots, \alpha_m; t) = \int_{J_t} z_1^{\alpha_1} \ldots z_m^{\alpha_m}\, dz_1 \ldots dz_m, \quad J_t = \left\{ Z \mid z_\nu \geq 0, \sum z_\nu \geq t \right\}.$$

Setting $z_\nu = tu_\nu$ to this integral yields

$$C(\alpha_1, \ldots, \alpha_m; t) = t^{\alpha_1 + \cdots + \alpha_m + m} \int_{J_1} z_1^{\alpha_1} \ldots z_m^{\alpha_m}\, dz_1 \ldots dz_m.$$

The remaining integral, $C(\alpha_1, \dots, \alpha_m)$ can be treated recursively:

$$C(\alpha_1, \dots, \alpha_m) = \int_0^1 z_m^{\alpha_m} \left(\int_{J'} z_1^{\alpha_1} \dots z_{m-1}^{\alpha_{m-1}} \, dz_1 \dots dz_{m-1} \right) dz_m,$$

$$J' = \left\{ (z_1, \dots, z_{m-1}) \mid z_\nu \geq 0, \ \sum_{\nu=1}^{m-1} z_\nu \leq 1 - z_m \right\}$$

$$= \int_0^1 z_m^{\alpha_m} C(\alpha_1, \dots, \alpha_{m-1}; 1 - z_m) \, dz_m$$

$$= C(\alpha_1, \dots, \alpha_{m-1}) \int_0^1 z_m^{\alpha_m} (1 - z_m)^{\alpha_1 + \dots + \alpha_{m-1} + m - 1} dz_m$$

$$= C(\alpha_1, \dots, \alpha_{m-1}) \frac{\Gamma(\alpha_m + 1) \Gamma(\alpha_1 + \dots + \alpha_{m-1} + m)}{\Gamma(\alpha_1 + \dots + \alpha_m + m + 1)}.$$

Since, finally,

$$C(\alpha_1) = \int_{0 \leq z_1 \leq 1} z_1^{\alpha_1} dz_1 = \frac{1}{\alpha_1 + 1} = \frac{\Gamma(\alpha_1 + 1)}{\Gamma(\alpha_1 + 2)},$$

we can see, by induction, that

$$C(\alpha_1, \dots, \alpha_m) = \frac{\Gamma(\alpha_1 + 1) \Gamma(\alpha_2 + 1) \dots \Gamma(\alpha_m + 1)}{\Gamma(\alpha_1 + \alpha_2 + \dots + \alpha_m + m + 1)}.$$

For our integral we must set: $m = r_1 + r_2$, $\alpha_i = 0$ $(0 < i \leq r_1)$, $\alpha_i = 1$ $(r_1 < i \leq r_1 + r_2)$: our domain, characterized by $\sum |x_\nu| \leq t$ has the volume

$$\mu = \frac{2^{r_1 - r_2} \pi^{r_2} t^n}{n!}.$$

This domain is convex, since for any z on the line between x and y:

$$\sum |x_\nu| \leq t, \quad \sum |y_\nu| \leq t, \quad z = \lambda x_v + (1 - \lambda) y_\nu \ (\nu = 1, \dots, n), \ 0 \leq \lambda \leq 1$$

$$|z_\nu| \leq \lambda |x_\nu| + (1 - \lambda) |y_\nu| \quad (\nu = 1, \dots, n)$$

$$\implies \sum_1^n |z_\nu| \leq \lambda t + (1 - \lambda) t = t$$

so that z also lies within the domain, to which, since it is also point-symmetric, the Theorem of Minkowski is applicable.

5. Approximation of the Class Number

We will use the above investigations to approximate the number of classes of ideals in an algebraic number field k of degree n. Let there exist, among the fields conjugate to k over $\mathbb{Q}$, r_1 real and r_2 complex number fields. To each $\beta \in k$ there corresponds a point (vector)

$$(\beta^{(1)}, \beta^{(2)}, \dots, \beta^{(n)}) \in V = \mathbb{R}^{r_1} \mathbb{C}^{r_2},$$

where the $\beta^{(\nu)}$ are conjugates of β in the order described in the last section. The sum of two elements of k corresponds to the sum of the corresponding vectors.

If $\mathfrak{a}$ an ideal of k, it can be represented in the form

$$\mathfrak{a} = \alpha_1 \mathbb{Z} + \alpha_2 \mathbb{Z} + \dots + \alpha_n \mathbb{Z}$$

with the ideal basis $\alpha_1, \ldots, \alpha_n$. The image of $\mathfrak{a}$ in V is spanned by the images of the α_i. To show that these form a lattice, we calculate the volume of the parallelepiped which they span, which is the absolute value of the determinant

$$\begin{vmatrix} \alpha_1^{(1)} & \alpha_1^{(2)} & \ldots & \alpha_1^{(n)} \\ \alpha_2^{(1)} & \alpha_2^{(2)} & \ldots & \alpha_2^{(n)} \\ \cdots & \cdots & \cdots & \cdots \\ \alpha_n^{(1)} & \alpha_n^{(2)} & \ldots & \alpha_n^{(n)} \end{vmatrix}$$

after the (r_1+i)-th column has been replaced by the column of its real parts, and the (r_1+r_2+i)-th column has been replaced by the imaginary parts of the (r_1+i)-th column for all $i = 1, 2, \ldots, r_2$. The volume is thus the absolute value of

$$\pm \left(i\frac{1}{2}\right)^{r_2} (D(\mathfrak{a}))^{\frac{1}{2}},$$

and remembering that $D(\mathfrak{a}) = (N\mathfrak{a})^2 D$, this equals

$$2^{-r_2} N\mathfrak{a} |D|^{\frac{1}{2}}.$$

Since the above term $\neq 0$, the image of $\mathfrak{a}$ is, in fact, a lattice in V with that loop volume.

To guarantee that a point of the lattice lies in the domain given by

$$\sum_{\nu=1}^{n} |\beta^{(\nu)}| \leq t$$

we need only choose t to satisfy

$$\frac{2^{r_1-r_2}\pi^{r_2}t^n}{n!} = 2^{n-r_2} N\mathfrak{a}|D|^{\frac{1}{2}} \implies t^n = \left(\frac{4}{\pi}\right)^{r_2} n!\, N\mathfrak{a}|D|^{\frac{1}{2}}.$$

Thus there exists a $\beta \neq 0$, $\beta \in \mathfrak{a}$, with

$$\left(\sum_{\nu=1}^{n} |\beta^{(\nu)}|\right)^n \leq \left(\frac{4}{\pi}\right)^{r_2} n!\, N\mathfrak{a}|D|^{\frac{1}{2}}.$$

But by the theorem of the geometric and arithmetic mean,

$$\prod_{\nu=1}^{n} |\beta^{(\nu)}| \leq \left(\frac{\sum |\beta^{(\nu)}|}{n}\right)^n,$$

and it follows that every ideal $\mathfrak{a}$ contains an element β such that

$$|N\beta| \leq \left(\frac{4}{\pi}\right)^{r_2} \frac{n!}{n^n} N\mathfrak{a}|D|^{\frac{1}{2}}.$$

As an incidental result we see that sign $D = (-1)^{r_2}$. For, the loop volume in the case of the unity ideal is $\pm\left(\frac{i}{2}\right)^{r_2} D^{\frac{1}{2}}$, and by squaring we see that $(-1)^{r_2}D > 0$.

For the theory it would have sufficed to prove the existence of an element $\beta \neq 0$ in $\mathfrak{a}$, and some constant C, dependent only upon k, such that

$$|N\beta| \leq C \cdot N\mathfrak{a}.$$

This weaker assertion can be proved more easily:

Let $\mathfrak{a}$ be an integral ideal, and $\omega_1, \omega_2, \ldots, \omega_n$ any linearly independent integers in k. Any number ξ of k can then be put in the form

$$\xi = x_1\omega_1 + x_2\omega_2 + \cdots + x_n\omega_n \quad (x_i \in \mathbb{Q}),$$

and then $N\xi$ is an n-th degree polynomial in the x_i. Thus

$$|N\xi| \leq C \cdot (\mathrm{Max}_i |x_i|)^n \quad \text{with some constant } C.$$

If we let each of the x_i take on all values $0, 1, \ldots, [\sqrt[n]{N\mathfrak{a}}]$, then ξ runs through more than $N\mathfrak{a}$ numbers in $\mathfrak{o}$, and obviously there exists among these a pair ξ_1, ξ_2:

$$\xi_1 \neq \xi_2, \quad \xi_1 - \xi_2 \in \mathfrak{a}$$

$$\Longrightarrow \eta = \xi_1 - \xi_2 = y_1\omega_1 + y_2\omega_2 + \cdots + y_n\omega_n \neq 0, \quad |y_i| \leq \sqrt[n]{N\mathfrak{a}}$$

$$|N\eta| \leq C\, N\mathfrak{a}, \quad \eta \neq 0,\ \eta \in \mathfrak{a}.$$

This inequality being homogenous, there also exists such a $\eta \in \mathfrak{a}$ for fractional ideals $\mathfrak{a}$.

¿From this result, we can, without difficulty, prove that the number h of classes of ideals of an algebraic number field of degree n is finite.

Let $\mathfrak{K}$ be any ideal class (or divisor class), and let $\mathfrak{a}$ be an ideal in the class $\mathfrak{K}^{-1}$ inverse to $\mathfrak{K}$. We can then find a $\beta \in \mathfrak{a}$, $\beta \neq 0$, with $|N\beta| \leq C \cdot N\mathfrak{a}$. Now, we have $\beta\mathfrak{o} \in \mathfrak{a} \Longrightarrow \beta\mathfrak{o} = \mathfrak{a}\mathfrak{b}$, $\mathfrak{b}$ integral. $\mathfrak{b}$ belongs to the class $\mathfrak{K}$, and since

$$|N\beta| = N\mathfrak{a} \cdot N\mathfrak{b} \leq C \cdot N\mathfrak{a}$$

we have: there exists an integral ideal $\mathfrak{b}$ in $\mathfrak{K}$ with

$$N\mathfrak{b} \leq C \quad (C \text{ can be set } = \left(\frac{4}{\pi}\right)^{r_2} \frac{n!}{n^n} |D|^{\frac{1}{2}}).$$

$N\mathfrak{b} \in \mathbb{Z}$, and it is immediately clear that to any integral norm there can be only a finite number of integral ideals in k (every valuation of $\mathbb{Q}$ having only finite many extensions onto k). Thus, the class number, h, is finite.

6. Examples

As an example we consider a cubic number field with discriminant $D < 0$. Since sign $D = (-1)^{r_2}$, we have $r_2 = 1$, $r_1 = 1$. The Minkowski constant,

$$C = \frac{4}{\pi} \frac{3!}{3} |D|^{\frac{1}{2}} = \frac{8}{9\pi} |D|^{\frac{1}{2}} < 0.283\, |D|^{\frac{1}{2}}.$$

For $|D| \leq 49$ we have $C < 2$, thus for cubic fields with $-49 < D < 0$ the class number $h = 1$; this holds for the cubic fields discussed earlier, with $D = -23$, -31, and -44.

For the cubic field with $D = -503$ we can set $C = 6.35$. It thus suffices to consider integral ideals, in fact only the prime ideals, with norm ≤ 6. We demonstrated earlier that these are all principal ideals; thus here too, $h = 1$.

We want to examine the cubic field with discriminant

$$D = -4027,$$

generated by a root of of the cubic equation

$$x^3 + 10x + 1 = 0.$$

The Minkowski bound for this case is < 18. A minimal basis is formed by $1, \alpha, \alpha^2$. That α is a unit is seen immediately from the equation. As in earlier cases, we factor the elements $\alpha + a$, $a \in \mathbb{Z}$, limiting ourselves to a residue system a modulo 17, say $-8, -7, \ldots, 0, \ldots, 8$. All prime ideals occuring here must be of the first degree, for the corresponding place is uniquely determined from the place of the base field by the condition

$$\alpha \to -a,$$

and $-a$ is the the residue class field of the base field. Thus,

$\alpha - 1 = \mathfrak{p}_2^2\mathfrak{p}_3$ for $N(\alpha-1) = 12 = 2\cdot 2\cdot 3$, and both prime ideals with norm 2 must be identical, the corresponding place being determined uniquely by $2 \to 0$ and $\alpha \to 1$.

$\alpha + 1 = \mathfrak{p}_2\mathfrak{p}_5$ for $N(\alpha+1) = 10 = 2\cdot 5$, and the prime ideal $\mathfrak{p}|2$ occuring here must be the same as above, because $1 \equiv -1 \pmod 2$.

$\alpha - 2 = \mathfrak{p}_{29}$

$\alpha + 2 = \mathfrak{p}_3^3$ for $N(\alpha+2) = 27 = 3\cdot 3\cdot 3$, and the prime ideal is uniquely determined by $3 \to 0$, $\alpha \to -2$.

Similar arguments yield:

$$\begin{array}{llll} \alpha - 3 = \mathfrak{p}_2\mathfrak{p}'_{29}, & \alpha + 3 = \mathfrak{p}_2^3\mathfrak{p}_7, & \alpha - 6 = \mathfrak{p}_{277}, & \alpha + 6 = \mathfrak{p}_{11}\mathfrak{p}_5^2, \\ \alpha - 4 = \mathfrak{p}_3\mathfrak{p}_5\mathfrak{p}_7, & \alpha + 4 = \mathfrak{p}_{103}, & \alpha - 7 = \mathfrak{p}_2\mathfrak{p}_3^2\mathfrak{p}_{23}, & \alpha + 7 = \mathfrak{p}_2^2\mathfrak{p}'_{103}, \\ \alpha - 5 = \mathfrak{p}_2^4\mathfrak{p}_{11}, & \alpha + 5 = \mathfrak{p}_2\mathfrak{p}_3\mathfrak{p}''_{29}, & \alpha - 8 = \mathfrak{p}_{593}, & \alpha + 8 = \mathfrak{p}_3\mathfrak{p}_{197}. \end{array}$$

The prime numbers 13 and 17 do not occur, and thus remain prime, and are principal ideals (divisors). All other prime numbers $p \leq 11$ show just one factor (1st degree), so that their decompositions must be

$$p = \mathfrak{p}_p\mathfrak{p}'_p \quad (\mathfrak{p}'_p \text{ of the } 2^{\text{nd}} \text{ degree}).$$

Denoting by $\mathfrak{a} \sim \mathfrak{b}$ that $\mathfrak{a}$ and $\mathfrak{b}$ lie in the same class (i.e. $\mathfrak{a}\mathfrak{b}^{-1} \sim 1$), we see, by comparing our decompositions:

$$\mathfrak{p}_2^6 = \frac{(\alpha-1)^3}{\alpha+2} \implies \mathfrak{p}_2^6 \sim 1$$

$$\mathfrak{p}_3\mathfrak{p}_2^2 \sim 1, \quad \mathfrak{p}_3 \sim \mathfrak{p}_2^4, \quad \mathfrak{p}_5 \sim \mathfrak{p}_2^5, \quad \mathfrak{p}_7 \sim \mathfrak{p}_2^3, \quad \mathfrak{p}_{11} \sim \mathfrak{p}_2^2.$$

All prime ideals $\mathfrak{p}_3, \mathfrak{p}_5, \mathfrak{p}_7, \mathfrak{p}_{11}$, are thus equivalent to a power of $\mathfrak{p}_2$, and since

$$\mathfrak{p}'_p = \frac{p}{\mathfrak{p}_p}$$

this also holds for the prime ideals $\mathfrak{p}'_3, \mathfrak{p}'_5, \mathfrak{p}'_7$, and $\mathfrak{p}'_{11}$. Because $\mathfrak{p}_2^6 \sim 1$ we have $h \leq 6$. One is led to believe that $h = 6$; one would have to show that neither $\mathfrak{p}_2^2$ or $\mathfrak{p}_2^3$ is a principal ideal. We will later see that this question is connected with the question whether α is the smallest unit.

In a similar manner we can, by comparing factorizations, find certain units, e.g. since

$$\frac{(\alpha+1)^2(\alpha-5)}{(\alpha+6)\frac{(\alpha-1)^3}{\alpha+2}} = \frac{\mathfrak{p}_2^2\mathfrak{p}_5^2\mathfrak{p}_2^4\mathfrak{p}_{11}}{\mathfrak{p}_{11}\mathfrak{p}_5^2\mathfrak{p}_2^6} = 1$$

and thus $\frac{(\alpha+1)^2(\alpha+2)(\alpha-5)}{(\alpha-1)^3(\alpha+6)}$ is a unit.

REMARK. Since the norm of an integral ideal with respect to the rational number field must be at least 1, the above investigations yield: there exists an integral ideal of k with

$$1 \leq N\mathfrak{b} \leq \left(\frac{4}{\pi}\right)^{r_2} \frac{n!}{n^n} \sqrt{|D|}.$$

This gives us an approximation for the discriminant:

$$|D| \geq \left(\frac{\pi}{4}\right)^{r_2} \frac{n^{2n}}{(n!)^2} \geq \left(\frac{\pi}{4}\right)^{n} \cdot \frac{n^{2n}}{(n!)^n} = a_n.$$

Then,

$$\frac{a_{n+1}}{a_n} = \frac{\pi}{4} \frac{(n+1)^{2n}}{(n!)^2} \cdot \frac{(n!)^2}{n^{2n}} = \frac{\pi}{4} \left(1 + \frac{1}{n}\right)^{2n} \geq \frac{\pi}{4} 2^2 = \pi.$$

This shows that the sequence of the a_n increases monotonically, and, since $a_2 = \frac{\pi^2}{4} > 1$, that $a_n > 1$ for $n \geq 2$. Thus for any algebraic number field of degree $n \geq 2$ over $\mathbb{Q}$ we have

$$|D| > 1.$$

This inequality, long conjectured last century, could finally be proved by Minkowski by the above considerations.

¿From this inequality it follows that in any algebraic number field of degree $n \geq 2$ there exist natural prime numbers that are ramified in that field (those primes that are factors of the discriminant). This theorem no longer holds if the base field is not the field of rationals, as is demonstrated by the field $k(i)$, with $i^2 = -1$, which is unramified over $k = \mathbb{Q}(\sqrt{5})$.

CHAPTER 13

THE UNIT THEOREM

The map of non-zero elements of the algebraic number field k into the divisors maps the units of the ring $\mathfrak{o}$ of integers onto the unity divisor. Thus, if ε is a unit in $\mathfrak{o}$ then

$$N_{k|\mathbb{Q}}\varepsilon = \pm 1$$

for ± 1 are the only rational units. If, on the other hand, $N_{k|\mathbb{Q}} = \pm 1$, $\varepsilon \in \mathfrak{o}$, then ε is a unit.

The units form a multiplicative group, E, which we want to examine more closely here. We will alway assume that the degree of k over $\mathbb{Q}$ is at least 2. r_1 is the number of real, r_2 the number of complex fields conjugate to k.

1. Upper Bound for the "Number" of Units

As it is easier to study an additive than a multiplicative group, we map E onto an additive group.

Let ε be any unit. We map ε onto the vector

$$(\varepsilon^{(1)}, \varepsilon^{(2)}, \dots \varepsilon^{(n)}) \quad (\varepsilon^{(1)} = \varepsilon),$$

with the conjugates of ε as components, ordered in the manner described previously. This map is an isomorphism. We reduce it to a homomorphism by replacing the components of the vectors by their absolute values:

$$(|\varepsilon^{(1)}|, |\varepsilon^{(2)}|, \dots, |\varepsilon^{(n)}|),$$

and finally eliminate the superfluous last r_2 components, which now, coincide pairwise with the previous r_2 components:

$$(|\varepsilon^{(1)}|, |\varepsilon^{(2)}|, \dots, |\varepsilon^{(r_1+r_2)}|),$$

Even here one component is superfluous, since $|N\varepsilon| = 1$, which implies

$$|\varepsilon^{(1)}|^{\delta_1} |\varepsilon^{(2)}|^{\delta_2} \dots |\varepsilon^{(r_1+r_2)}|^{\delta_{r_1+r_2}} = 1, \qquad \delta = \begin{cases} 1, & 0 < i \le r_1 \\ 2, & r_1 < i \le r_1 + r_2. \end{cases}$$

Thus, we can eliminate the last component. To at last get to an additive group we take the logarithm, and set $\delta_i \log |\varepsilon^{(i)}| = l_i(\varepsilon)$ for $i = 1, 2, \dots, r_1 + r_2$.

$$L \colon \varepsilon \to L(\varepsilon) = (l_1(\varepsilon), l_2(\varepsilon), \dots, l_r(\varepsilon)),$$

where $r = r_1 + r_2 - 1$, maps the unit group E homomorphically onto the additive group $L(E)$ of vectors of a vector space V of dimension $r = r_1 + r_2 - 1$.

Assertion. $L(E)$ is a lattice in V, and the kernel of L is a finite cyclic group (hence composed of roots of unity).

The proof is based upon the boundedness of the inverse image of a compact set of V under L. Let K be such a compact set of V, and $\varepsilon \in L^{-1}(K)$, then $l_i(\varepsilon)$ is bounded from above and below, for all i, and because

$$l_{r+1}(\varepsilon) = -l_1(\varepsilon) - l_2(\varepsilon) - \cdots - l_r(\varepsilon),$$

$l_{r+1}(\varepsilon)$ is also bounded. (If $r_1 + r_2 = 1$, hence $r = 0$, this statement remains true.) This implies, however, that

$$|\varepsilon^{(i)}| \leq M \quad \text{for } i = 1, 2, \ldots, r, r+1, \ldots, n.$$

That $L^{-1}(L(E) \cap K))$ is a finite set follows from the general theorem: The set of all algebraic integers of degree $\leq n$ whose conjugates are bounded by a fixed constant is finite. For, such a number satisfies an algebraic equation of degree $\leq n$, and the boundedness of the conjugates implies the boundedness of the rational integral coefficients of the defining equation, so that there can be only a finite number of such equations.

Thus, $L(E)$ is a lattice in V of dimension $\leq r$. The kernel of L is a finite group $(L^{-1}(0))$. But, a finite multiplicative group of elements of a field must consist of only roots of unity, and be cyclic.

Let w be the order of the kernel of L, which then consists of all w-th roots of unity. Since in any field the numbers ± 1 belong to this kernel, w is always an even number. In real fields $w = 2$, for ± 1 are the only roots of unity. For most non-real number fields we also have $w = 2$.

Let $\varepsilon_1, \varepsilon_2, \ldots, \varepsilon_s$ $(s \leq r)$ be units of k such that their images $L(\varepsilon_1), \ldots, L(\varepsilon_s)$ generate the lattice $L(E)$. Let ε be an arbitrary unit in E. Then,

$$L(\varepsilon) = x_1 L(\varepsilon_1) + x_2 L(\varepsilon_2) + \cdots + x_s L(\varepsilon_s), \quad x_i \text{ rational integers,}$$

$$\Longrightarrow L(\varepsilon) = L(\varepsilon_1^{x_1} \varepsilon_2^{x_2} \ldots \varepsilon_s^{x_s}).$$

If two units have the same image in V, their difference is an element of the kernel. Thus every unit can be represented in the form

$$\varepsilon = \zeta \varepsilon_1^{x_1} \varepsilon_2^{x_2} \ldots \varepsilon_s^{x_s}, \quad \text{where } \zeta^w = 1,$$

and the rational integers $x_1, x_2, \ldots, x_s$ are uniquely determined by ε (even by $L(\varepsilon)$). The unit group is the direct product of a cyclic group of order w and a free group with s free generators. It remains to be shown that $s = r$.

2. Lower Bound for the "Number" of Units

In the following we will again designate with V the n-dimensional vector space introduced in the proof of the finiteness of the class number, i.e., the space with r_1 real coordinates, and r_2 complex coordinates and their conjugates:

$$x = (x_1, x_2, \ldots, x_n), \quad n = r_1 + 2r_2.$$

We again define $Nx = x_1, x_2, \ldots, x_n$. By identifying elements $\alpha \in k$ with vectors

$$(\alpha^{(1)}, \alpha^{(2)}, \ldots, \alpha^{(n)})$$

with conjugates $\alpha^{(i)}$ of α, k is embedded in V.

Fix $x \in V$ with $\frac{1}{2} \leq |Nx| \leq 1$, and consider the lattice

$$x\mathfrak{o} \subset V, \quad \mathfrak{o} = \text{ring of integers of } k.$$

If $\mathfrak{o} = \omega_1\mathbb{Z} + \omega_2\mathbb{Z} + \cdots + \omega_n\mathbb{Z}$ (i.e., if the ω_i form a minimal basis), then $x\mathfrak{o} = x\omega_1\mathbb{Z} + x\omega_2\mathbb{Z} + \cdots + x\omega_n\mathbb{Z}$.

$x\mathfrak{o}$ is a lattice, whose loop volume equals the absolute, value of the determinant

$$\begin{vmatrix} x_1\omega_1^{(1)} & x_2\omega_1^{(2)} & \dots & x_n\omega_1^{(n)} \\ x_1\omega_2^{(1)} & x_2\omega_2^{(2)} & \dots & x_n\omega_2^{(n)} \\ \dots & \dots & \dots & \dots \\ x_1\omega_n^{(1)} & x_2\omega_n^{(2)} & \dots & x_n\omega_n^{(n)} \end{vmatrix} = x_1 x_2 \dots x_n \begin{vmatrix} \omega_1^{(1)} & \omega_1^{(2)} & \dots & \omega_1^{(n)} \\ \omega_2^{(1)} & \omega_2^{(2)} & \dots & \omega_2^{(n)} \\ \dots & \dots & \dots & \dots \\ \omega_n^{(1)} & \omega_n^{(2)} & \dots & \omega_n^{(n)} \end{vmatrix}.$$

after each pair of complex conjugate columns is replaced by it real and imaginary parts. As in previous calculations, this yields the lattice loop volume

$$2^{-r_2}|D|^{\frac{1}{2}}|Nx|.$$

Thus, the lattice loop volume is bounded for all possible x.

Let T be a compact, point symmetric, convex subset of V with sufficiently large volume. Then, for any possible x, at least one non-zero point of the lattice lies in T, i.e. to each such x there exists a $\gamma \neq 0$, $\gamma \in \mathfrak{o}$, $x\gamma \in T$. The points of T, having bounded coordinates, have bounded norms:

$$x\gamma \in T \implies |Nx\, N\gamma| \leq M \implies |N\gamma| \leq \frac{M}{Nx} \leq 2M.$$

Consider, for every x, the corresponding γ and corresponding principal ideal $\gamma\mathfrak{o}$. We have

$$N(\gamma\mathfrak{o}) = |N\gamma| \leq 2M,$$

and since $\gamma\mathfrak{o}$ is an integral ideal, there can only be a finite number of such $\gamma\mathfrak{o}$, say $\gamma_1\mathfrak{o}, \gamma_2\mathfrak{o}, \dots, \gamma_t\mathfrak{o}$. Thus, if γ corresponds to x, then $\gamma\mathfrak{o} = \gamma_i\mathfrak{o}$ for some $i \in 1, 2, \dots, t$, i.e. there exists a unit such that $\gamma = \gamma_i\varepsilon$. Thus, for any $x \in V$ with $\frac{1}{2} \leq |Nx| \leq 1$ we can find an $i \in 1, 2, \dots, t$ with $\gamma_i\varepsilon x \in T$:

$$\varepsilon x \in \frac{1}{\gamma_1}T \cup \frac{1}{\gamma_2}T \cup \dots \cup \frac{1}{\gamma_t}T.$$

Since the points of this set have bounded coordinates, we see that to each x with $\frac{1}{2} \leq |Nx| \leq 1$ there exists a unit ε such that all coordinates of εx are bounded (uniformly in x).

As there is nothing further to prove in case $r = 0$ (E is then a cyclic group of order w),[1] we can assume that $r = r_1 + r_2 - 1 \geq 1$. We choose an i with $1 \leq i \leq r + 1 = r_1 + r_2$ and an $x = (x_1, x_2, \dots, x_n)$ all of whose coordinates except x_i (and x_{i+r_2}, if $i > r_1$) are very large, and x_i (and x_{i+r_2}) sufficiently small so that $Nx = 1$. There then exists a unit $\varepsilon_i \in E$ such that $\varepsilon_i x$ has bounded coordinates,

$$\begin{aligned} &|\varepsilon_i^{(j)}| < 1 \quad \text{for } 1 \leq j \leq r_1 + r_2,\ j \neq i \\ \implies &l_j(\varepsilon_i) < 0 \quad \text{for } j \neq i. \end{aligned}$$

Assertion. $L(\varepsilon_1), L(\varepsilon_2), \dots, L(\varepsilon_r)$ are independent vectors of the lattice $L(E)$.

We must prove that the determinant

$$\begin{vmatrix} l_1(\varepsilon_1) & l_2(\varepsilon_1) & \dots & l_r(\varepsilon_1) \\ l_1(\varepsilon_2) & l_2(\varepsilon_2) & \dots & l_r(\varepsilon_2) \\ \dots & \dots & \dots & \dots \\ l_1(\varepsilon_r) & l_2(\varepsilon_r) & \dots & l_r(\varepsilon_r) \end{vmatrix} \neq 0.$$

[1] If $r_1 + r_2 = 1$, then either $r_1 = 1$, $r_2 = 0$ and $k = \mathbb{Q}$, or $r_1 = 0$, $r_2 = 1$ and k is imaginary quadratic. In both cases, E is fnite and thus cyclic. *Editor's note.*

All elements of the determinant except the principal diagonal are negative, and

$$\sum_{\nu=1}^{r+1} l_\nu(\varepsilon_i) = 0$$

$$\Longrightarrow \sum_{\nu=1}^{r} l_\nu(\varepsilon_i) = -l_{r+1}(\varepsilon_i) > 0$$

for $i = 1, 2, \ldots, r$, i.e. the row sums are positive. The assertion then immediately follows from the following general theorem of Minkowski:

If (a_{ij}) is a real r-row square matrix, with the properties:

1.) $a_{ij} < 0$ for $j \neq i$,
2.) $\sum_{\nu=1}^{r} a_{i\nu} > 0$ for $i = 1, 2, \ldots, r$,

then $|(a_{ij})| \neq 0$. (Minkowski even shows that $|(a_{ij})| > 0$).

For, if $|(a_{ij})| = 0$, then the system of equations

$$\sum_{\nu=1}^{r} a_{i\nu} x_\nu = 0 \quad (i = 1, 2, \ldots, r)$$

has a non-trivial solution, x_j $(j = 1, 2, \ldots, r)$. Let

$$|x_i| = \operatorname{Max}(|x_1|, |x_2|, \ldots, |x_r|).$$

Because of the homogeneity of the equations, we can assume $x_i = 1$. But then $|x_j| \leq 1$ for all j, which yields a contradiction in the i-th equation:

$$0 = \sum_{\nu} a_{i\nu} x_\nu = a_{ii} + \sum_{\nu \neq i} a_{i\nu} x_\nu \geq a_{ii} + \sum_{\nu \neq i} a_{iv} > 0.$$

Then independence of the lattice vectors $L(\varepsilon_i)$ is demonstrated. Together with the previously shown, this proves that the lattice $L(E)$ is r-dimensional. The group of units E is the direct product of a finite cyclic group and a free abelian group with r generators. A system of units, $\varepsilon_1, \varepsilon_2, \ldots, \varepsilon_r$, whose images under the map L generate the lattice $L(E)$ are termed *base units*. The absolute value of the determinant $l_j(\varepsilon_i)$ is called the *regulator* of the field k in analytic number theory; it is independent of the choice of base units.

EXAMPLES. 1.) $r = 0$. a) $r_1 = 1$, $r_2 = 0$. k is the rational number field. $E = \{+1, -1\}$.

b) $r_1 = 0$, $r_2 = 1$. k is an imaginary quadratic number field. E consists of the roots of unity of k. Let $k = \mathbb{Q}(\sqrt{m})$ with negative integer m without square factors. If $m \equiv 2, 3 \pmod 4$ then $1, \sqrt{m}$ is a minimal basis. If $m = -1$ (Gaussian number field), then E contains the 4th roots of unity, $\pm i, \pm 1$; otherwise only ± 1. If $m \equiv 1 \pmod 4$ then $1, \frac{1+\sqrt{m}}{2}$ is a minimal basis; for $m = -3$, E consists of the 6th roots of unity, otherwise it contains only ± 1.

2.) $r = 1$. a) $r_1 = 2$, $r_2 = 0$. k is a real quadratic number field. As in any real field, the only roots of unity are ± 1. Aside from this factor of ± 1, all units are powers of some single base unit.

b) $r_1 = 1$, $r_2 = 1$. k is of the third degree and $D < 0$, Here too, ± 1 are the only roots of unity, and, up to this factor, E consists of all powers of a fixed element.

c) $r_1 = 0$, $r_2 = 2$. $n = 4$, k is a totally imaginary biquadratic number field. Up to the factor of the cyclic group of roots of unity contained in k, E again is generated by a single base unit.

3. Actual Calculation of Base Units

Certain algorithms have been developed for the calculation of the base units of quadratic number field, and these are to be found in the literature. We will here carry out the determination of base units for a special type of real cubic field, k, characterized by

$$r_1 = 1, \quad r_2 = 1, \quad \text{i.e. } r = 1.$$

Then, $n = 3$, and $D < 0$. The only roots of unity are ± 1, and the group of units is

$$E = \{\pm\varepsilon^\nu\}$$

with some base unit ε, which we now want to calculate.

We may assume that $\varepsilon > 1$, since, if ε is a base unit, so are $\frac{1}{\varepsilon}$ and $-\varepsilon$. We seek a lower bound for ε. Let its conjugates be[2]

$$u^2 = \varepsilon, \; u^{-1}\varepsilon^{iv}, \; u^{-1}e^{-iv} \quad (0 \le v \le \pi)$$

and let Δ be the discriminant of the equation for ε. Then,

$$\begin{aligned} \Delta^{\frac{1}{2}} &= (u^2 - u^{-1}e^{iv})(u^2 - u^{-1}e^{-iv})(u^{-1}e^{iv} - u^{-1}e^{-iv}) \\ &= 2i(u^2 + u^{-3} - 2\cos v)\sin v. \end{aligned}$$

If we set $2\xi = u^3 + u^{-3}$, $x = \cos v$, then we get:

$$|\Delta|^{\frac{1}{2}} = 4(\xi - \cos v)\sin v,$$

which, for a given u, has a maximum where $\xi\cos v - \cos^2 v + \sin^2 v = 0$ or

$$-g(x) = \xi x - 2x^2 + 1 = 0, \quad |x| \le 1.$$

This maximum yields

$$|\Delta| \le 16(\xi^2 - 2\xi x + x^2)(1 - x^2),$$

and by applying the conditions $\xi x = 2x^2 - 1$, $\xi^2 x^2 = 4x^4 - 4x^2 + 1$, we have

$$\begin{aligned} |\Delta| &\le 16(\xi^2 + 1 - x^2 - x^4) \\ &= 4u^6 + 24 + 4(u^{-6} - 4x^2 - 4x^4). \end{aligned}$$

But, $g(1) = 1 - \xi < 0$ (because $u > 1$, $\xi = \frac{u^3+u^{-3}}{2} > 1$), and $g\left(-\frac{1}{2u^3}\right) = \frac{3}{4}(u^{-6} - 1) < 0$, it follows that $g(x) = 0$ has one root > 1, and that the desired root, with $|x| \le 1$ is $< \frac{1}{2u^3}$,

$$x^2 > \frac{1}{4u^6} \implies u^{-6} - 4x^2 < 0 \implies u^{-6} - 4x^2 - 4x^4 < 0.$$

Thus, $|\Delta| < 4u^6 + 24$, or

$$|\Delta| < 4\varepsilon^3 + 24.$$

The discriminant Δ is a square multiple of the field discriminant D, so that we certainly have

$$|D| < 4\varepsilon^3 + 24.$$

This approximation of the base unit is valuable help in its calculation.

---

[2]In this formula, u is just the positive square root of ε. It is not an element of k. *Editor's note.*

EXAMPLE. We consider the previously mentioned cubic number field, generated by a real root α of the equation

$$\alpha^3 + 10\alpha + 1 = 0; \quad D = -4027.$$

The inequality derived above yields, for the base unit here,

$$\varepsilon > 10.$$

α itself is a unit, and therefore so is $\beta = -\frac{1}{\alpha}$, which satisfies

$$\beta^3 - 10\beta^2 - 1 = 0 \implies 10 < \beta < 11.$$

β is, except possibly for its sign, a power of a base unit. But, this is possible only if β is itself such a base unit. But then, α is also a base unit and

$$E = \{\pm\alpha^\nu\}.$$

(The same holds true, of course, if one associates α with an imaginary root of the equation. Our restriction to real roots served the simplification of the notation of the proof.)

Now it is also easy to show that this field has class number

$$h = 6.$$

We had previously shown that

$$\mathfrak{p}_2^6 = \frac{(\alpha-1)^3}{\alpha+2}, \quad \text{and} \quad \mathfrak{p}_2 \mid 2,$$

and all that remained to be proved was that neither $\mathfrak{p}_2^3$ nor $\mathfrak{p}_2^2$ is a principal ideal.

But

$$\mathfrak{p}_2^3 = \gamma,\ \gamma \in k \implies \pm\alpha^\nu \frac{(\alpha-1)^3}{\alpha+2} = \gamma^2$$

for some ν and one of the signs. But then, one of the numbers

$$\pm\alpha^\delta \frac{\alpha-1}{\alpha+2} \quad (\delta = 0, 1)$$

would necessarily be a square in k, and then certainly a square modulo all prime numbers. To derive a contradiction we consider the prime ideal $\mathfrak{p}_{29}$, with $\mathfrak{p}_{29} \mid 29$, and defined by

$$\alpha \to 2 \pmod{29} \implies \alpha - 1 \to 1,\ \alpha + 2 \to 4,\ -1 \to -1.$$

The numbers 1, 4, and -1 are squares modulo 29, but 2 is not; hence δ would have to be 0. Since $\frac{\alpha-1}{\alpha+2}$ is negative, we immediately see by considering the absolute value that the negative sign would have to be chosen. Thus, only the number

$$-\frac{\alpha-1}{\alpha+2}$$

is left to be considered. The contradiction is found by considering the prime ideal $\mathfrak{p}_7$, with $\mathfrak{p}_7 \mid 7$, defined by $\alpha \to -3$. For then,

$$-\frac{\alpha-1}{\alpha+2} \to -4, \quad -4 \text{ not a square modulo } 7.$$

$\mathfrak{p}_2^3$ is not a principal ideal.

In a similar manner the supposition $\mathfrak{p}_2^2 = \gamma$ leads to a contradiction. For, this would imply that

$$\pm\alpha^\nu \frac{(\alpha-1)^3}{\alpha+2} = \gamma^3$$

from which it would follow that one of the numbers $\frac{\alpha^\nu}{\alpha+2}$ $(\nu = 0, 1, -1)$ is a cube. We consider such prime numbers p, modulo which not all prime numbers are cubes:

$$p \equiv +1 \pmod 3.$$

The considerations modulo $\mathfrak{p}_7$ show that

$$\alpha \to -3, \ \alpha + 2 \to -1, \ \alpha^\nu \to (-3)^\nu.$$

The number of cubes is one third the number of non-zero residue classes, and thus $(-3)^\nu$ can be a cube modulo 7 only for $\nu = 0$. Since -1 is a cube modulo 7, the problem is reduced to whether $\alpha+2$ is a cube in k. To answer this, we factor $\alpha+14$:

$$\alpha + 14 = \mathfrak{p}_3\mathfrak{p}_{31}^2$$

where $\mathfrak{p}_{31}$ is defined by $\alpha \to -14 \implies \alpha + 2 \to -12$.

Because $2 = 4^3 \pmod{31}$ the numbers 1, 2, 4, 8, -15, -1, -2, -4, -8, and 15 are cubes modulo 31. But, these are as many cubes as can exist modulo 31, and -12 does not occur, hence $\alpha + 2$ is no cube.

We have thus demonstrated that for this field the class number $h = 6$.

The method used in the above example to determine the base units fails for fields with small discriminants. For such fields one can use the following procedure. One lists all cubic equations

$$x^3 + a_1x^2 + a_2x - 1 = 0,$$

with integral coefficients a_1, a_2 and which have a real root between 1 and 1.5. This list will be finite, as the coefficients lie between fixed bounds. For each of these equations one calculates the roots between 1 and 1.5, and the corresponding field discriminant. This will lead to a list of cubic fields and their base units.

It should still be mentioned that for totally real cubic fields ($r_1 = 3$, $r_2 = 0$), an approximation of the two base units is arrived at more easily by use of the discriminant than by the method in our example. The Minkowski lattice point theorem is applied, and the regulator is approximated.

Papers

by Emil Artin

AXIOMATIC CHARACTERIZATION OF FIELDS BY THE PRODUCT FORMULA FOR VALUATIONS

EMIL ARTIN AND GEORGE WHAPLES

Introduction. The theorems of class field theory are known to hold for two kinds of fields: algebraic extensions of the rational field and algebraic extensions of a field of functions of one variable over a field of constants. We shall refer to these fields as number fields and function fields, respectively. For class field theory, the function fields must indeed be restricted to those with a Galois field as field of constants; however, we make this restriction only in §5, and until then consider fields with an arbitrary field of constants.

In proving these theorems, the product formula for valuations plays an important rôle. This formula states that, for a suitable set of inequivalent valuations $|\ |_{\mathfrak{p}}$,

$$\prod_{\mathfrak{p}} |\alpha|_{\mathfrak{p}} = 1$$

for all numbers $\alpha \neq 0$ of the field. For fields of the types mentioned, this product formula is easy to prove. After reviewing this proof (§1), we shall show (§2) that, conversely, the number fields and function fields are characterized by their possession of a product formula. Namely, we prove that if a field has a product formula for valuations, and if one of its valuations is of suitable type, then it is either a function field or a number field.

This shows that the theorems of class field theory are consequences of two simple axioms concerning the valuations, and suggests the possibility of deriving these theorems directly from our axioms. We do this in the later sections of this paper for the generalized Dirichlet unit theorem, the theorem that the class number is finite, and certain others fundamental to class field theory. This axiomatic method has the decided advantage of uniting the two cases; also, it simplifies the proofs. For example, we avoid the use of either ideal theory or the Minkowski theory of lattice points. Thus these two theories are unnecessary to class field theory, since they are needed only to prove the unit theorem.

1. Preliminaries on valuations.

If k is any field, then a function $|\alpha|$, defined for all $\alpha \in k$, is called a valuation of K if:

An address delivered by Professor Artin before the Chicago meeting of the Society on April 23, 1943, by invitation of the Program Committee; received by the editors February 3, 1945.

(1) $|\alpha|$ *is a real number not less than* 0, *and* $|\alpha|=0$ *only if* $\alpha=0$,

(2) $$|\alpha\beta| = |\alpha||\beta|,$$

(3) $$|\alpha+\beta| \leqq |\alpha| + |\beta|.$$

We call a valuation nonarchimedean if in addition to (3) it satisfies

(3′) $$|\alpha+\beta| \leqq \max(|\alpha|, |\beta|).$$

Note that the assumption that $|\alpha|$ is a real number eliminates the possibility of certain valuations discussed in various recent papers.

The theory of a field with respect to one given valuation is supposed to be known by the reader and shall be called the local theory. We review the most important facts. The valuation $|\alpha|=1$ for all $\alpha\neq 0$ is called the trivial valuation. Two nontrivial valuations $|\alpha|_1$ and $|\alpha|_2$ are called equivalent when $|\alpha|_1<1$ implies $|\alpha|_2<1$, and it is easy to show[1] that there is a positive real number ρ such that $|\alpha|_1^{\rho}=|\alpha|_2$ for all $\alpha\in k$. If $|\alpha|$ is a nonarchimedean valuation, then $|\alpha|^{\rho}$ is an equivalent valuation for any $\rho>0$. If $|\alpha|$ is archimedean and ρ positive, then $|\alpha|^{\rho}$ will be a valuation only for sufficiently small values of ρ. However, we shall in the remainder of this paper use the word "valuation" to mean any function $|\alpha|^{\rho}$ where ρ is any positive number and $|\alpha|$ is a true valuation.

A set of equivalent and nontrivial valuations of a field k is called a prime divisor of that field, and denoted by letters like p, $\mathfrak{p}$, $\mathfrak{P}$, q, $\mathfrak{q}$, $\mathfrak{Q}$, $\cdots$. If $\mathfrak{p}$ is a prime divisor, $|\alpha|_{\mathfrak{p}}$ stands for a particular, fixed valuation chosen from this set. The sign $\|\alpha\|_{\mathfrak{p}}$ will later be used to stand for another valuation of the same set.

If R is a subfield of k then each set $\mathfrak{p}$ of equivalent valuations of k is also a set of equivalent valuations of R. If these valuations are nontrivial on R then they define a prime divisor p of R, and $\mathfrak{p}$ is said to divide p: $\mathfrak{p}|p$. One p may be divisible by several $\mathfrak{p}$ of k. By well known methods,[2] the field k can be extended to the field $k_{\mathfrak{p}}$ which is completed with respect to the valuations of $\mathfrak{p}$. If R_p is the corresponding completion of R then R_p is a subfield of k and the degree $n(\mathfrak{p})=(k_{\mathfrak{p}}/R_p)$ is called the local degree. If k itself is a finite extension of R of degree n it is easy to prove[3] the inequality

[1] See van der Waerden [7, pp. 254–255], or Artin [1]. Numbers in brackets refer to the references cited at the end of the paper.

[2] van der Waerden [7, p. 250].

[3] To prove this, assume first that $k=R(\alpha)$, where α is a root of a polynomial $f(x)$, irreducible in R of degree n. Let $P(x)$ be the polynomial, irreducible in R_p of degree $n_{\mathfrak{p}}$, with root α. Since (van der Waerden [7, p. 264]) an extension field of R_p can be evaluated in only one way by a divisor $\mathfrak{p}$ of p, it follows that different divisors

$$\sum_{\mathfrak{p}|p} n(\mathfrak{p}) \leqq n \tag{4}$$

for each p of R.

In case $\mathfrak{p}$ is a discrete valuation, $n(\mathfrak{p})=e(\mathfrak{p})f(\mathfrak{p})$ where $e(\mathfrak{p})$ is the ramification number and $f(\mathfrak{p})$ the degree of the residue class field of k over that of R.

We proceed now to study a finite set of nontrivial inequivalent valuations $|\ |_1, |\ |_2, \cdots, |\ |_n$.

LEMMA 1. *If* $|\ |_1$ *and* $|\ |_2$ *are two nontrivial, inequivalent valuations of* k, *then there is a* $\gamma \in k$ *with* $|\gamma|_1<1$ *and* $|\gamma|_2>1$.

PROOF. Since the valuations are inequivalent there is an α with $|\alpha|_1<1$ and $|\alpha|_2 \geqq 1$ and a β with $|\beta|_1 \geqq 1$ and $|\beta|_2<1$. Take $\gamma=\alpha/\beta$.

LEMMA 2. *If* $|\ |_1, |\ |_2, \cdots, |\ |_n$ *are nontrivial and inequivalent there is an* $\alpha \in k$ *such that* $|\alpha|_1>1$ *and* $|\alpha|_\nu<1$ *for* $\nu=2, \cdots, n$.

PROOF. The lemma is true for $n=2$ by Lemma 1. We use induction, assuming that we have found a β such that $|\beta|_1>1$ and $|\beta|_\nu<1$ for $\nu=2, \cdots, n-1$. Choose γ so that $|\gamma|_1>1$ and $|\gamma|_n<1$. There are two cases:

Case 1. If $|\beta|_n \leqq 1$ let $\alpha=\beta^r\gamma$. Then $|\alpha|_1>1$ and, for r sufficiently large, $|\alpha|_2, |\alpha|_3, \cdots, |\alpha|_n$ are all less than 1.

Case 2. If $|\beta|_n>1$ let

$$\alpha = \frac{\beta^r}{\beta^r+1}\gamma$$

so that

$$|\alpha|_\nu = \frac{|\beta|_\nu^r |\gamma|_\nu}{|\beta^r+1|_\nu} \leqq \frac{|\beta|_\nu^r}{1-|\beta|_\nu^r}|\gamma|_\nu, \qquad \nu = 2, \cdots, n-1,$$

$$|\alpha|_n \leqq \frac{|\beta|_n^r}{|\beta|_n^r-1}|\gamma|_n.$$

Now $|\alpha|_\nu<1$ for r large; namely $\lim_{r\to\infty}|\beta|_\nu^r=0$ and $|\alpha|_n<1$ since

---

$\mathfrak{p}_1, \mathfrak{p}_2, \cdots$ of p will lead to different irreducible polynomials $P(x)$. Since $f(x)$ is divisible by the product of the polynomials $P(x)$, this proves the inequality for this case. If several elements have to be adjoined to R in order to get k, we prove the theorem by repeated application of the simple case.

This proof shows also that in case of an inseparable extension k one can not expect to replace the inequality by an equality. If this can be done in a special case, it is a noteworthy property of the particular field. In §3 we shall find a class of fields with this property.

$$\lim_{r\to\infty} \frac{|\beta|_n^r}{|\beta|_n^r - 1} = 1.$$

For $\nu=1$ we find

$$|\alpha|_1 \geqq \frac{|\beta|_1^r}{1+|\beta|_1^r} |\gamma|_1,$$

so for large r, $|\alpha|_1>1$.

LEMMA 3. *If any n nontrivial inequivalent valuations of k are given, then for any positive ϵ there is an α such that*

$$|\alpha - 1|_1 \leqq \epsilon, \quad |\alpha|_\nu \leqq \epsilon \quad \textit{for} \quad \nu > 1.$$

PROOF. Choose β, by Lemma 2, so that $|\beta|_1>1$ and $|\beta|_\nu<1$ for $\nu>1$ and take

$$\alpha = \frac{\beta^r}{1+\beta^r}\cdot$$

Then

$$|\alpha - 1|_1 = \frac{1}{|1+\beta^r|_1} \leqq \frac{1}{|\beta|^r - 1} \leqq \epsilon$$

for r sufficiently large. For $\nu>1$,

$$|\alpha|_\nu = \frac{|\beta|_\nu^r}{|1+\beta|_\nu^r} \leqq \frac{|\beta|_\nu^r}{1-|\beta|_\nu^r} \leqq \epsilon$$

for r sufficiently large.

THEOREM 1 (APPROXIMATION THEOREM). *If we are given any n nontrivial inequivalent valuations $|\ \ |_\nu$ of k, an element α_ν of k for each valuation, and an $\epsilon>0$, then we can find an element α of k such that*

$$|\alpha - \alpha_\nu|_\nu \leqq \epsilon \quad \textit{for} \quad \textit{each} \quad \nu = 1, 2, \cdots, n.$$

PROOF. Let M be the maximum of the numbers $|\alpha_i|_j$ for all combinations of i and j and choose β_i ($i=1, 2, \cdots, n$) such that

$$|1 - \beta_i|_i < \frac{\epsilon}{nM}, \qquad |\beta_i|_\nu < \frac{\epsilon}{nM} \quad \text{for} \quad \nu \neq i.$$

Let

$\alpha = \beta_1\alpha_1 + \beta_2\alpha_2 + \cdots + \beta_n\alpha_n$; then $|\alpha - \alpha_i|_i < \epsilon$ for each i.

COROLLARY. *If* $|\ |_1, |\ |_2, \cdots, |\ |_n$ *are nontrivial and inequivalent then a relation*

$$|x|_1^{\nu_1} |x|_2^{\nu_2} \cdots |x|_n^{\nu_n} = 1$$

is true for all $x \in k$, $x \neq 0$, *if and only if all* $\nu_i = 0$.

PROOF. If any $\nu_i \neq 0$, an x for which $|x|_i$ is sufficiently large and the other $|x|_\nu$ for $\nu \neq i$ are sufficiently near 1 gives a contradiction.

2. **The product formula.** Our corollary precludes the possibility that a finite number of valuations can be interrelated in a field. Such an interrelation may nevertheless happen for an infinite number of valuations. In case of the ordinary function fields and number fields that is not only the case but this fact may even be used to derive all the properties of these fields on a common basis.

We shall assume for our field k:

AXIOM 1. *There is a set* $\mathfrak{M}$ *of prime divisors* $\mathfrak{p}$ *and a fixed set of valuations* $|\ |_\mathfrak{p}$, *one for each* $\mathfrak{p} \in \mathfrak{M}$, *such that, for every* $\alpha \neq 0$ *of* k, $|\alpha|_\mathfrak{p} = 1$ *for all but a finite number of* $\mathfrak{p} \in \mathfrak{M}$ *and*

$$\prod_\mathfrak{p} |\alpha|_\mathfrak{p} = 1,$$

where this product is extended over all $\mathfrak{p} \in \mathfrak{M}$.

If this axiom is satisfied, $\mathfrak{M}$ can contain only a finite number of archimedean divisors: for $|1+1|_\mathfrak{p} > 1$ at all archimedean $\mathfrak{p}$. Suppose that Axiom 1 is satisfied and that $\mathfrak{M}$ contains no archimedean divisors at all; consider the set k_0 of all α for which $|\alpha|_\mathfrak{p} \leqq 1$ at all $\mathfrak{p} \in \mathfrak{M}$. Let α and β be two elements of k_0. It follows at once that $-\alpha$, $\alpha\beta$, and $\alpha+\beta$ are also in the set. If $\alpha \in k_0$ and $\alpha \neq 0$, the product formula gives at once $|\alpha|_\mathfrak{p} = 1$ for all $\mathfrak{p}$. It now follows that α^{-1} is in k_0. Thus k_0 forms a subfield of k, called the field of constants. It consists of 0 and those elements of k which satisfy $|\alpha|_\mathfrak{p} = 1$ for all $\mathfrak{p}$. It may also be defined as the largest subfield of k for which all $\mathfrak{p}$ reduce to the trivial valuation. If $\mathfrak{M}$ contains archimedean divisors, then there is no field of constants.

We associate with our set $\mathfrak{M}$ of valuations $\mathfrak{p}$ a certain space of vectors $\mathfrak{a}$ with one component $\alpha_\mathfrak{p}$ for each divisor $\mathfrak{p}$. The component $\alpha_\mathfrak{p}$ may range freely over the $\mathfrak{p}$-adic completion $k_\mathfrak{p}$ of k. If $\mathfrak{a}$ is such a vector we shall for brevity write $|\mathfrak{a}|_\mathfrak{p}$ instead of $|\alpha_\mathfrak{p}|_\mathfrak{p}$. The idèles of Chevalley[4] are special cases of these vectors; for an idèle we must have $\alpha_\mathfrak{p} \neq 0$ for all $\mathfrak{p}$ and $|\mathfrak{a}|_\mathfrak{p} = 1$ for all but a finite number of $\mathfrak{p}$.

[4] See Chevalley [3, 4].

Our field k may be considered a subset of this space inasmuch as $\alpha \in k$ may also be considered as the vector whose $\mathfrak{p}$-coordinate is the element α of $k_{\mathfrak{p}}$.

With each idèle $\mathfrak{a}$ we associate the product

$$V(\mathfrak{a}) = \prod_{\mathfrak{p}} | \mathfrak{a} |_{\mathfrak{p}}$$

and think of it as something measuring the size of $\mathfrak{a}$. In a moment we shall see that it may be interpreted as a "volume."

For elements α of k the product formula yields

$$V(\alpha) = 1$$

so that for all idèles $\mathfrak{a}$ we get

$$V(\alpha\mathfrak{a}) = V(\mathfrak{a}).$$

If we select real numbers $x_{\mathfrak{p}} > 0$ for each $\mathfrak{p}$ and take care that $x_{\mathfrak{p}} \neq 1$ for a finite number of $\mathfrak{p}$ only, then we call the set of vectors $\mathfrak{c}$ satisfying

$$| \mathfrak{c} |_{\mathfrak{p}} \leqq x_{\mathfrak{p}} \quad \text{for all } \mathfrak{p}$$

a parallelotope of dimensions $x_{\mathfrak{p}}$.

We shall find later that every valuation is either archimedean or discrete. If this is true then there is an element $\alpha_{\mathfrak{p}}$ in $k_{\mathfrak{p}}$ whose value is maximal and not greater than $x_{\mathfrak{p}}$ so that it is no restriction of generality to start with a given idèle $\mathfrak{a}$ and to construct all vectors $\mathfrak{c}$ satisfying

$$| \mathfrak{c} |_{\mathfrak{p}} \leqq | \mathfrak{a} |_{\mathfrak{p}}.$$

We talk in this case of a parallelotope of size $\mathfrak{a}$. The product $V(\mathfrak{a})$ may then be interpreted as its volume.

Next we introduce the "order" of a given set of elements. It is a notion that shall unite different types of fields. If k has an archimedean valuation we mean by order the number of elements. Otherwise k has a field k_0 of constants: we let q stand for an arbitrarily selected but fixed number greater than 1 when the number of elements of k_0 is infinite, and for the number of elements of k_0 when this number is finite. By order of a set we mean in this case the number q^s where s is the number of elements in our set that are linearly independent with respect to k_0. Should k_0 contain q elements and our set be closed under addition and under multiplication by elements of k_0 then q^s is the number of elements in the set.

In the next section we shall be interested in the order of the set of elements α of k that are contained in a given parallelotope of size $\mathfrak{a}$. We denote this order by $M(\mathfrak{a})$. If $\theta \neq 0$ is in k than $M(\theta\mathfrak{a}) = M(\mathfrak{a})$.

Indeed multiplication by θ transforms the parallelotope of size $\mathfrak{a}$ into the parallelotope of size $\theta\mathfrak{a}$ and does not change the order.

In the next section it will be shown that $V(\mathfrak{a})$ and $M(\mathfrak{a})$ are related; namely that they are of the same order of magnitude.

If $\mathfrak{p}$ is a nonarchimedean prime divisor, the elements $\alpha \in k$ for which $|\alpha|_\mathfrak{p} \leqq 1$ form a ring $\mathfrak{o}_\mathfrak{p}$, called the ring of $\mathfrak{p}$-integers (or local integers). The elements α for which $|\alpha|_\mathfrak{p} < 1$ form a prime ideal in this ring and we denote this ideal by the same symbol $\mathfrak{p}$ as the prime divisor. If the residue class field $\mathfrak{o}_\mathfrak{p}/\mathfrak{p}$ is of finite order then we call this order the norm of $\mathfrak{p}$ and denote it by $N\mathfrak{p}$. We can talk of the order also in case of a constant field k_0 since k_0 may be considered as subfield of the field $\mathfrak{o}_\mathfrak{p}/\mathfrak{p}$. Thus, if f is the degree of $\mathfrak{o}_\mathfrak{p}/\mathfrak{p}$ over k_0, and if f is finite, we put $N\mathfrak{p} = q^f$.

Axiom 2. *The set $\mathfrak{M}$ of Axiom* 1 *contains at least one prime* $\mathfrak{q}$, *which is of one of the following two types*:

1. *Discrete, with a residue class field of finite order* $N\mathfrak{q}$.
2. *Archimedean, with a completed field $k_\mathfrak{q}$ which is either the real or the complex field.*[5]

As mentioned before, there are an infinity of equivalent valuations belonging to one prime divisor $\mathfrak{p}$. One of them, $|\alpha|_\mathfrak{p}$, is singled out by our Axiom 1. For primes $\mathfrak{p}$ satisfying Axiom 2 we shall define another one that is singled out by inner properties. In case 1 of Axiom 2 we put (for $\alpha \neq 0$)

$$\|\alpha\|_\mathfrak{p} = \frac{1}{N\mathfrak{p}^\nu}$$

where ν is the ordinal number of α at $\mathfrak{p}$. In case 2 we take $\|\alpha\|_\mathfrak{p}$ to be ordinary absolute value when $k_\mathfrak{p}$ is the real field and the square of ordinary absolute value when $k_\mathfrak{p}$ is the complex field. Note that in the latter case $\|\alpha\|_\mathfrak{p}$ is not a true valuation. We call $\|\alpha\|_\mathfrak{p}$ the normed valuation at $\mathfrak{p}$.

Theorem 2. *In case of the following special fields k we can construct a set $\mathfrak{M}$ of valuations such that our axioms hold, and the second one even holds for all $\mathfrak{p}$ of $\mathfrak{M}$*:

1. *Any finite algebraic number field (that is, a finite extension of the field of rational numbers).*
2. *Any field of algebraic functions over any given field k_1 (that is, a finite extension of the field $k_1(z)$ where z is transcendental with respect to k_1).*

[5] It is well known (Ostrowski [5]) that we could drop this condition on the completed field.

In case 2, the constant field k_0 of k with respect to $\mathfrak{M}$ consists of all elements of k that are algebraic with respect to k_1.

The proof is contained in the following chain of statements:

LEMMA 4. *Let k be a field for which Axiom 1 holds and R a subfield consisting not exclusively of constants of k. Let $\mathfrak{N}$ be the set of those nontrivial divisors p of R that are divisible by some $\mathfrak{p}$ of $\mathfrak{M}$. Then Axiom 1 holds in R for this set $\mathfrak{N}$.*

PROOF. Let p be any divisor of $\mathfrak{N}$ and a an element of R such that $|a|_p>1$. Then $|a|_{\mathfrak{p}}>1$ for all $\mathfrak{p}$ that divide p. Because of Axiom 1 there can be only a finite number of $\mathfrak{p}|p$. Let us now define

$$|b|_p = \prod_{\mathfrak{p}|p} |b|_{\mathfrak{p}} \quad \text{for all} \quad b \in R$$

and we have a set of valuations $|\ |_p$ for which Axiom 1 holds.

LEMMA 5. *Let k be a field for which Axiom 1 holds and K a finite algebraic extension of k. Let $\mathfrak{N}$ be the set of all divisors $\mathfrak{P}$ of K that divide some $\mathfrak{p}$ of $\mathfrak{M}$. Then Axiom 1 holds in K for some subset $\mathfrak{N}'$ of $\mathfrak{N}$.*

(It would not be difficult to show now that $\mathfrak{N}'=\mathfrak{N}$, but it is better to postpone this and other details until the next section.)

PROOF. 1. Let $\mathrm{A}\neq 0$ be an element of K and $f(x)=0$ the equation for A with coefficients in k and with highest coefficient 1. If $\mathfrak{p}$ is a nonarchimedean valuation for which all coefficients in $f(x)$ have a value not greater than 1 and $\mathfrak{P}$ a divisor of $\mathfrak{p}$, then $|\mathrm{A}|_{\mathfrak{P}}\leqq 1$ or else no cancellation could take place between the highest term in $f(\mathrm{A})=0$ and the others. So $|\mathrm{A}|_{\mathfrak{P}}\leqq 1$ for all but a finite number of $\mathfrak{P}$. For the same reason $|1/\mathrm{A}|_{\mathfrak{P}}\leqq 1$ for all but a finite number of $\mathfrak{P}$. Therefore $|\mathrm{A}|_{\mathfrak{P}}\neq 1$ for only a finite number of $\mathfrak{P}$.

2. Let $F(x)$ be a polynomial in k that has the generators of K among its roots ($F(x)$ need not be irreducible). If K' is the splitting field of $F(x)$ we may first prove Lemma 5 for K' instead of K and then descend to the subfield K by use of Lemma 4. This shows that we may already assume that K is the splitting field of a polynomial $F(x)$ in k.

The algebraic structure of such a field is well known. If $\mathfrak{G}$ is the group of all its automorphisms σ and if we construct for any $\mathrm{A}\in K$ the product

$$\prod \mathrm{A}^{\sigma} = \alpha$$

then α is invariant under $\mathfrak{G}$. Since we have to consider also the inseparable case we do not know that α is in k. But there is always a positive integer m such that

$$\prod_{\sigma} (A^{\sigma})^{m} = a$$

is in k whatever A may be. Because of the product formula in k we get

$$\prod_{\mathfrak{p}} | a |_{\mathfrak{p}} = 1.$$

Now select, for each $\mathfrak{p}$, one divisor $\mathfrak{P}$ of K which divides $\mathfrak{p}$ and define $| \ |_{\mathfrak{P}}$ in such a way that $|b|_{\mathfrak{P}} = |b|_{\mathfrak{p}}$ for all b in k. Then

$$| a |_{\mathfrak{p}} = \prod_{\sigma} | A^{\sigma} |_{\mathfrak{P}}^{m}.$$

If we consider the expression $|A^{\sigma}|_{\mathfrak{P}}$ as function of A, it is clearly a valuation of K that belongs to a divisor $\mathfrak{P}'$ which divides $\mathfrak{p}$ and may be equal to or different from $\mathfrak{P}$. If we change our notation slightly we obviously get

$$| a |_{\mathfrak{p}} = \prod_{\mathfrak{P}'} | A |_{\mathfrak{P}'}',$$

where $\mathfrak{P}'$ runs through some divisors of $\mathfrak{p}$ and where $| \ |_{\mathfrak{P}'}'$ is a certain valuation belonging to $\mathfrak{P}'$.

If we substitute this in our product-formula we get

$$\prod_{\mathfrak{p}} \prod_{\mathfrak{P}'} | A |_{\mathfrak{P}'}' = 1$$

and this proves Lemma 5.

Before we proceed with our next lemma let us consider the special field $R = k_1(z)$ of Theorem 2. If p is a nontrivial valuation of R that reduces to the trivial one on k_1 then p is nonarchimedean and we distinguish two cases:

1. If $|z|_p \leqq 1$ then $|f(z)|_p \leqq 1$ for every polynomial in z. Let $p(z)$ be a polynomial of lowest degree such that $|p(z)|_p < 1$. If $g(z)$ is another polynomial with $|g(z)|_p < 1$ then we divide:

$$g(z) = p(z)h(z) + r(z),$$

where the degree of $r(z)$ is lower than that of $p(z)$. From

$$r(z) = g(z) - p(z)h(z)$$

we get $|r(z)|_p < 1$; hence $r(z) = 0$.

Now let $\phi(z)$ be any element of R and put

$$\phi(z) = p(z)^{\nu} \cdot \psi(z),$$

where neither numerator nor denominator of $\psi(z)$ is divisible by $p(z)$. Then $|\psi(z)|_p=1$ so

$$|\phi(z)|_p = |p(z)|_p^\nu = c^\nu, \quad \text{where} \quad c = |p(z)|_p < 1.$$

$p(z)$ is obviously irreducible.

In order to find the normed valuation $\| \|_p$ in this case we have to determine the degree of the residue class field (mod $p(z)$). It is the degree f of $p(z)$ so that $Np=q^f$ and

$$\|\phi(z)\|_p = q^{-\nu f}.$$

2. If $|z|_p>1$ we replace z by $y=1/z$. Then $|y|_p<1$ and we have our previous case. The polynomial in y of lowest degree is y itself, so that there is only one prime divisor p of this kind. We denote it by p_∞.

Let $\phi(z)=g(z)/h(z)$ where $g(z)$ and $h(z)$ are polynomials of degrees m and n. Then

$$\phi(z) = y^{n-m}\cdot\frac{g_1(y)}{h_1(y)}$$

where $g_1(y)$ and $h_1(y)$ are polynomials not divisible by y. Hence

$$\|\phi(z)\|_{p_\infty} = q^{m-n}.$$

A product formula connecting all these valuations or a subset of them can be written in the form

$$\prod_p \|\phi(z)\|_p^{\lambda(p)} = 1,$$

where $\lambda(p)$ are constants not less than 0. If we substitute for $\phi(z)$ the irreducible polynomials $p(z)$, then only two factors can possibly be different from 1: the valuations at the p belonging to $p(z)$ and at p_∞. This gives

$$q^{-f\lambda(p_\infty)}\cdot q^{f\lambda(p)} = 1$$

or $\lambda(p)=\lambda(p_\infty)$. So all $\lambda(p)$ are equal and may therefore be assumed to be equal to 1. In order to show that this product formula holds we put

$$V(\phi(z)) = \prod_p \|\phi(z)\|_p.$$

It is obvious that $V(\phi(z)\cdot\psi(z))=V(\phi(z))\cdot V(\psi(z))$ and that a similar rule holds for quotients.

We have just seen that $V(p(z))=1$ for any irreducible polynomial; it follows that $V(\phi(z))=1$ for any element $\phi(z)\neq 0$ of R.

In the same fashion we can discuss the field R of rational numbers.

It is well known that all valuations are either the single archimedean p_∞ for which $\|a\|_{p_\infty}$ is the ordinary absolute value or the p-adic valuations of R where p is a prime number. The normed valuation $|a|_p$ in this latter case is given by $1/p^\nu$, if ν is the ordinal number of a. Just as before we consider a hypothetical product formula

$$\prod_p \|a\|_p^{\lambda(p)} = 1.$$

Substituting for a a prime number p we get

$$\left(\frac{1}{p}\right)^{\lambda(p)} \cdot p^{\lambda(p_\infty)} = 1$$

or $\lambda(p) = \lambda(p_\infty)$. The numbers $\lambda(p)$ are therefore equal and may be considered equal to 1. The same method as before shows that the product formula really holds.

LEMMA 6. *Axiom 1 holds in the case of the field R of rational numbers and that of $R = k_1(z)$. If R is the rational field, $\mathfrak{M}$ is the set of all valuations; if $R = k_1(z)$, it is the set of all valuations that are trivial on k_1. The product formula itself takes on the form*

$$\prod_p \|a\|_p = 1$$

or a power of it and there is no other relation between these valuations.

Lemmas 5 and 6 already show that Axiom 1 holds for the fields mentioned in Theorem 2. That all valuations of $\mathfrak{M}$ satisfy Axiom 2 follows from the fact that this is true in R and consequently in a finite extension k.

It remains to prove the statement about the field k_0 of constants. If $\mathfrak{p}$ is trivial on k_1 it is also trivial on an algebraic extension of k_1. Hence we need only show that any constant c of k_0 is algebraic with respect to k_1. If on the contrary c were transcendental with respect to k_1 then from the equation c satisfied with respect to $k_1(z)$ it follows that z would be algebraic with respect to $k_1(c)$. Since $k_1(c)$ is in k_0, this would mean that z is in k_0. So all of k would be in k_0, contradicting the fact that no $\mathfrak{p}$ of $\mathfrak{M}$ is trivial on k.

More detailed information about the fields of Theorem 2 will follow from the next section.

3. Characterization of fields by the product formula. In this section we assume k to be any field for which the Axioms 1 and 2 hold and are going to prove that k is of the type described in Theorem 2.

For any prime $\mathfrak{p}$ that satisfies Axiom 2 we shall have to distinguish the valuation $|\alpha|_{\mathfrak{p}}$ of Axiom 1 and the equivalent normed valuation $\|\alpha\|_{\mathfrak{p}}$. We define the real number $\rho(\mathfrak{p})>0$ by

$$|\alpha|_{\mathfrak{p}} = \|\alpha\|_{\mathfrak{p}}^{\rho(\mathfrak{p})}.$$

By R we mean the following subfield of k:

1. If $\mathfrak{M}$ has archimedean valuations, R is the rational field. By $\|a\|_{p_\infty}$ we mean the ordinary absolute value in R.

2. In the other case k has a field k_0 of constants and cannot contain any algebraic extension of k_0 since our valuations are trivial on k_0 and would be trivial on that extension. Let z be any element of k not in k_0; then $R=k_0(z)$ is a transcendental extension of k_0. By integers we mean in this case the polynomials in z. By $\|a\|_{p_\infty}$ we mean the one valuation we found in proving Lemma 6 that has $\|z\|_{p_\infty}>1$.

In both cases we mean by $\mathfrak{p}_\infty$ any divisor of $\mathfrak{M}$ that divides p_∞. Since the product formula, when applied to elements of R, must reduce to the formula of Theorem 2, our set $\mathfrak{M}$ always contains at least one $\mathfrak{p}_\infty$. The other primes of $\mathfrak{M}$ shall be called finite. For elements a of R the valuations $|a|_{\mathfrak{p}_\infty}$ and $\|a\|_{p_\infty}$ are equivalent. We define the real numbers $\lambda(\mathfrak{p}_\infty)>0$ by

$$|a|_{\mathfrak{p}_\infty} = \|a\|_{p_\infty}^{\lambda(\mathfrak{p}_\infty)} \text{ for all } a \in R.$$

LEMMA 7. *Let $\mathfrak{q}$ be one of the primes satisfying Axiom 2 and $\mathfrak{S}$ be a set of elements of k of an order $M>1$. Let x be an upper bound for $|\alpha|_{\mathfrak{q}}$ for all α of $\mathfrak{S}$: $|\alpha|_{\mathfrak{q}}\leqq x$. Then there is an element θ of k with the following properties:*

(1) *θ is either a difference of two elements of $\mathfrak{S}$ or, in case there is a field of constants, a linear combination of elements of $\mathfrak{S}$ with coefficients in k_0.*

(2) *$\theta\neq 0$.*

(3) *$|\theta|_{\mathfrak{q}}\leqq A_{\mathfrak{q}}x/M^{\rho(\mathfrak{q})}$ where $A_{\mathfrak{q}}$ is a constant depending only on $\mathfrak{q}$.*

PROOF. 1. $\mathfrak{q}$ archimedean, $k_{\mathfrak{q}}$ real. In this case we may treat k as a subfield of the real field. We have

$$\|\alpha\|_{\mathfrak{q}} \leqq x^{1/\rho(\mathfrak{q})}$$

for each of the M elements α of $\mathfrak{S}$. Divide the interval from $-x^{1/\rho(\mathfrak{q})}$ to $x^{1/\rho(\mathfrak{q})}$ into $M-1$ equal parts. Two of the α's must be in the same compartment so their difference θ satisfies

$$\|\theta\|_{\mathfrak{q}} \leqq \frac{2x^{1/\rho(\mathfrak{q})}}{M-1} \leqq \frac{4x^{1/\rho(\mathfrak{q})}}{M},$$

hence

$$|\theta|_{\mathfrak{q}} \leqq \frac{4^{\rho(\mathfrak{q})}x}{M^{\rho(\mathfrak{q})}}.$$

2. $\mathfrak{q}$ archimedean, $k_{\mathfrak{q}}$ complex. Treating k as a subfield of the complex field, $\|\alpha\|_{\mathfrak{q}}^{1/2} = |\alpha|_{\mathfrak{q}}^{1/2\rho(\mathfrak{q})}$ is the ordinary distance from the origin to the point α and is less than $x^{1/2\rho(\mathfrak{q})}$. Writing $\alpha = \xi + i\eta$ for each $\alpha \in \mathfrak{S}$ we know that $\mathfrak{S}$ is in the square $|\xi| \leqq x^{1/2\rho(\mathfrak{q})}$ $|\eta| \leqq x^{1/2\rho(\mathfrak{q})}$. Divide this square into N^2 small squares by dividing each side into N equal parts, where $N < M^{1/2} \leqq N+1$. Then some two α's are in the same subdivision so their difference θ satisfies:

$$\|\theta\|_{\mathfrak{q}}^{1/2} \leqq \frac{2^{3/2}x^{1/2\rho(\mathfrak{q})}}{N} \leqq \frac{2^{5/2}x^{1/2\rho(\mathfrak{q})}}{M^{1/2}}$$

so

$$|\theta|_{\mathfrak{q}} \leqq \frac{(2^{5/2})^{2\rho(\mathfrak{q})}x}{M^{\rho(\mathfrak{q})}}.$$

3. $\mathfrak{q}$ discrete. Let α_1 be an α for which $|\alpha|_{\mathfrak{q}}$ is maximum. This exists since $\mathfrak{q}$ is discrete and $|\alpha|_{\mathfrak{q}} \leqq x$. Then for each $\alpha \in \mathfrak{S}$, $|\alpha/\alpha_1|_{\mathfrak{q}} \leqq 1$.

Choose r so that $N\mathfrak{q}^r < M \leqq N\mathfrak{q}^{r+1}$. If the number of elements in the residue class field is finite then the local theory shows easily that $\mathfrak{o}_{\mathfrak{q}}$ contains at most $N\mathfrak{q}^r$ residue classes mod $\mathfrak{q}^r$. Hence two of the $M > N\mathfrak{q}^r$ elements of $(1/\alpha_1)\mathfrak{S}$ are in the same residue class and their difference θ/α_1 has at least the ordinal number r. Should there be a field k_0, let f be the degree of $\mathfrak{o}_{\mathfrak{q}}/\mathfrak{q}$ over k_0, so that $N\mathfrak{q} = q^f$. Then there are at most rf elements of $\mathfrak{o}_{\mathfrak{q}}$ that are linearly independent mod $\mathfrak{q}^r$. Taking more than rf of our elements α/α_1 that are independent considered as elements of k (possible since $M > N\mathfrak{q}^r$) we can find a linear combination $\theta/\alpha_1 \neq 0$ of them that is congruent to 0 (mod $\mathfrak{q}^r$) and hence has at least the ordinal number r. In both cases we get

$$\left\|\frac{\theta}{\alpha_1}\right\|_{\mathfrak{q}} \leqq \frac{1}{N\mathfrak{q}^r} = \frac{N\mathfrak{q}}{N\mathfrak{q}^{r+1}} \leqq \frac{N\mathfrak{q}}{M}, \qquad \|\theta\|_{\mathfrak{q}} \leqq \frac{N\mathfrak{q}\cdot\|\alpha_1\|_{\mathfrak{q}}}{M},$$

or

$$|\theta|_{\mathfrak{q}} \leqq \frac{N\mathfrak{q}^{\rho(\mathfrak{q})}|\alpha_1|_{\mathfrak{q}}}{M^{\rho(\mathfrak{q})}} \leqq \frac{N\mathfrak{q}^{\rho(\mathfrak{q})}\cdot x}{M^{\rho(\mathfrak{q})}}.$$

LEMMA 8. *Let M be the order of the set of elements $\alpha \in k$ that is contained in a parallelotope of dimensions $x_{\mathfrak{p}}$. If $\mathfrak{q}$ is a prime satisfying Axiom 2 we can find a constant $B_{\mathfrak{q}}$ depending only on $\mathfrak{q}$ such that either $M = 1$ (if our set contains only $\alpha = 0$) or*

$$M \leqq B_{\mathfrak{q}}\left(\prod_{\mathfrak{p}} x_{\mathfrak{p}}\right)^{1/\rho(\mathfrak{q})}.$$

PROOF. Assume $M>1$. By Lemma 7 there is a $\theta\neq 0$ satisfying

$$|\theta|_{\mathfrak{q}} \leqq \frac{A_{\mathfrak{q}}x_{\mathfrak{q}}}{M^{\rho(\mathfrak{q})}}.$$

For the other $\mathfrak{p}$ of $\mathfrak{M}$ we estimate θ directly and get

$$|\theta|_{\mathfrak{p}} \leqq \begin{cases} x_{\mathfrak{p}} \text{ at any nonarchimedean } \mathfrak{p}, \\ 4^{\rho(\mathfrak{p})}\cdot x_{\mathfrak{p}} \text{ at any archimedean } \mathfrak{p}. \end{cases}$$

Substituting in the product formula $\prod_{\mathfrak{p}}|\theta|_{\mathfrak{p}}=1$ we get (if $D_{\mathfrak{q}}$ is a certain constant):

$$1 \leqq \frac{D_{\mathfrak{q}}\cdot\prod_{\mathfrak{p}}x_{\mathfrak{p}}}{M^{\rho(\mathfrak{q})}},$$

hence the lemma.

LEMMA 9. *If $\alpha_1, \alpha_2, \cdots, \alpha_l$ are linearly independent with respect to the subfield R and if y is a given nonzero integer of R, we can construct a certain set $\mathfrak{S}$ of elements α with the following properties:*

1. $|\alpha|_{\mathfrak{p}} \leqq a_{\mathfrak{p}} = \max_{i=1,\ldots,l}(|\alpha_i|_{\mathfrak{p}})$ *for every finite* $\mathfrak{p}$.
2. $|\alpha|_{\mathfrak{p}_\infty} \leqq B\cdot|y|_{\mathfrak{p}_\infty}$ *with a certain constant B that can be easily estimated.*
3. *If there is a field of constants k_0 then $\mathfrak{S}$ is closed under addition and under multiplication by elements of k_0, so $\mathfrak{S}$ may be considered as a vector space over k_0.*
4. *The order of $\mathfrak{S}$ is greater than $\|y\|^l_{p_\infty}$.*

PROOF. Let $\mathfrak{S}$ consist of all α of the form

$$\alpha = \nu_1\alpha_1 + \nu_2\alpha_2 + \cdots + \nu_l\alpha_l$$

where the ν_i range over all integers of R that satisfy

$$\|\nu_i\|_{p_\infty} \leqq \|y\|_{p_\infty}.$$

This settles at once property 3 and implies $|\nu_i|_{\mathfrak{p}_\infty} \leqq |y|_{\mathfrak{p}_\infty}$ for each $\mathfrak{p}_\infty$ and consequently property 2. Property 1 holds since $|\nu_i|_{\mathfrak{p}}\leqq 1$ for all finite $\mathfrak{p}$. Property 4 is clear if p_∞ is archimedean; if not then assume $\|y\|_{p_\infty}=q^d$ so that y is a polynomial of degree d. Each ν_i ranges over all polynomials of degree not greater than d. This gives for $\mathfrak{S}$ a vector space of $(d+1)l$ dimensions and our statement is obvious.

LEMMA 10. *The degree n of k over R is finite; every $\mathfrak{p}$ of M satisfies Axiom 2; and the inequality*

$$n \leqq \frac{1}{\rho(\mathfrak{p})} \cdot \sum_{\mathfrak{p}_\infty} \lambda(\mathfrak{p}_\infty)$$

holds for each p.

PROOF. Apply Lemma 8 to the set of Lemma 9. We get the inequality

$$\|y\|_{p_\infty}^{l} \leqq E \cdot \prod_{\mathfrak{p}_\infty} |y|_{\mathfrak{p}_\infty}^{1/\rho(\mathfrak{q})} = E \cdot \|y\|_{p_\infty}^{(1/\rho(\mathfrak{q}))\Sigma_{\mathfrak{p}_\infty}\lambda(\mathfrak{p}_\infty)},$$

where E is a certain constant that depends on the constants in the previous lemmas. Since $\|y\|_{p_\infty}$ takes on arbitrarily large values we get

$$l \leqq \frac{1}{\rho(\mathfrak{q})} \sum_{\mathfrak{p}_\infty} \lambda(\mathfrak{p}_\infty).$$

This proves that n is finite. None of our valuations $\mathfrak{p}$ is trivial on R or else it would be trivial on the finite extension k. Let p be the divisor of R that is divisible by $\mathfrak{p}$. The local theory shows now (since p is nontrivial) that $\mathfrak{p}$ satisfies Axiom 2. In our previous inequality we can therefore assume $l=n$ and take for $\mathfrak{q}$ each prime $\mathfrak{p}$ of $\mathfrak{M}$.

Let r be a positive real number and let us replace each valuation $|\alpha|_\mathfrak{p}$ by its rth power $|\alpha|_\mathfrak{p}^r$. This would be a new set of valuations for which Axioms 1 and 2 would hold again. The numbers $\lambda(\mathfrak{p}_\infty)$ would then be replaced by $r\lambda(\mathfrak{p}_\infty)$. This shows that it is no restriction of generality to assume that

$$\sum_{\mathfrak{p}_\infty} \lambda(\mathfrak{p}_\infty) = n.$$

Then Lemma 10 gives

$$\rho(\mathfrak{p}) \leqq 1$$

for every $\mathfrak{p}$.

Assume now that $\mathfrak{p}|p$, where p is a nonarchimedean divisor of R, and let us compare $\|a\|_\mathfrak{p}$ and $\|a\|_p$ for elements a of R. The ordinal number of a in k is $e(\mathfrak{p})$ times the ordinal number of a measured in R; we also have $N\mathfrak{p} = (Np)^{f(\mathfrak{p})}$. $e(\mathfrak{p})$ is the ramification number and $f(\mathfrak{p})$ the degree of the residue class fields. So

$$\|a\|_\mathfrak{p} = \|a\|_p^{e(\mathfrak{p})f(\mathfrak{p})} = \|a\|_p^{n(\mathfrak{p})}.$$

For an archimedean $\mathfrak{p}$ this equality follows directly from the definitions. Hence we have

$$| a |_{\mathfrak{p}} = \|a\|_p^{n(\mathfrak{p})\rho(\mathfrak{p})} \text{for all } \mathfrak{p} \text{ and all } a \in R.$$

We note in particular $\lambda(\mathfrak{p}_\infty) = n(\mathfrak{p}_\infty)\rho(\mathfrak{p}_\infty)$ so that

$$\sum_{\mathfrak{p}_\infty} n(\mathfrak{p}_\infty)\rho(\mathfrak{p}_\infty) = n.$$

Now we apply the product formula to an $a \in R$:

$$1 = \prod_{\mathfrak{p}} | a |_{\mathfrak{p}} = \prod_p \Bigl(\prod_{\mathfrak{p}|p,\, \mathfrak{p} \in \mathfrak{M}} \|a\|_p^{n(\mathfrak{p})\rho(\mathfrak{p})} \Bigr) = \prod_p \|a\|_p^{\nu(p)},$$

where

$$\nu(p) = \sum_{\mathfrak{p}|p,\, \mathfrak{p} \in \mathfrak{M}} n(\mathfrak{p})\rho(\mathfrak{p}).$$

But Lemma 6 shows that all $\nu(p)$ are equal and the special case $p = p_\infty$ shows finally

$$n = \sum_{\mathfrak{p}|p,\, \mathfrak{p} \in \mathfrak{M}} n(\mathfrak{p})\rho(\mathfrak{p}).$$

If we compare this with $\rho(\mathfrak{p}) \leqq 1$ and $\sum_{\mathfrak{p}|p} n(\mathfrak{p}) \leqq n$ we find:
(1) all $\mathfrak{p} | p$ are in $\mathfrak{M}$,
(2) all $\rho(\mathfrak{p}) = 1$,
(3) $\sum_{\mathfrak{p}|p} n(\mathfrak{p}) = n$.
Thus we have proved:

THEOREM 3. *If k is a field that satisfies the Axioms 1 and 2 it is an extension of a finite degree n either of the rational field R or of the field $R = k_0(z)$ of rational functions over its field of constants k_0. All valuations satisfy Axiom 2. $\mathfrak{M}$ consists of all extensions of the well known valuations of R. Replacing if necessary all valuations in the product formula by the same power we can assume that they are all normed (that is, $|\alpha|_{\mathfrak{p}} = \|\alpha\|_{\mathfrak{p}}$). We have $\sum_{\mathfrak{p}|p} n_{\mathfrak{p}} = n$ for all p of R.*

Let now α be an element of k and $\mathfrak{p}_1, \mathfrak{p}_2, \cdots, \mathfrak{p}_r$ those among the finite primes for which $\|\alpha\|_{\mathfrak{p}_i} > 1$. Let $p_1, p_2, \cdots, p_l$ be the primes of R that have the $\mathfrak{p}_\nu$ as divisors. Construct an integer in R whose absolute value at each p_i is sufficiently small and $a\alpha$ will now be less than or equal to 1 at all finite primes. This shows that any set $\alpha_1, \alpha_2, \cdots, \alpha_l$ of elements that are linearly independent with respect to R can be replaced by a set of elements which are integral at all finite primes. This will be useful in the next section.

4. Parallelotopes. We still make the same assumptions as in the previous chapter so that we can assume Theorem 3. Thus we will take $|\alpha|_{\mathfrak{p}} = \|\alpha\|_{\mathfrak{p}}$ and $\rho(\mathfrak{p}) = 1$. In Lemma 9 we may assume $l = n$.

THEOREM 4. *There are two positive constants C and D such that for all idèles* $\mathfrak{a}$ *we have*

$$CV(\mathfrak{a}) < M(\mathfrak{a}) \leqq \max (1, DV(\mathfrak{a})).$$

PROOF. If we apply Lemma 8 for one particular $\mathfrak{q}$ we get the right half of the inequality.

If we replace $\mathfrak{a}$ by $\alpha\mathfrak{a}$ then $V(\mathfrak{a})$ and $M(\mathfrak{a})$ remain unchanged. Select $\alpha = \alpha_1 y$ where α_1 and y are selected as follows:

Theorem 1 shows that there is an α_1 such that

$$4B \leqq \|\alpha_1\mathfrak{a}\|_{\mathfrak{p}_\infty} \leqq 5B \quad \text{for all} \quad \mathfrak{p}_\infty,$$

where B is the constant of Lemma 9. We choose such an α_1 and then select an integer y of R in such a way that $\|\alpha_1 y\mathfrak{a}\|_\mathfrak{p} \leqq 1$ at all finite $\mathfrak{p}$. This shows that it is sufficient to prove our theorem for all idèles $\mathfrak{a}$ satisfying

$$4B\|y\|_{\mathfrak{p}_\infty} \leqq \|\mathfrak{a}\|_{\mathfrak{p}_\infty} \leqq 5B\|y\|_{\mathfrak{p}_\infty} \quad \text{for all} \quad \mathfrak{p}_\infty$$

and $\|\mathfrak{a}\|_\mathfrak{p} \leqq 1$ at all finite $\mathfrak{p}$, where y is an integer of R. Using this integer y, we now apply Lemma 9, taking $a_\mathfrak{p} = 1$ at each finite $\mathfrak{p}$ and constructing a set $\mathfrak{S}$ of elements α of k with the following properties:

1. $\|\alpha\|_\mathfrak{p} \leqq 1$ for all finite $\mathfrak{p}$.
2. $\|\alpha\|_{\mathfrak{p}_\infty} \leqq B\|y\|_{\mathfrak{p}_\infty}$.
3. In case there is a field k_0, $\mathfrak{S}$ is a vectorspace over k_0.
4. The order of $\mathfrak{S}$ is greater than $\|y\|_{p_\infty}^n$ and we have

$$\|y\|_{p_\infty}^n = \|y\|_{p_\infty}^{\Sigma_{\mathfrak{p}_\infty}\lambda(\mathfrak{p}_\infty)} = \prod_{\mathfrak{p}_\infty} \|y\|_{\mathfrak{p}_\infty} \geqq \prod_{\mathfrak{p}_\infty} \frac{1}{5B}\|\mathfrak{a}\|_{\mathfrak{p}_\infty}.$$

So the order is greater than $C\prod_{\mathfrak{p}_\infty}\|\mathfrak{a}\|_{\mathfrak{p}_\infty}$ where C is a certain constant not equal to 0.

We distinguish two cases:

1. Order of a set means number. Consider the set $\mathfrak{o}$ of all integers of k (that is, all elements that are integers for every finite $\mathfrak{p}$) and the subset $\{\mathfrak{a}\}$ of all integers β satisfying $\|\beta\|_\mathfrak{p} \leqq \|\mathfrak{a}\|_\mathfrak{p}$ for all finite $\mathfrak{p}$. This subset $\{\mathfrak{a}\}$ forms an additive group which is the intersection of all the groups $\{\mathfrak{a}\}_\mathfrak{p}$ of integers satisfying $\|\beta\|_\mathfrak{p} \leqq \|\mathfrak{a}\|_\mathfrak{p}$ for only this particular $\mathfrak{p}$. The local theory shows that the index of $\mathfrak{o}$ mod $\{\mathfrak{a}\}_\mathfrak{p}$ is at most $1/\|\mathfrak{a}\|_\mathfrak{p}$. So the index of $\{\mathfrak{a}\}$ in the group of all integers is at most $N = \prod_{\mathfrak{p}\text{ fin}}(1/\|\mathfrak{a}\|_\mathfrak{p})$. If we consider now our set $\mathfrak{S}$ modulo $\{\mathfrak{a}\}$ we get at most N residue classes. So one residue class contains more than

$$\frac{C \cdot \prod \|\mathfrak{a}\|_{\mathfrak{p}_\infty}}{N} = CV(\mathfrak{a})$$

elements. If we select one special element of this residue class and subtract it from each of the others, we get more than $CV(\mathfrak{a})$ elements γ of $\{\mathfrak{a}\}$. As such they satisfy $\|\gamma\|_{\mathfrak{p}} \leqq \|\mathfrak{a}\|_{\mathfrak{p}}$ for all finite p. At a $\mathfrak{p}_\infty$ we get $\|\gamma\|_{\mathfrak{p}_\infty} \leqq 4B\|y\|_{\mathfrak{p}_\infty} \leqq \|\mathfrak{a}\|_{\mathfrak{p}_\infty}$. (The factor 4 instead of the expected 2 must be used since one of the valuations may be the square of a true valuation.) So we have found more than $CV(\mathfrak{a})$ elements in our parallelotope of size $\mathfrak{a}$.

2. There is a constant field k_0. We define $\{\mathfrak{a}\}$ and $\{\mathfrak{a}\}_{\mathfrak{p}}$ as before. Assume that m is the dimension of the vector space $\mathfrak{S}$ over k_0. The residue classes of the integers mod $\{\mathfrak{a}\}_{\mathfrak{p}}$ also form a vector space over k_0; let $d(\mathfrak{p})$ denote its dimension. Then $q^{d(\mathfrak{p})} \leqq 1/\|\mathfrak{a}\|_{\mathfrak{p}}$. Let $\mathfrak{S}_1$ be the intersection of $\mathfrak{S}$ and $\{\mathfrak{a}\}_{\mathfrak{p}_1}$. Starting with a basis for the space $\mathfrak{S}_1$, we need at most $d(\mathfrak{p}_1)$ vectors to complete it to a basis of $\mathfrak{S}$. So the dimension of $\mathfrak{S}_1$ is at least $m-d(\mathfrak{p}_1)$. Repeating this process for all finite $\mathfrak{p}$ and calling $\mathfrak{T}$ the intersection of $\mathfrak{S}$ and $\{\mathfrak{a}\}$, we see that its dimension is at least $m-\sum_{\mathfrak{p}\text{ fin}} d(\mathfrak{p})$. So the order of $\mathfrak{T}$ is at least

$$q^m \cdot \prod_{\mathfrak{p}\text{ fin}} \frac{1}{q^{d(\mathfrak{p})}} \geqq q^m \cdot \prod_{\mathfrak{p}\text{ fin}} \|\mathfrak{a}\|_{\mathfrak{p}}.$$

Since the order q^m of $\mathfrak{S}$ is $C\prod_{\mathfrak{p}_\infty}\|\mathfrak{a}\|_{\mathfrak{p}_\infty}$ we find that the order of $\mathfrak{T}$ is greater than $CV(\mathfrak{a})$. That the elements γ of $\mathfrak{T}$ satisfy $\|\gamma\|_{\mathfrak{p}} \leqq \|\mathfrak{a}\|_{\mathfrak{p}}$ follows for a finite $\mathfrak{p}$ from the fact that they are in $\{\mathfrak{a}\}$. For $\mathfrak{p}_\infty$, since they are in $\mathfrak{S}$,

$$\|\gamma\|_{\mathfrak{p}_\infty} \leqq B\|y\|_{\mathfrak{p}_\infty} \leqq \|\mathfrak{a}\|_{\mathfrak{p}_\infty}.$$

COROLLARY. *If* $V(\mathfrak{a}) \geqq 1/C$ *then there is a* β *in* k *such that*

$$1 \leqq \|\beta\mathfrak{a}\|_{\mathfrak{p}} \leqq V(\mathfrak{a}) \quad \textit{for all} \quad \mathfrak{p}.$$

PROOF. The field elements in the parallelotope of size $\mathfrak{a}$ form a set of order greater than 1 so there is an $\alpha \neq 0$ such that $\|\alpha\|_{\mathfrak{p}} \leqq \|\mathfrak{a}\|_{\mathfrak{p}}$ for all $\mathfrak{p}$. Put $\beta = 1/\alpha$; then $1 \leqq \|\beta\mathfrak{a}\|_{\mathfrak{p}}$. Now for each $\mathfrak{q}$

$$\|\beta\mathfrak{a}\|_{\mathfrak{q}} = \frac{V(\beta\mathfrak{a})}{\prod_{\mathfrak{p}\neq\mathfrak{q}}\|\beta\mathfrak{a}\|_{\mathfrak{p}}} \leqq V(\beta\mathfrak{a}) = V(\mathfrak{a}).$$

LEMMA 11. *Let* $\mathfrak{a}$ *be any idèle and* $\mathfrak{q}$ *a fixed prime; then there is a* β *in* k *such that*

$$1 \leqq \|\beta\mathfrak{a}\|_{\mathfrak{p}} \leqq N\mathfrak{q}/C \quad \textit{for} \quad \mathfrak{p} \neq \mathfrak{q};$$

$$(C/N\mathfrak{q})V(\mathfrak{a}) \leqq \|\beta\mathfrak{a}\|_{\mathfrak{q}} \leqq V(\mathfrak{a}).$$

For an archimedean prime $\mathfrak{q}$ we mean by $N\mathfrak{q}$ the number 1. C is the constant of the preceding corollary.

PROOF. If we replace in $\mathfrak{a}$ the component $\alpha_{\mathfrak{q}}$ by a suitable $\alpha_{\mathfrak{q}}'$ and leave all other components unchanged we can achieve that the new idèle $\mathfrak{a}'$ satisfies

$$1/C \leqq V(\mathfrak{a}') \leqq N\mathfrak{q}/C.$$

Then we determine the β of our corollary and get

$$1 \leqq \|\beta\mathfrak{a}\|_{\mathfrak{p}} \leqq V(\mathfrak{a}') \leqq N\mathfrak{q}/C \quad \text{for} \quad \mathfrak{p} \neq \mathfrak{q}$$

and

$$1 \leqq \|\beta\mathfrak{a}'\|_{\mathfrak{q}} \leqq V(\mathfrak{a}').$$

Now

$$\|\beta\mathfrak{a}\|_{\mathfrak{q}} = \frac{V(\beta\mathfrak{a})}{V(\beta\mathfrak{a}')}\cdot\|\beta\mathfrak{a}'\|_{\mathfrak{q}} = \frac{V(\mathfrak{a})}{V(\mathfrak{a}')}\|\beta\mathfrak{a}'\|_{\mathfrak{q}}.$$

Hence

$$(C/N\mathfrak{q})V(\mathfrak{a}) \leqq V(\mathfrak{a})/V(\mathfrak{a}') \leqq \|\beta\mathfrak{a}\|_{\mathfrak{q}} \leqq V(\mathfrak{a}).$$

Let now U be the multiplicative group of all absolute units, that is, the set of all ζ of k satisfying $\|\zeta\|_{\mathfrak{p}}=1$ for all $\mathfrak{p}$. In case there is a constant field k_0, our group consists of the elements not equal to 0 of k_0. In case order means number of elements, U must be a finite group since it is contained in the parallelotope of size 1; so U consists in this case of all roots of unity of k and is a finite cyclic group.

We select a finite non-empty set S of primes $\mathfrak{p}$ that contains at least all archimedean primes. By $\mathfrak{a}_S$ we mean the idèles satisfying $\|\mathfrak{a}_S\|_{\mathfrak{p}}=1$ for all $\mathfrak{p}$ not in S. An element ϵ_S of k that belongs to $\mathfrak{a}_S$ is called an S-unit.

Let $\mathfrak{p}_1, \mathfrak{p}_2, \cdots, \mathfrak{p}_s$ be the primes of S. If ϵ_S is an S-unit and we know the s positive numbers $\|\epsilon_S\|_{\mathfrak{p}_1}, \|\epsilon_S\|_{\mathfrak{p}_2}, \cdots, \|\epsilon_S\|_{\mathfrak{p}_s}$, then we know $\|\epsilon_S\|_{\mathfrak{p}}$ for all $\mathfrak{p}$, so $\|\epsilon_S\|$ is known except for a factor in U. Let us call two S-units equivalent if they differ only by a factor in U. The product formula gives

$$\prod_{\nu=1}^{s}\|\epsilon_S\|_{\mathfrak{p}_\nu} = 1$$

and shows that it suffices to know the $s-1$ numbers $\|\epsilon_S\|_{\mathfrak{p}_1}, \|\epsilon_S\|_{\mathfrak{p}_2}, \cdots, \|\epsilon_S\|_{\mathfrak{p}_{s-1}}$. (Should $s=1$ then ϵ_S is already in U as the product formula shows.)

It is more convenient to take the logarithms of our numbers so we map the unit ϵ_S onto the following vector $v(\epsilon_S)$ of an ordinary space R_{s-1} of $s-1$ dimensions:

$$v(\epsilon_S) = (\log\|\epsilon_S\|_{\mathfrak{p}_1}, \log\|\epsilon_S\|_{\mathfrak{p}_2}, \cdots, \log\|\epsilon_S\|_{\mathfrak{p}_{s-1}}).$$

We have then for two units ϵ_S and η_S the relation

$$v(\epsilon_S\eta_S) = v(\epsilon_S) + v(\eta_S).$$

So the maps $v(\epsilon_S)$ form an additive group of vectors in R_{s-1}.

The product formula gives

$$\log \|\epsilon_S\|_{\mathfrak{p}_s} = -\sum_{\nu=1}^{s-1} \log \|\epsilon_S\|_{\mathfrak{p}_\nu}.$$

Let us consider a bounded region in R_{s-1} that gives bounds for $\log \|\epsilon_S\|_{\mathfrak{p}_\nu}$ $(\nu=1, 2, \cdots, s-1)$, say

$$-K \leqq \log \|\epsilon_S\|_{\mathfrak{p}_\nu} \leqq K \quad (\nu = 1, 2, \cdots, s-1).$$

Then we get for $\log \|\epsilon_S\|_{\mathfrak{p}_s}$ the bounds $-(s-1)K \leqq \log \|\epsilon_S\|_{\mathfrak{p}_s} \leqq (s-1)K$.

In case all the $\mathfrak{p}_\nu$ of S are discrete this gives only a finite number of possibilities for the ordinal number at each $\mathfrak{p}_\nu$; hence only a finite number of units inequivalent mod U. If there are archimedean primes in S then all ϵ_S of our region are contained in a parallelotope, so their order is finite. But order means number in this case. So we have proved:

Lemma 12. *There are only a finite number of vectors $v(\epsilon_S)$ in a bounded region of R_{s-1}.*

The following lemma is well known; we repeat its proof here for the convenience of the reader.

Lemma 13. *Let G be an additive group of vectors in an ordinary euclidean n-space R_n, such that no bounded region of R_n contains an infinite number of vectors of G. Assume that we can find m but not more vectors of G that are linearly independent with respect to real numbers. Then these m vectors may be selected in such a fashion that any vector of G is a linear combination of them with integral coefficients. In other words: G is a lattice of dimension m.*

Proof. The proof is by induction according to m.

Let $v_1, v_2, \cdots, v_m$ be a maximal set of independent vectors and G_0 be the subgroup of G contained in the subspace spanned by the vectors $v_1, v_2, \cdots, v_{m-1}$. Because of induction we may already assume that any vector in G_0 is a linear integral combination of $v_1, v_2, \cdots, v_{m-1}$.

Consider the subset $\mathfrak{S}$ of all v of G of the form

$$v = x_1v_1 + x_2v_2 + \cdots + x_{m-1}v_{m-1} + x_mv_m$$

with real coefficients $x_1, x_2, \cdots, x_m$ that satisfy

$$0 \leqq x_i < 1 \qquad \text{for } i = 1, 2, \cdots, m-1$$

and

$$0 \leqq x_m \leqq 1.$$

It is a bounded set. Let v'_m be a vector of S with the smallest possible $x_m \neq 0$, say

$$v'_m = \xi_1 v_1 + \xi_2 v_2 + \cdots + \xi_m v_m.$$

Starting now with any vector v of G we can select integral coefficients $y_1, y_2, \cdots, y_m$ in such a way that

$$v' = v - y_m v'_m - y_1 v_1 - y_2 v_2 - \cdots - y_{m-1} v_{m-1}$$

is in $\mathfrak{S}$ and the coefficient of v_m is even less than ξ_m. So this coefficient of v_m is 0, that is, v' is in G_0. So v' is an integral linear combination of $v_1, v_2, \cdots, v_{m-1}$ and therefore v is an integral linear combination of $v_1, v_2, \cdots, v_{m-1}$ and v'_m.

THEOREM 5. *The vectors $v(\epsilon_S)$ form a lattice of at most $s-1$ dimensions. The ϵ_S themselves form* mod U *a free abelian group with at most $s-1$ generators.*

5. **A more restrictive axiom.** If we wish to derive stronger theorems as for instance that of the existence of enough units, we must replace Axiom 2 by a stronger axiom. So we assume from now on that we have in k besides Axiom 1 also

AXIOM 2a. *There is at least one prime in $\mathfrak{M}$ that is either archimedean, with the real field or the complex field as its completed field, or else discrete with residue class field having only a finite number of elements.*

Since Axiom 2 is a consequence of Axiom 2a we can assume all the results we derived thus far and thus we see that k is either a number field or else a function-field where k_0 has only a finite number of elements. We see immediately that Axiom 2a holds for all primes of $\mathfrak{M}$.

LEMMA 14. *To any integer M there are only a finite number of primes $\mathfrak{p}$ with $N\mathfrak{p} \leqq M$.*

PROOF. Since there are only a finite number of archimedean primes we are concerned only with the nonarchimedean ones. Consider $M+1$ integers α_ν of R and let $\mathfrak{p}$ be a prime with $N\mathfrak{p} \leqq M$. Two α_i are in the same residue class, say α_1 and α_2; hence $|\alpha_1 - \alpha_2|_{\mathfrak{p}} < 1$. So our $\mathfrak{p}$'s are contained among the primes for which one of the differences $\alpha_i - \alpha_k$ $(i \neq k)$ has an absolute value $|\alpha_i - \alpha_k|_{\mathfrak{p}} < 1$. Because of Axiom 1 our lemma holds.

Now let S be again a finite and non-empty set of primes.

LEMMA 15. *There is a constant E such that to any idèle $\mathfrak{a}_S$ and any prime $\mathfrak{q}$ of S we can find an S-unit ϵ_S such that*

$$\|\epsilon_S \mathfrak{a}_S\|_{\mathfrak{p}} \leqq E \quad \textit{for all } \mathfrak{p} \neq \mathfrak{q} \textit{ of } S.$$

PROOF. Select β according to Lemma 11. Then $1 \leqq \|\beta \mathfrak{a}_S\|_{\mathfrak{p}} \leqq N\mathfrak{q}/C$ for $\mathfrak{p} \neq \mathfrak{q}$. So

$$1 \leqq \|\beta\|_{\mathfrak{p}} \leqq N\mathfrak{q}/C \quad \text{for all } \mathfrak{p} \text{ not in } S.$$

If $\|\beta\|_{\mathfrak{p}} \neq 1$, then $\|\beta\|_{\mathfrak{p}} \geqq N\mathfrak{p}$. Since there are only a finite number $\mathfrak{p}_1, \mathfrak{p}_2, \cdots, \mathfrak{p}_l$ of primes with $N\mathfrak{p}_i \leqq N\mathfrak{q}/C$ we get $\|\beta\|_{\mathfrak{p}} = 1$ for all $\mathfrak{p} \neq \mathfrak{p}_1, \mathfrak{p}_2, \mathfrak{p}_l$ and $\mathfrak{p}$ not in S. Since $\mathfrak{p}_1, \mathfrak{p}_2, \cdots, \mathfrak{p}_l$ are discrete we get only a finite number of possibilities for each $\|\beta\|_{\mathfrak{p}_i}$.

Assume that $\beta_1, \beta_2, \cdots, \beta_r$ already realize any possible distribution of values for $\|\beta\|_{\mathfrak{p}_i}$. Then to any of our β there is a β_k with $\|\beta\|_{\mathfrak{p}} = \|\beta_k\|_{\mathfrak{p}}$ for all $\mathfrak{p}$ not in S, or $\beta = \beta_k \epsilon_S$. Substituting back we get

$$\|\beta_k \epsilon_S \mathfrak{a}_S\|_{\mathfrak{p}} \leqq N\mathfrak{q}/C \quad \text{for} \quad \mathfrak{p} \neq \mathfrak{q}.$$

So $\|\epsilon_S \mathfrak{a}_S\|_{\mathfrak{p}} \leqq E$ for all $\mathfrak{p} \neq \mathfrak{q}$ of S where

$$E = \max_{\nu=1,2,\cdots,r;\ \mathfrak{p}\in S} \left(\frac{N\mathfrak{q}}{C\|\beta_\nu\|_{\mathfrak{p}}} \right).$$

Now select an $\mathfrak{a}_S$ so that $\|\mathfrak{a}_S\|_{\mathfrak{p}} > E$ for all $\mathfrak{p}$ of S. If ϵ_S is the corresponding unit then $\|\epsilon_S\|_{\mathfrak{p}} < 1$ for all $\mathfrak{p} \neq \mathfrak{q}$ of S.

Assume now that $\mathfrak{p}_1, \mathfrak{p}_2, \cdots, \mathfrak{p}_s$ are all the primes in S. Then $\mathfrak{q}$ could be any of the primes $\mathfrak{p}_i$. We get in this fashion s S-units $\epsilon_1, \epsilon_2, \cdots, \epsilon_s$, where ϵ_i satisfies $\|\epsilon_i\|_{\mathfrak{p}_k} < 1$ for $k \neq i$. Because of the product formula we also get

$$\|\epsilon_i\|_{\mathfrak{p}_i} > 1.$$

The first $s-1$ of these S-units are mapped onto vectors

$$v_i = (a_{i1}, a_{i2}, \cdots, a_{i,s-1}), \qquad i = 1, 2, \cdots, s-1,$$

where $a_{ik} = \log \|\epsilon_i\|_{\mathfrak{p}_k}$. Then $a_{ii} > 0$ and $a_{ik} < 0$ for $i \neq k$, but $\sum_{\nu=1}^{s-1} a_{i\nu} = \sum_{\nu=1}^{s-1} \log \|\epsilon_i\|_{\mathfrak{p}_\nu} = -\log \|\epsilon_i\|_{\mathfrak{p}_s} > 0$.

We prove now that the vectors v_i are linearly independent, that is, that the homogeneous equations

$$\sum_{\nu=1}^{s-1} x_\nu a_{\nu k} = 0, \qquad k = 1, 2, \cdots, s-1,$$

have only the trivial solution. To that effect it suffices to show that the homogeneous equations

$$\sum_{\nu=1}^{s-1} a_{i\nu} y_\nu = 0, \qquad i = 1, 2, \cdots, s-1,$$

have only the trivial solution.

Assume indeed that $y_1, y_2, \cdots, y_{s-1}$ is a non-trivial solution and that y_i has the greatest absolute value. It is no restriction to assume $y_i > 0$ so that $y_i \geqq y_j$ for all $j \neq i$. Since $a_{ij} < 0$ we get $a_{ij} y_i \leqq a_{ij} y_j$. Now

$$\begin{aligned} a_{i1}y_1 + a_{i2}y_2 + \cdots + a_{in}y_n &\geqq a_{i1}y_i + a_{i2}y_i + \cdots + a_{in}y_i \\ &\geqq (a_{i1} + a_{i2} + \cdots + a_{in})y_i. \end{aligned}$$

The left side of the inequality should be 0 but on the right side both factors are positive.

This proves:

THEOREM 6 (UNIT THEOREM).[6] *If Axioms* 1 *and* 2a *hold then the dimension mentioned in Theorem* 5 *is precisely* $s-1$, *so the S-units form* mod U *a free abelian group with* $s-1$ *generators.*

Another consequence of Axiom 2a is the following: If we go back to Lemma 11 and select in it for $\mathfrak{q}$ one of the primes of S then the inequalities show just as in the proof of Lemma 15 that $\|\beta\mathfrak{a}\|_\mathfrak{p} = 1$ for all $\mathfrak{p}$ with $N\mathfrak{p} > N\mathfrak{q}/C$ and that outside of S there are only a finite number of possibilities for the value distribution of $\|\beta\mathfrak{a}\|_\mathfrak{p}$. Assume that the idèles $\mathfrak{a}_1, \mathfrak{a}_2, \cdots, \mathfrak{a}_m$ realize any possible case; then there is always an i such that $\|\beta\mathfrak{a}\|_\mathfrak{p} = \|\mathfrak{a}_i\|_\mathfrak{p}$ for all p not in S or $\beta\mathfrak{a} = \mathfrak{a}_i \cdot \mathfrak{a}_S$. This proves:

THEOREM 7 (FINITENESS OF CLASS NUMBER). *There is a finite set of idèles* $\mathfrak{a}_1, \mathfrak{a}_2, \cdots, \mathfrak{a}_m$ *such that any idèle* $\mathfrak{a}$ *is of the form*

$$\mathfrak{a} = \alpha \mathfrak{a}_i \mathfrak{a}_S$$

for a suitable i, $\alpha \in k$ *and* $\mathfrak{a}_S$.

We mention the special case, important for class field theory:

THEOREM 8. *If the set* S *is big enough then* $\mathfrak{a} = \alpha\mathfrak{a}_S$ *for all idèles* $\mathfrak{a}$.

PROOF. Add to the previous set S also the primes $\mathfrak{p}$ where any $\|\mathfrak{a}_i\|_\mathfrak{p} \neq 1$.

[6] The unit theorem in this form is due to Hasse. It is proved by Chevalley in [4].

References

1. E. Artin, *Über die Bewertungen algebraischer Zahlkörper*, Journal für Mathematik vol. 167 (1932) pp. 157–159.

2. C. Chevalley, *Sur la théorie du corps de classes dans les corps finis et les corps locaux*, Journal of College of Sciences, Tokyo, 1933, II, part 9.

3. ———, *Généralization de la théorie de corps de classes pour les extensions infinies*, Journal de mathématiques pures et appliquées (9) vol. 15 (1936) pp. 359–371.

4. ———, *La théorie du corps de classes*, Ann. of Math. vol. 41 (1940) pp. 394–418.

5. A. Ostrowski, *Über einige Lösungen der Funktionalgleichung* $\phi(x)\phi(y)=\phi(xy)$, Acta Math. vol. 41 (1918) pp. 271–284.

6. ———, *Untersuchung in der arithmetische Theorie der Körper*, Parts I, II, and III, Math. Zeit. 39 (1935) pp. 269–321.

7. B. L. van der Waerden, *Moderne Algebra*, vol. 1, 2d ed., Berlin, 1937.

8. G. Whaples, *Non-analytic class field theory and Grunwald's theorem*, Duke Math. J. vol. 9 (1942) pp. 455–473.

Indiana University and
University of Pennsylvania

A NOTE ON AXIOMATIC CHARACTERIZATION OF FIELDS

E. ARTIN AND G. WHAPLES

Since publication of our paper, *Axiomatic characterization of fields by the product formula for valuations*,[1] we have found that the fields of class field theory can be characterized by somewhat weaker axioms; we can drop the assumption, in Axiom 1, that $|\alpha|_{\mathfrak{p}}=1$ for all but a finite number of $\mathfrak{p}$, replacing it by the assumption that the product of all valuations converges absolutely to the limit 1 for all α.

Our original proof can be adapted to the new axiom with a few modifications, which we shall describe here. In §2, we keep Axiom 1 for reference and introduce:

AXIOM 1*. *There is a set $\mathfrak{M}$ of prime divisors $\mathfrak{p}$ and a fixed set of valuations $|\ |_{\mathfrak{p}}$, one for each $\mathfrak{p}\in\mathfrak{M}$, such that, for every $\alpha\neq 0$ of k, the product $\prod_{\mathfrak{p}}|\alpha|_{\mathfrak{p}}$ converges absolutely to the limit* 1. (*That is, the series* $\sum_{\mathfrak{p}}\log|\alpha|_{\mathfrak{p}}$ *converges absolutely to* 0.)

We must then omit the statement that there are only a finite number of archimedean primes, since this does not follow immediately from 1*; instead of it, we use the fact that $\sum_{\mathfrak{p}_\infty}\rho(\mathfrak{p}_\infty)$ and $\sum_{\mathfrak{p}_\infty}\lambda(\mathfrak{p}_\infty)$ converge absolutely. These quantities are defined on p. 480; the convergence follows from the fact that the product over all $\mathfrak{p}_\infty$ of $|1+1|_{\mathfrak{p}_\infty}$ must converge absolutely. Also, we must temporarily broaden the definition of "parallelotope" so as to permit a parallelotope to be defined by any valuation vector $\mathfrak{a}$ for which $\prod_{\mathfrak{p}}|\mathfrak{a}|_{\mathfrak{p}}$ converges absolutely (rather than restricting $\mathfrak{a}$ to be an idèle). In the statement of Axiom 2 we must replace "Axiom 1" by "Axiom 1*," Theorem 2, however, is left unchanged, together with Lemmas 4, 5, and 6, which are needed only to prove it; this theorem shows that the fields of class field theory really satisfy Axiom 1, so that at the end of the whole proof we shall find that Axiom 1 is a consequence of Axioms 1* and 2.

In §3, k is assumed to be any field for which Axioms 1* and 2 hold. Lemma 8 holds under assumption of Axiom 1*, for our slightly more general parallelotopes; in its proof we have only to note, in case of archimedean primes, that the product $\prod_{\mathfrak{p}_\infty}4^{\rho(\mathfrak{p}_\infty)}$ converges absolutely. In Lemma 9, property 2 must be replaced by:

2*. $|\alpha|_{\mathfrak{p}_\infty}\leq B_{\mathfrak{p}_\infty}|y|_{\mathfrak{p}_\infty}$, *with a set of constants $B_{\mathfrak{p}_\infty}$ for which $\prod_{\mathfrak{p}_\infty}B_{\mathfrak{p}_\infty}$ converges absolutely.*

Received by the editors December 9, 1945.

[1] Bull. Amer. Math. Soc. vol. 51 (1945) pp. 469–492.

To prove existence of these constants, let, at each $\mathfrak{p}_\infty$, $M_{\mathfrak{p}_\infty}$ be the maximum of $|\alpha_i|_{\mathfrak{p}_\infty}$ for $i=1 \cdots l$; then $\prod_{\mathfrak{p}_\infty} M_{\mathfrak{p}_\infty}$ converges to a finite limit. Take $B_{\mathfrak{p}_\infty}=M_{\mathfrak{p}_\infty} l^{\lambda(\mathfrak{p}_\infty)}$; since $\sum_{\mathfrak{p}_\infty}\lambda(\mathfrak{p}_\infty)$ was absolutely convergent, our conclusion follows.

Lemma 10 holds as stated, although the set of $\mathfrak{p}_\infty$ is not now known to be finite. But as soon as we have proved that n is finite, it follows from Theorem 2 that our original Axiom 1 holds, so no further changes are necessary. (The theorems about parallelotopes in §4 hold only for parallelotopes defined by ideal elements.)

It is easy to construct an example of a field which satisfies Axiom 1* but does not satisfy Axiom 1 (nor, of course, Axiom 2). Let $k=R(x, z)$ be the set of all rational functions of x and z over the rational field. Let $k_0=R(x)$, consider k as the set $k_0(z)$ of all rational functions of z with k_0 as constant field, and denote by $\mathfrak{M}_0$ the set of all divisors which are trivial on k_0. We construct $\mathfrak{M}_0$, and define the set of normed valuations, exactly as in the proof of Lemma 6 of our original paper (pp. 477–479). Let $V_0(A)=\prod\|A\|_{p_0}$ where the product is taken over all $p_0\in\mathfrak{M}_0$; by Lemma 6, $V_0(A)=1$ for all $A\in k$.

Now let $x_1=x+z$, $x_2=x+2z, \cdots, x_i=x+iz, \cdots$; let $k_i=R(x_i)$ and for each i construct the sets $\mathfrak{M}_i$ of divisors p_i by repeating exactly the above process with k_0 replaced by k_i. The products $V_i(A)$ are all equal to 1. These sets $\mathfrak{M}_i$ are by no means disjoint; for example one can easily see that the irreducible polynomial z defines the same valuation in each $\mathfrak{M}_i$. However, it is unnecessary to explore these duplications in detail; we shall need only the facts that the valuations $p_{i\infty}$ and $p_{j\infty}$ are inequivalent for $i\neq j$, and are not equivalent to any of the finite p_ν. Namely, $x_i=x+iz=x_j+(i-j)z$ has value 1 at $p_{i\infty}$, but value $q>1$ at all $p_{j\infty}$ with $j\neq i$. And z has value $q>1$ at all $p_{i\infty}$, but has value $\leqq 1$ at all finite p_ν.

To construct our example, let ϵ_ν $(\nu=0, 1, 2, \cdots)$ be an infinite sequence of positive numbers whose sum is finite. Form the product

$$\prod \|A\|_{p_i}^{\epsilon_i}$$

over all $p_i\in\mathfrak{M}_i$, all i, and in this product unite each set of equivalent valuations into a single valuation. The exponents insure the convergence of the infinite products involved in this step. To show that the whole product is absolutely convergent for each $A\in k$, write A in the form $A=g(x, z)/h(x, z)$ where g and h are polynomials with rational coefficients. If N and M are the maximum degrees in x and z, respectively, for both numerator and denominator, then A can be written in the form $g_i(z)/h_i(z)$, where numerator and denominator are polynomials in z with coefficients in k_i, and are of degree at most $N+M$ in z. It follows from this that, for fixed A, the number of factors of $V_i(A)$ which are greater than 1 (or which are less than 1) is bounded, and their size is bounded; and this bound is uniform for all i. Hence the exponents ϵ_i insure absolute convergence. Finally, we note that our product, applied to z, contains an infinity of factors different from 1.

Taking the product over sets $\mathfrak{M}_0$ and $\mathfrak{M}_1$ only gives an example in which Axiom 1 is satisfied but Axiom 2 is not; for the field of constants with respect to $\mathfrak{M}_0 \cup \mathfrak{M}_1$ is the rational field $k_0 \cap k_1$.

To get an example of a field possessing a valuation satisfying Axiom 2, but such that this valuation cannot be contained in any set $\mathfrak{M}$ satisfying Axiom 1, take the p-adic closure of either the rational field or any of the fields $k_0(z)$ of our original paper, with p any of the divisors of Lemma 6. Because of Theorem 3, such an $\mathfrak{M}$ cannot exist.

Indiana University and
University of Pennsylvania

A Characterization of the Field of Real Algebraic Numbers*

Emil Artin in Hamburg

In the following, R will denote the field of rational numbers, P the field of real algebraic numbers, and Ω the field of all algebraic numbers. One has

$$\Omega = P(i) \qquad (i = \sqrt{-1}). \tag{1}$$

Instead of i one can use any complex number. We now ask about all subfields K of Ω for which Ω over K is finite, i.e. for which in analogy to (1) there is a number α such that

$$\Omega = K(\alpha). \tag{2}$$

We find

THEOREM. *If Ω is finite over K and $K \neq \Omega$, then $[\Omega : K] = 2$. More exactly:*

$$\Omega = K(i). \tag{3}$$

Moreover, K is conjugate to P within Ω, i.e. there is an automorphism σ of Ω which takes K to P.

Naturally an algebraic characterization of a field can only be given up to isomorphism. Thus, P is essentially characterized as being the "only" proper subfield of Ω for which Ω/K is finite.

If K is a field for which (2) holds, then Ω/K is a Galois extension since Ω is algebraically closed. We now proceed stepwise:

1. Suppose that Ω is of prime degree p over K and thus cyclic. Since Ω is algebraically closed, the irreducible polynomials over K have degree either p or 1. In particular, every polynomial over K of degree $p-1$ or less must split as a product of linear factors. Since the p'th roots of unity satisfy such a polynomial, they must lie in K. Since Ω is cyclic over K, this shows that the number α in (2) can be taken to have the form $\alpha = \sqrt[p]{a}$, where a belongs to K.

However, the p^2 roots of unity cannot all belong to K. If they did, consider the polynomial $x^{p^2} - a$. It's linear factors, $x - \epsilon_\nu \sqrt[p^2]{a}$, where ϵ_ν is a p^2 root of unity, cannot lie in K. Otherwise, would follow that $\sqrt[p^2]{a}$ was in K and so $\alpha = \sqrt[p]{a}$ would be in K, contrary to the assumption about α. A factor of degree p of $x^{p^2} - a$ would have a constant term of the form $\epsilon \sqrt[p]{a}$, where ϵ is a p^2 root of unity. This is immediate from the form of its linear factors. If ϵ were in K then, once again, $\sqrt[p]{a}$ would be in K, contrary to assumption.

---

*Kennzeichnung der Körper der reellen algebraischen Zahlen, Hamb. Abh., Bd. 3 (1924), 319–323.

Let ϵ be a primitive p^2 root of unity and ζ be a primitive p^3 root of unity. Since ϵ is not in K and Ω/K has degree p, ϵ generates Ω, thus

$$\Omega = K(\epsilon). \tag{4}$$

Over K, ζ must satisfy an irreducible polynomial of degree p which is a factor of $x^{p^3}-1$. Thus, its coefficients must lie in $R(\zeta)$. This polynomial is also irreducible over the intersection k of K with $R(\zeta)$. Since ζ is a root of this polynomial, $R(\zeta)/k$ has degree p. The degree of k over R is therefore $p(p-1)$. Now, one can see immediately that p must be 2. If p is odd, $R(\zeta)/R$ is cyclic. Thus, there is only one subfield whose degree over R is $p(p-1)$, namely the field of p^2 roots of unity, $R(\epsilon)$. This would mean, however, that ϵ was in k and thus in K. This contradicts what has been shown above.

Thus, $p = 2$ and so, because of (4), $\Omega = K(i)$.

2. Now, suppose that the degree of Ω over K is arbitrary. We set $K_1 = K(i)$ and assert that $K_1 = \Omega$. If this were not the case, Ω/K_1 would have degree n different from 1. Let $\mathfrak{g}$ be a subgroup of prime order p of the Galois group of Ω/K_1. Ω is of degree p over the fixed field K_2 of $\mathfrak{g}$. According to Section 1, $\Omega = K_2(i)$. However, K_2 contains K_1 which contains i. It follows that $K_2 = \Omega$, contrary to the fact that Ω/K_2 has degree p. Thus, $K_1 = \Omega$, as asserted, and the first part of the theorem is proven, in particular (3).

3. In K, -1 cannot be written as the sum of two squares. Otherwise, $-1 = a^2 + b^2$ with both $a \neq 0$ and $b \neq 0$ since if either were equal to zero, then $\sqrt{-1}$ would be in K contrary to (3). Now, consider the equation

$$(x^2 - a)^2 + b^2 = 0\ . \tag{5}$$

None of its four roots $\pm\sqrt{a \pm ib}$ belong to K since otherwise $a \pm ib$ would be in K and thus i itself would be in K. The left hand side of (5) therefore factors over K into the product of two quadratic factors. The constant terms are, according to the various cases, $\pm a \pm ib$ or $\pm\sqrt{a^2+b^2} = \pm i$. This is impossible, since otherwise i would be in K.

4. Each number a in K is either a square or the negative of a square in K.

In any case, the number $\sqrt{a}$ is in $\Omega = K(i)$ and therefore can be written in the form $\sqrt{a} = b + ic$, where b and c are in K. Square both sides and compare the coefficients of i. One finds that $bc = 0$. If $c = 0$, then $a = c^2$, and if $b = 0$, then $a = -c^2$.

5. Every sum of arbitrarily many squares in K is itself a square.

Consider the sum of two squares. It follows from Step 4 that $a_1^2 + a_2^2 = \pm b^2$. For $b = 0$ there is nothing to prove. If $b \neq 0$, then $(a_1/b)^2 + (a_2/b)^2 = \pm 1$. By Step 3, the plus sign must hold.

The general result now follows by complete induction.

6. In K, -1 cannot be written as a sum of squares.

This follows from Step 5, since if -1 were a sum of squares it would have to be a square and so $\sqrt{-1}$ would be in K.

7. Let K^* is a finite dimensional number field contained in K. At least one conjugate of K^* is contained in the real numbers.

If no conjugate of K^* is real, then -1 would be totally positive. It would follow, contrary to Step 6, that -1 is a sum of squares in K^* and so also in K.

(See E. Landau, Über die Zerlegung total positiver Zahlen in Quadrate, Göttinger Nachrichten 1919 and C. Siegel, Darstellung total positiver Zahlen durch Quadrate, Mathematische Zeitschrift, Bd 11.)[1]

8. The set of numbers in K is countable, say $K = \{\alpha_1, \alpha_2, \dots\}$. Set $K_1 = K(\alpha_1), K_2 = K_1(\alpha_2), \dots, K_\nu = K_{\nu-1}(\alpha_\nu), \dots$, etc. In this way we construct a tower of fields $K_1, K_2, K_3, \dots$ whose union is K.

According to Section 7, each field K_ν has a least one real conjugate K'_ν. Let σ_ν denote an isomorphism which takeK_ν to K'_ν. There is some freedom in the choice of K'_ν and σ_ν. Since K_ν is of finite degree over R, there are only finitely many choices for each ν.

We now give the following recursive procedure:

If $\sigma_1, \sigma_2, \dots, \sigma_{\nu-1}$ are already determined, choose, if possible, σ_ν and K'_ν as an extension of $\sigma_{\nu-1}$, i.e. σ_ν restricted to $K_{\nu-1}$ should be $\sigma_{\nu-1}$. If the choice of a real K'_ν is impossible, let σ_ν be any isomorphism of K_ν with a real conjugate K'_ν. Now change the whole series of fields $K'_1, K'_2, \dots, K'_{\nu-1}$ and isomorphisms $\sigma_1, \sigma_2, \dots, \sigma_{\nu-1}$ by defining σ_i to be the restriction of σ_ν to K_i and K'_i to be the image of σ_i. The new sequence of isomorphisms $\sigma_1, \sigma_2, \dots, \sigma_{\nu-1}, \sigma_\nu$ are then, by construction, extensions of one another.

We observe now, for fixed r, the mapping σ_r of K_r to the real field K'_r. At first, it looks like at each new step σ_r will be changed. This, however, is not the case. After finitely many steps σ_r remains fixed from then on. If at a certain step σ_r changes, it cannot at any later stage return to its original designation. A return to its original designation would mean that we could have preserved that designation earlier, at the next stage, contrary to our procedure. Since for fixed r there are only finitely many choices for σ_r, the choice of σ_r must eventually remain fixed.

We have, finally, a tower of real fields $K'_1, K'_2, \dots$ and a sequence of isomorphisms $\sigma_1, \sigma_2, \dots$ such that σ_ν restricted to $K_{\nu-1}$ coincides with $\sigma_{\nu-1}$.

Let K' denote the union of the fields K'_ν and σ the mapping of K to K' given by $\sigma(\alpha) = \sigma_r(\alpha)$ if α is in K_r. One sees immediately that σ is an isomorphism of K with K'.

For a and b in K set $\sigma(a + ib) = \sigma(a) + i\sigma(b)$. An easy calculation shows that for α and β in Ω we have $\sigma(\alpha + \beta) = \sigma(\alpha) + \sigma(\beta)$ and $\sigma(\alpha\beta) = \sigma(\alpha)\sigma(\beta)$. From $\alpha = a + ib$ and $\sigma(\alpha) = \sigma(a) + i\sigma(b) = 0$ it follows that $\sigma(a) = 0$ and $\sigma(b) = 0$, since both $\sigma(a)$ and $\sigma(b)$ are real. Consequently, $a = 0$ and $b = 0$ and so $\alpha = 0$. Thus, $\sigma(\alpha) = \sigma(\beta)$ implies $\alpha = \beta$.

The extended map σ is thus an isomorphism between $\Omega = K(i)$ and $K'(i)$. Since Ω is algebraically closed, so is $K'(i)$. Since Ω is algebraic over $K'(i)$ it follows that $\Omega = K'(i)$. This isomorphism σ is thus an automorphism of Ω.

Since K' is real and $K'(i) = \Omega$ it follows that K' is the whole field of real algebraic numbers, i.e. $K' = P$. Thus, we have found an automorphism of Ω which takes K onto P (moreover, as one easily sees, this automorphism is essentially unique), whereby all parts of our theorem have been proven.

Since Galois theory remains valid in the simple cases before us, one easily thinks through the implications of our theorem for the structure of the Galois group.

---

[1]The modern reader will note that this follows from the theorem that over a number field, a quadratic form in five or more variables which is indefinite at each infinite place must have a non-trivial zero. Simply apply this to the sum of five squares to conclude that over a number field with no real conjugate, -1 is a sum of four squares. *Editor's Note.*

THEOREM. *Every finite subgroup of the group* $\mathfrak{G}$ *of all automorphisms of* Ω *is of order* 2. *The elements of order* 2 *constitute a single conjugacy class in* $\mathfrak{G}$. *An element* τ *of this class commutes only with* e *and itself.*

THE MATHEMATICAL SEMINAR, HAMBURG, JUNE 1924

The Algebraic Construction of Real Fields*

Emil Artin and Otto Schreier in Hamburg

With his paper "Algebraische Theorie der Körper"[1] E. Steinitz has opened up wide areas of algebra to the abstact mode of treatment; his path-breaking investigation is largely responsible for the strong development of modern algebra since his work appeared. However, there still exist many branches of algebra which have not benefitted from the abstract method, among them, perhaps, real algebra and certain parts of algebraic number theory. We mention, for example, the theorem of Sturm about the number of real roots of a polynomial equation, the theory of units in an algebraic number field, class field theory, and the reciprocity laws.

In order to be able to handle real algebra abstractly, one must necessarily ask, to begin with, which properties can be used to characterize real fields, for example the field of of all real numbers and the field of all real algebraic numbers, among other fields. One would attempt to describe these properties by simple axioms. Such axioms would have to satisfy various demands. In the first place, they must be consistent with the idea of "real" in the usual sense. For example, an algebraic number field would only be called real if was isomorphic to a subfield of the real numbers. Moreover, they should make it possible to prove, purely algebraically, the existence of a maximally extended class of real fields which would include, as a special case, the field of all real algebraic numbers. In these abstract real fields it should be possible to prove all the theorems of real algebra.

It is, in fact, possible to give such a characterization of real fields. It would be tempting to proceed from the idea of an ordered field. From the standpoint of higher algebra which deals with arbitrary fields, it would be preferable to have a definition which only uses addition and multiplication yet carries with it the possibility of ordering the field under consideration. One might also expect of such a definition that it would lead more easily to an algebract existence proof of real fields.

The basic property of real fields that we are looking for is the following: If a sum of squares in the field is zero, then the individual summands must be zero. Equivalently, -1 cannot be written as a sum of squares. This condition is suggested by recent investigation of one of the authors[2] in which the field of real algebraic numbers is characterized by algebraic properties. That one can now build up real algebra in a completely abstract manner will be demonstrated in this article.

* *Algebraische Konstruction reeller Körper*, Hamb. Abh., Bd. 5 (1926), 85–99.

[1] Crelle, Bd. 137 (1910), pp. 167–309.

[2] E. Artin, *Kennzeichnung des Körper der reellen algebraischen Zahlen*, Hamb. Abh., Bd. 3 (1924), 319–323.

In the following work[3] the fruitfulness of this approach will be demonstrated. Through these methods the question of when a field element can be written as a sum of squares will be answered, and therewith a solution given to Hilbert's problem about positive definite rational functions.

I. Definition and Principal Properties of Real Fields

A field will be called real if -1 cannot be written as a sum of squares

A real field must have characteristic zero since in a field of characteristic p, -1 is the sum of $p-1$ copies of 1^2.

A field K is called ordered if a property "positive" (≥ 0) is given for its elements which satisfies the following requirements

1. *For each element $a \in K$ exactly one of the following conditions holds*

$$a = 0, \quad a > 0, \quad -a > 0.$$

2. *If $a > 0$ and $b > 0$, then $a + b > 0$ and $ab > 0$.*

If $-a > 0$ we say that a is negative. In an ordered field we define a size relationship by

$$a > b \text{ (or } b < a) \text{ if } a - b > 0.$$

One can show without difficulty that the usual axioms of order must hold.

The absolute value of a, $|a|$, is defined to be the non-negative element of the pair $\{a, -a\}$. The following rules hold

$$|ab| = |a|\ |b|, \quad |a + b| \leq |a| + |b|.$$

Also, one sees that $|a|^2 = a^2$. Thus, a square is always non-negative. In particular, $1 = 1^2 > 0$, and $-1 < 0$. Thus, -1 cannot be a sum of squares. Consequently, every ordered field is real. *A field P is called real closed if P is real and every proper algebraic extension of P is not real.*

THEOREM 1. *Every real closed field can be ordered in one and only one way.*

If P is real closed, we will show that for every non-zero element $a \in P$, either a is a square or $-a$ is a square, these cases being mutually exclusive. We also show that the sum of squares is a square.

From these assertions, the theorem follows at once. One sets $a > 0$ when a is a square different from zero. This defines an order on P and it is the only order possible because a square is non-negative in any ordering.

If $\gamma \in P$ is not a square, then $P(\sqrt{\gamma})$ is a proper algebraic extension of P and therefore not real. Thus, we have an equation

$$-1 = \sum_{\nu=1}^{n} (\alpha_\nu \sqrt{\gamma} + \beta_\nu)^2$$

or

$$-1 = \gamma \sum_{\nu=1}^{n} \alpha_\nu^2 + \sum_{\nu=1}^{n} \beta_\nu^2 + 2\sqrt{\gamma} \sum_{\nu=1}^{n} \alpha_\nu \beta_\nu,$$

where α_ν and β_ν belong to P. The last term must vanish, since otherwise $\sqrt{\gamma} \in P$. On the other hand, the first term cannot vanish, since otherwise P would not be real. From this we can conclude, to begin with, that γ cannot be a sum of squares

[3] E. Artin, Über die Zerlegung definiter Funktionen in Quadrate.

since if it were, -1 would be a sum of squares in P. We've shown that if γ is not a square, then γ is not a sum of squares. Put positively, every sum of squares is a square.

Now, we have

$$-\gamma = \frac{1 + \sum_{\nu=1}^{n} \beta_\nu^2}{\sum_{\nu=1}^{n} \alpha_\nu^2}.$$

Both numerator and denominator are sums of squares and thus squares, which shows that $-\gamma = c^2$, where $c \in P$. Consequently, for every $\gamma \in P$ either $\gamma = b^2$ or $-\gamma = c^2$. If $\gamma \neq 0$ then both conditions cannot hold simultaneously since otherwise $-1 = (c/b)^2$ which is impossible.

Based on Theorem 1, in what follows we will consider every real closed field to be ordered.

THEOREM 2. *In a real closed field, every polynomial of odd degree has at least one root.*

The theorem is trivial for polynomials of degree one. Assume it is true for every every polynomial of odd degree less than n where $n > 1$ is an odd integer. Let $f(x)$ be a polynomial of odd degree n. If $f(x)$ is reducible over the real closed field P, it must have a factor of odd degree less than n, and thus a root in P. The assumption that $f(x)$ is irreducible will lead to a contradiction. Let α be a root in some extension field of P. Since $P(\alpha)$ is not a real field, one must have an equation

$$-1 = \sum_{\nu=1}^{n} \phi_\nu(\alpha)^2, \tag{1}$$

where the $\phi_\nu(x)$ are polynomials of degree at most $n-1$. From (1) we derive an identity

$$-1 = \sum_{\nu=1}^{n} \phi_\nu(x)^2 + f(x)g(x) \tag{2}$$

The sum of the $\phi_\nu(x)^2$ is of even degree since the leading coefficients are squares and cannot add up to zero. Furthermore, the degree of the sum must be positive since otherwise equation (1) would yield a contradiction. Thus, $g(x)$ must have odd degree $\leq n-2$ and therefore a root a in P. Substituting $x = a$ in equation (2) yields

$$-1 = \sum_{\nu=1}^{n} \phi_\nu(a)^2$$

Thus, we have arrived at a contradiction, since the $\phi_\nu(a)$ lie in P.

THEOREM 3. *A real closed field is not algebraically closed. On the other hand, the field obtained by adjoining i $(= \sqrt{-1})$ is algebraically closed.*

The first half is trivial, since the equation $x^2 + 1 = 0$ has no root in any real field. The second half follows directly from

THEOREM 3A. *Suppose K is an ordered field with the following two properties. Every positive element has a square-root in K, and every polynomial of odd degree has a root in K. Then, the field obtained by adjoining i to K is algebraically closed.*

To begin with, we remark that every element in $K(i)$ has a square-root and therefore every quadratic equation is solvable. To show the existence of a square root, suppose $a+bi$ is in $K(i)$ where a and b are in K. Then, $\sqrt{a^2+b^2}$ is also in K and, further, $|\sqrt{a^2+b^2}| \geq |a|$. Thus,

$$c_1 = \left|\sqrt{\frac{a+|\sqrt{a^2+b^2}|}{2}}\right| \quad \text{and} \quad c_2 = \left|\sqrt{\frac{-a+|\sqrt{a^2+b^2}|}{2}}\right|$$

also belongs to K and $(c_1 + ic_2 \text{ sign } b)^2 = a + ib$.

To show that every irreducible polynomial over $K(i)$ has a root we can, following Gauss, proceed as follows. Suppose it has been proven that every polynomial with coefficients in K and without repeated roots, whose degree is divisible by 2^{m-1} but not by 2^m, has a root in $K(i)$ (by assumption, this is true for $m=1$). Let $f(x)$ be a polynomial of degree n with distinct roots, where $n = 2^m q$ with q odd. Let $\{\alpha_1, \alpha_2, \ldots, \alpha_n\}$ be the roots of $f(x)$ in some extension field of K. We choose a c in K so that all the quantities $\alpha_j\alpha_k + c(\alpha_j + \alpha_k)$ for $1 \leq j < k \leq n$ are distinct.[4] Evidently, these quantities satisfy a polynomial of degree $n(n-1)/2$ with coefficients in K. By the induction assumption, at least one of them must lie in $K(i)$, $\alpha_1\alpha_2 + c(\alpha_1+\alpha_2)$ say. A consequence of the condition imposed on c is that $K(\alpha_1\alpha_2, \alpha_1+\alpha_2) = K(\alpha_1\alpha_2 + c(\alpha_1+\alpha_2))$ which is contained in $K(i)$. It follows that α_1 and α_2 satisfy a quadratic equation over $K(i)$, and so are in $K(i)$.

With the help of Galois theory a proof can be given along the following lines. Since every polynomial of odd degree over K has a root in K, it follows that K only has extensions of even degree. Let L be a Galois extension of K of degree $2^m q$, where $m \geq 1$ and q is odd. Let G be the Galois group of L over K and H a 2-Sylow subgroup of G. If E is the fixed field of H, then the degree of E over K is q, which is odd. Consequently, $q = 1$ and $E = K$. This implies that G is a 2-group and that L can be obtained from K by a series of quadratic extensions. From what has be said previously, this shows that L is contained in $K(i)$, which is what we wanted to prove.

THEOREM 4. *Let Ω be an algebraically closed field of characteristic zero. Suppose that P a proper subfield such that Ω is a simple extension of P. Then, P is a real closed field.*

Suppose $\Omega = P(\xi)$. ξ cannot be transcendental over P. Otherwise $x^2 - \xi$ would not have a root in Ω, and so Ω would not be algebraically closed. Thus, Ω is a finite extension of P. From here on the proof exactly follows the procedure given in the paper of E. Artin referred to in footnote 2.

In characteristic zero,[5] the real closed fields are identical with those which become algebraically closed through a simple algebraic extension.

THEOREM 5. *Let $f(x)$ be a polynomial with coefficients in a real closed field P and suppose a and b are elements of P such that $f(a) < 0$ and $f(b) > 0$. Then there is at least one element c in P between a and b, such that $f(c) = 0$.*

Since $P(i)$ is algebraically closed, polynomials in $P[x]$ factor into a product of linear and quadratic irreducible polynomials. An irreducible quadratic polynomial

[4]This is possible since we assumed that $f(x)$ has no repeated roots.

[5]This assumption can be avoided, a point to which we will return.

$x^2 + px + q$ is always positive because we can write $x^2 + px + q = (x + \frac{p}{2})^2 + (q - \frac{p^2}{4})$. The first term is always non-negative and the second term must be positive, otherwise the polynomial would not be irreducible. Thus, a change in sign of the polynomial can only occur if one of the linear factors changes sign. This can only occur if there is a zero of $f(x)$ in the interval between a and b.

THEOREM 6. *In a real closed field, the theorems of real algebra hold. For example: Uniform continuity of polynomials in every interval $a \leq x \leq b$. Rolle's theorem. The mean value theorem of differential calculus. The theorem of Sturm about the number of zeros in an interval. Every rational function whose denominator does not vanish for $a \leq x \leq b$ takes on a maximum and a minimum value, and, to be sure, assumes these values among the points a, b, ξ_j, where ξ_j run through the zeros of the derviative of our function in the interval under consideration. All the zeros of the polynomial $x^n + a_1x^{n-1} + \cdots + a_n$ have absolute value less than $1 + |a_1| + |a_2| + \cdots + |a_n|$.*

The proofs, based on Theorem 5, follow word for word as in the usual case. One can compare this with the corresponding sections of H. Weber's *Lehrbuch der Algebra* I (in particular, Sections 35, 91, 112, and 114 of the second edition).

II. Existence and Uniqueness Theorems

We now turn our attention to the existence of certain real closed extensions of real fields, as well as real closed subfields of algebraically closed fields.

THEOREM 7. *Let K be a real field and Ω an algebraically closed field containing K. There exists (at least one) real closed field P between K and Ω such that $\Omega = P(i)$.*[6]

Let Λ be the set of real fields lying between K and Ω. Order Λ by inclusion. Λ is non-empty since K is in Λ. It is easy to check that the conditions of Zorn's lemma apply. Let P be a maximal element in Λ. From the definition, it is clear that P is a real closed field. Ω is algebraic over P, since if ξ in Ω is transcendental over P, it is easy to see that $P(\xi)$ is a real field, contradicting the maximality of P. By Theorem 3, $P(i)$ is algebraically closed. It follows that $\Omega = P(i)$.

A special case of Theorem 7 which is an immediate consequence, deserves a special formulation

THEOREM 7A. *Every real field K has at least one real closed extension field which is algebraic over K.*

For the proof we need only take Ω to be the algebraic closure of K in Theorem 7.

THEOREM 7B. *Every real field can be ordered in at least one way.*

This follows immediately from Theorems 1 and 7A.

If Ω is any algebraically closed field of characteristic zero, we can take K to be the field of rational numbers in Theorem 7. Thus,

THEOREM 7C. *Every algebraically closed field of characteristic zero contains at least one real closed subfield P such that $\Omega = P(i)$.*

[6]For simplicity, we use Zorn's lemma in the first part of the proof. The original uses the well ordering principle. *Editor's note.*

For ordered fields, Theorem 7 can be considerably strengthened.

THEOREM 8. *Let K be an ordered field. There exists one, and, up to isomorphism, only one algebraic, real closed extension field P of K whose ordering extends that of K. The only automorphism of P which leaves K fixed is the identity.*

We first prove,

LEMMA. *Let K be an ordered field and let $\bar{K}$ be the field obtained from K by adjoining the square roots of all positive elements of K. $\bar{K}$ is a real field.*

It clearly suffices to show the non-existence of equations of the form

$$-1 = \sum_{\nu=1}^{n} c_\nu \xi_\nu^2, \tag{3}$$

where the c_ν are positive elements of K and the ξ_ν are in $\bar{K}$. Assume there is such an equation. Naturally, only finitely many square-roots of positive elements, $\sqrt{a_1}.\sqrt{a_2},\ldots,\sqrt{a_r}$, can appear in the ξ_ν. Among all equations of the form (3), choose one in which the number r is least. (To be sure, $r \geq 1$ since it is impossible for an equation of the form (3) to exist over K). Each ξ_ν can be written in the form $\eta_\nu + \zeta_\nu\sqrt{a_r}$, where η_ν and ζ_ν are in $K(\sqrt{a_1},\sqrt{a_2},\ldots,\sqrt{a_{r-1}})$. Thus we have

$$-1 = \sum_{\nu=1}^{n} c_\nu \eta_\nu^2 + \sum_{\nu=1}^{n} c_\nu a_\nu \zeta_\nu^2 + 2\sqrt{a_r}\sum_{\nu=1}^{n} c_\nu \eta_\nu \zeta_\nu. \tag{4}$$

If the last sum in (4) vanishes, we are left with an equation of the same form as (3), but involving less than r square-roots. If the last sum doesn't vanish, then $\sqrt{a_r}$ is in $K(\sqrt{a_1},\sqrt{a_2},\ldots,\sqrt{a_{r-1}})$. In both cases, we have a contradicition to the definition of r.

After these preparations, we can now give the proof of Theorem 8. Let P be a real closed, algebraic extension of $\bar{K}$. Such an extension exists by Theorem 7A since the Lemma shows that $\bar{K}$ is real. P is also an algebraic extension of K and the order of P is a continuation of the order of K since every positive element of K is a square in $\bar{K}$ and thus in P.

Now, let P^* be another algebraic, real closed extension of K whose order extends that of K. Let $f(x)$ be a, not necessarily irreducible, polynomial with coefficients in K. Sturm's theorem allows us to decide, already in K, how many roots $f(x)$ has in P. We only have to evaluate a Sturm sequence[7] for $f(x) = x^n + a_1x^{n-1} + \cdots + a_n$ at the points $\pm(1 + |a_1| + \cdots + |a_n|)$ (see Theorem 6). Thus, $f(x)$ has the same number of roots in P^* as it has in P. In particular, a polynomial with coefficients in K which has a root in P must also have a root in P^* and vice-versa. Let $\alpha_1,\alpha_2,\ldots,\alpha_r$ be the roots of $f(x)$ in P and $\beta_1^*,\beta_2^*,\ldots,\beta_r^*$ be the roots of $f(x)$ in P^*. Further, let ξ be chosen so that $K(\xi) = K(\alpha_1,\alpha_2,\ldots,\alpha_r)$ and let $F(x)$ be the irreducible polynomial for ξ over K. $F(x)$ has a root ξ in P, so $F(x)$ has at least one root η^* in P^*. The isomorphism which leaves K fixed, and takes ξ to η^* must map the roots of $f(x)$ in $K(\xi)$ to the roots of $f(x)$ in $K(\eta^*)$. Since $K(\eta^*)$ in a subfield of P^*, and $K(\xi) = K(\alpha_1,\alpha_2,\ldots,\alpha_r)$, we must have $K(\eta^*) = K(\beta_1^*,\beta_2^*,\ldots,\beta_r^*)$. It follows that $K(\alpha_1,\alpha_2,\ldots,\alpha_r)$ and $K(\beta_1^*,\beta_2^*,\ldots,\beta_r^*)$ are isomorphic over K. In order to show that P and P^* are isomorphic over K,

[7]The reader can consult N. Jacobson, Basic Algebra I, Section 5.2, Second Edition, W. H. Freeman and Co. *Editor's note.*

we remark that such an isomorphism would have to preserve the order since this is determined by whether or not an element is a square. We therefore define the following mapping from P to P^*. Let α be an element of P and $p(x)$ its irreducible polynomial over K. Let $\alpha_1 < \alpha_2 < \cdots < \alpha_r$ be all the roots of $p(x)$ in P; in particular, let $\alpha = \alpha_k$. Let $\alpha_1^* < \alpha_2^* < \cdots < \alpha_r^*$ be the roots of $p(x)$ in P^*. Set $\sigma(\alpha) = \alpha_k^*$. Clearly, σ is one to one and it leaves the elements of K fixed. It must be shown that σ is an isomorphic mapping. To this end, let $f(x)$ be any polynomial with coefficients in K and $\gamma_1, \gamma_2, \ldots, \gamma_s$ be its roots in P and $\gamma_1^*, \gamma_2^*, \ldots, \gamma_s^*$ be its roots in P^*. Further, let $g(x)$ be the polynomial whose roots are the square-roots of the quantities $\gamma_i - \gamma_j$ for $1 \leq i, j \leq s$. Let $\delta_1, \delta_2, \ldots, \delta_t$ b the roots of $g(x)$ in P and $\delta_1^*, \delta_2^*, \ldots, \delta_t^*$ be the roots of $g(x)$ in P^*. According to what has been shown above, the fields $G = K(\gamma_1, \ldots, \gamma_s, \delta_1, \ldots, \delta_t)$ and $G^* = K(\gamma_1^*, \ldots, \gamma_s^*, \delta_1^*, \ldots, \delta_t^*)$ are isomorphic over K. Let τ be such an isomorphism. The map τ takes each γ to a γ^* and each δ to a δ^*. Choose the notation so that $\tau(\gamma_k) = \gamma_k^*$ and $\tau(\delta_h) = \delta_h^*$. Now, if $\gamma_k < \gamma_l$ in P, then $\gamma_l - \gamma_k = \delta_h^2$ for some index h. Applying τ we find $\gamma_l^* - \gamma_k^* = (\delta_h^*)^2$, which implies $\gamma_k < \gamma_l$. Thus, τ maps the roots of $f(x)$ in P to the roots of $f(x)$ in P^* and preserves their order. This must also be true for the roots of the irreducible factors of $f(x)$, which implies that $\tau(\gamma_k) = \sigma(\gamma_k)$ for $k = 1, 2, \ldots, s$. For any two elements α and β in P, choose $f(x)$ to be a polynomial with coefficients in K which has $\alpha, \beta, \alpha\beta, \alpha + \beta$ among its roots. Then, $\sigma(\alpha+\beta) = \tau(\alpha+\beta) = \tau(\alpha)+\tau(\beta) = \sigma(\alpha)+\sigma(\beta)$, and similarly $\sigma(\alpha\beta) = \sigma(\alpha)\sigma(\beta)$. Thus, σ is an isomorphic mapping of P onto P^* and it is, to be sure, the only one which leaves K elementwise fixed. If we consider the case where $P = P^*$, this proves the assertion about automorphisms of P which leave K fixed.

We now want to investigate ordered fields which satisfy either the Archimedean axiom or a certain generalization of it.

Let G be an ordered field, and K a subfield. An element α of G is said to be "infinitely large with respect to K" if $c < |\alpha|$ for all elements c in K, or, on the other hand, "infinitely small with respect to K" if $0 < |\alpha| < c$ for all positive elements c in K.

Note that α is infinitely large with respect to K if and only if $1/\alpha$ is infinitely small with respect to K.[8]

We call an ordered field containing K Archimedean with respect to K if it contains no elements which are infinitely large (or infinitely small) with respect to K.

In the case where K is the field of rational numbers, we omit the phrase "with respect to K".

If K_1 is Archimedean with respect to K_2 and K_2 is Archimedean with respect to K_3, then K_1 is Archimedean with respect to K_3.

A field A between K and G is called maximal Archimedean with respect to K if it is Archimedean with respect to K and no extension of A within G has this property.

[8]It might be good to have an example in mind. Let K be an ordered field and let $K(x)$ be the field of rational functions over K. Let's call $f(x)$ positive if the first non-vanishing coefficient of its Laurent expansion in positive in K. This defines an order on $K(x)$. Then, x is positive and infinitely small with respect to K and $1/x$ is infinitely large with respect to K. *Editor's note.*

It is easy to see using the well-ordering principle[9] that there is always at least one intermediate extension A which is maximal Archimedean with respect to K (see the proof of Theorem 7). In the same way it can be shown that if B is an intermediate extension which is Archimedean with respect to K, then A can be chosen to contain B.

For real closed fields we can go further and prove

THEOREM 9. *Let P be a real closed field and let K be a subfield of P. Then, all subfields of P which are maximal Archimedean with respect to K are isomorphic. Moreover, they are all real closed.*

Let A be a subfield of P which is maximal Archimedean with respect to K. We first prove that any element γ of P which is algebraic over A is already in A. Every element of $A(\gamma)$ is algebraic over A and so, by Theorem 6, it has absolute value less than an element of A. Thus, $A(\gamma)$ contains no element which is infinitely large with respect to A. Thus, $A(\gamma)$ is Archimedean over A. Since A is Archimedean over K, it follows that $A(\gamma)$ is Archimedean over K. From the maximality of A, we must have $A(\gamma) = A$.

Now, every polynomial of odd degree with coefficients in A has at least one root in P (see Theorem 2). By what has just been shown, that root must lie in A. In the same way one sees that the square root of every positive element of A must lie in A. By Theorem 3a, $A(i)$ is algebraically closed and therefore no proper algebraic extension of A is real. This shows that A is real closed.

Now, let Γ be the set of elements in P which, with respect to K, are not infinitely large. Obviously, Γ is a ring. Further, let $\mathfrak{u}$ be the set of elements in Γ consisting of 0 and the elements of P which are infinitely small with respect to K. $\mathfrak{u}$ is a subset of Γ and, in fact, $\mathfrak{u}$ is a prime ideal in Γ. This follows since the difference of two elements of P which are infinitely small with respect to K is again infinitely small with respect to K. Morever, the product of two elements is infinitely small with respect to K if and only if at least one of the factors is. Let $\mathfrak{A}$ denote the domain of residue classed of Γ modulo $\mathfrak{u}$. $\mathfrak{A}$ is a field since if $\gamma \not\equiv 0 \pmod{\mathfrak{u}}$, γ is not infinitely small with respect to K and so $1/\gamma$ is not infinitely large with respect to K. i.e. $1/\gamma$ is in Γ. A non-zero residue class consists entirely of positive or entirely of negative elements of Γ. We will call a non-zero residue class positive if all the elements in it are positive. This puts an order on the field $\mathfrak{A}$ since all the requirements of the property "positive" are satisfied. In every residue class there is at most one element of K.[10] The residue classes represent by elements of K form a field $\mathfrak{K}$ which is isomorphic to K. $\mathfrak{A}$ is Archimedean with respect to $\mathfrak{K}$ since Γ contains no element which is infinitely large with respect to K, and so $\mathfrak{A}$ contains no element which is infinitely large with respect to $\mathfrak{K}$.

We now assert: Every subfield A of P which is maximal Archimedean with respect to K is a complete set of representatives for $\Gamma/\mathfrak{u} = \mathfrak{A}$. Thus, the map which takes an element of A to its coset modulo $\mathfrak{u}$ is an isomorphism. This will constitute a complete proof of Theorem 9. To begin with, the fact that A is Archimedean wih respect to K shows A belongs to Γ and also that every residue class modulo $\mathfrak{u}$ contains at most one element of A. Also, every residue class contains at least one element of A. Suppose, on the contrary, that there is a residue class R which

[9] Or Zorn's lemma. *Editor's note.*

[10] Suppose a and b are in K. If $a \equiv b \pmod{\mathfrak{u}}$, then $a - b \in \mathfrak{u}$ and so $a = b$. Otherwise $a - b$ is infinitely small with respect to K, which is impossible since $a - b$ is in K.

contains no element of A. According to what has already been proven, every element of R is transcendental over A. Let t be in R. We want to show: $A(t)$ is Archimedean with respect to A and thus also with respect to K. This will yield a contradiction to the maximality of A. Every element of $A(t)$ has the form $f(t)/g(t)$, where $f(t)$ and $g(t)$ are polynoimials with coefficients in A. Since t is in Γ, so is $f(t)$, so it is sufficient to show that $g(t)$ is not infinitely small with respect to K. To this end, we determine two elements a and b in A such that $a < t < b$ and such that $g(x)$ has no zero (in P) on the interval $a \leq x \leq b$. This is always possible (as we now show). If $g(x)$ has no zero (in P) which is greater than t, take for b any element of A which is greater than t. Otherwise, let b' be the smallest zero of $g(x)$ which is greater than t. Since b' is algebraic over A, it belongs to A, and so $b' \not\equiv t \pmod{\mathfrak{u}}$.[11] Since $b' - t$ is not infinitely small with respect to K, there exists a positive c in K such that $0 < c < b' - t$. Set $b = b' - c$ and we have $t < b < b'$. Moreover, $g(x)$ doesn't vanish on $t \leq x \leq b$. The element a can be found in an analogous manner. By Theorem 6, $|g(x)|$ has a positive minimum μ on $a \leq x \leq b$, and $\mu = |g(\xi)|$, where ξ is either a or b or a zero of the derivative of $g(x)$. In each case, ξ belongs to A and thus, so does $|\mu|$. Therefore, $\mu \leq |g(t)|$, which shows that $g(t)$ is not infinitely small with respect to A. This completes the proof.

III. Examples and Applications

In this last section we want to give several applications to algebraic number fields and also to give examples of real fields, which in many respects cast new light on the results we have obtained. We begin with a lemma which will be very useful in applications.

LEMMA. *If K is the quotient field of an ordered domain R, then K can be ordered in one and only one way which extends the order on R.*

Suppose K is ordered in the desired manner. Let $a = b/c$ be an arbitrary element of K, where b and c belong to R and $c \neq 0$. Then, $a > 0, a = 0$, or $a < 0$ implies that $bc > 0, bc = 0$, or $bc < 0$. Thus, the order on R uniquely determines the order on K. Conversely, one sees immediately that the stipulation $a > 0$ when $bc > 0$, in fact, defines an order of the desired type on K.

In particular, the field of rational numbers can be ordered in only one way since, obviously, only the natural order is possible for the ring of rational integers. Using this and Theorem 8 we find

THEOREM 8A. *There is, up to isomorphism, one and only one real closed field which is algebraic over the field of rational numbers, namely the field of, in the usual sense,* **real**[12] *algebraic numbers.*[13]

We prove further

THEOREM 10. *A real, absolutely algebraic field K is always isomorphic to a* **real** *algebraic number field. Every ordering of K corresponds uniquely to an isomorphic mapping of K to a* **real** *algebraic number field in which the order of K is taken to the natural order of the corresponding* **real** *number field. Different orderings of K*

[11] Recall that R is assumed to contain no element of A. *Editor's note.*

[12] "Real" in the usual sense will hereafter be denoted by boldface letters.

[13] This is already proven the paper of E. Artin referred to in footnote 2), albeit not purely algebraically.

lead to the same **real** *field if and only if the orderings correspond by means of an automorphism of K*

Namely, suppose that K is a real, absolutely algebraic field. K can be ordered in some manner (see Theorem 7B). Let P be a real absolutely closed algebraic extension of K whose order extends that of K (see Theorem 8). Then, P is isomorphic to the field P^* of all **real** algebraic numbers. Therefore, K is isomorphic to a subfield K^* of P^*. Since an isomorphism between P and P^* must preserve their orders, the induced isomorphism from K to K^* must take the order of K to the natural order of K^*. Conversely, an isomorphism of K with **real** number field K^* gives rise to an order on K. One just carries the natural order on K^* back to K via the isomorphism.

The assertion about uniquenes follows from two remarks. An automorphism of a **real** number field which preserves the natural order must be the identity. Moreover, two different **real** number fields cannot have an isomorphism between them which preserves the natural order.[14]

As opposed to Theorem 8A, the following holds for transcendental fields

THEOREM 11. *Let Ω be an algebraically closed field of characteristic zero which is not absolutely algebraic. Then there are two*[15] *real closed subfields P_1 and P_2 which are not isomorphic and for which $\Omega = P_1(i) = P_2(i)$. If Ω has transcendence degree over the rational numbers $\leq \mathfrak{c}$ (here, $\mathfrak{c}$ is the cardinality of the continuum), the fields P_1 and P_2 can be chosen to be Archimedean.*

For the proof, let R be the field of rational numbers and $\mathfrak{R}$ the field of all **real** numbers. We suppose chosen a transcendence basis $\mathfrak{B}$ for $\mathfrak{R}$ over R, i.e. a set with the following two properties: 1. Every element of $\mathfrak{B}$ is transcendental over the field generated by the remaining elements of $\mathfrak{B}$ over R, and 2. $\mathfrak{R}$ is algebraic over $R(\mathfrak{B})$. (The existence of such a basis is proven in the usual way.)

Now, suppose Ω is algebraically closed of characteristic zero and of trancendence degree $\mathfrak{t}$ (> 0) over R. To begin with, let's suppose that $\mathfrak{t} \leq \mathfrak{c}$. Inside $\mathfrak{B}$ choose two non-equal subsets $\mathfrak{B}_1$ and $\mathfrak{B}_2$, both with cardinality $\mathfrak{t}$. Choose a in $\mathfrak{B}_1$ such that a is not in $\mathfrak{B}_2$. Now, let Ω_1 and Ω_2 be the algebraic closures of $R(\mathfrak{B}_1)$ and $R(\mathfrak{B}_2)$ inside the complex numbers. Ω is isomorphic to Ω_1 and Ω_2. Let P_1 and P_2 be the subsets of real numbers in Ω_1 and Ω_2 respectively. P_1 and P_2 are real closed since, obviously, $\Omega_1 = P_1(i)$ and $\Omega_2 = P_2(i)$, but they are certainly not isomorphic. Namely, suppose given an isomorphic mapping of P_1 onto P_2. It must leave R fixed, and preserve the order on P_1 and P_2 (see Theorem 1). This, however, is impossible since P_2 has no number which generates the same Dedekind cut in R as the number a which belongs to P_1. Because of the isomorphisms between Ω_1, Ω_2, and Ω, it follows that Ω has two subfields isomorphic to P_1 and P_2 respectively which together with i generate Ω. This proves our assertion when $\mathfrak{t} \leq \mathfrak{c}$.

For arbitrary $\mathfrak{t} > 0$ we proceed as follows. Let $\mathfrak{X}$ be a transcendence basis for Ω over R. Order $\mathfrak{X}$ in some fashion.[16] We now order the monomials formed with

[14] Suppose K^* is a **real** number field and that σ is an automorphism, not the identity. There is an element α in K^* such that $\sigma(\alpha) \neq \alpha$. We may assume $\sigma(\alpha) > \alpha$. Let r be a rational number such that $\sigma(\alpha) > r > \alpha$. Then, $\alpha - r < 0$, but $\sigma(\alpha - r) = \sigma(\alpha) - r > 0$. Thus, σ does not preserve the order. The second remark follows from similar considerations. *Editor's note.*

[15] Even infinitely many.

[16] "Order" is meant in the sense of set theory, not in the sense of an order on a field.

elements of $\mathfrak{X}$. Let $x_1, x_2, \ldots, x_n$ be elements of $\mathfrak{X}$, where the numbering respects the order in $\mathfrak{X}$. Then, we will say $x_1^{a_1} x_2^{a_2} \ldots x_n^{a_n}$ is less than $x_1^{b_1} x_2^{b_2} \ldots x_n^{b_n}$ if the first non-vanishing difference $b_j - a_j$ is positive. If $f(x)$ is an element of the polynomial ring $R[\mathfrak{X}]$, we stipulate that $f(x)$ is positive if the smallest monomial that occurs in $f(x)$ has a positive coefficient. This gives an order on $R[\mathfrak{X}]$, and this order extends to $R(\mathfrak{X})$ according to our Lemma. Let P_1 be the real closed extension of $R(\mathfrak{X})$ in Ω which preserves the order on $R(\mathfrak{X})$ we have just defined. Obviously, $\Omega = P_1(i)$, and the maximal Archimedean subfield of P_1 has the type of the field of all **real** algebraic numbers.

We now order $R[\mathfrak{X}]$, and thereby $R(\mathfrak{X})$, in another way. Let y be an element of $\mathfrak{X}$ and let $\mathfrak{X}'$ be $\mathfrak{X}$ with y removed. We order the monomials in $R[\mathfrak{X}']$ in the same way as before. If $f(y, x')$ is an element of the polynomial ring $R[\mathfrak{X}]$, let $g(y)$ be the coefficient of the smallest occuring monomial in the x'. We will say that $f(y, x')$ is positive if $g(e)$ is a positive real number; here, e denotes the base of the natural logaritms. We define P_2 to be a real closed field between $R(\mathfrak{X})$ and Ω which preserves this new order. Once again, we have $\Omega = P_2(i)$. However, P_2 contains an Archimedean field of transcendence degree 1, namely $R(y)$. It follows that P_1 and P_2 cannot be isomorphic (see Theorem 9).

In a real closed field, the zeros of a polynomial with coefficients in the field can always be separated. The following simple example shows that this need not be the case in an ordered field which is not real closed. Let $R(x)$ be the field of rational functions over the field R of rational numbers. We order $R(x)$ so that it satisfies the requirement that x is positive and infinitely small with respect to R, i.e. we say a polynomial is positive if the coefficient of the smallest occuring power of x is positive. Let P be the real closed subfield of the algebraic closure of $R(x)$ which preserves the given order on $R(x)$. In P, the equation $(y^2 - x)^2 - x^3$ has two positive roots $\sqrt{x(1 \pm \sqrt{x})}$. These two roots cannot be separated in $R(x)$. This example shows that the uniqueness part of Theorem 8 must be carried through without relying on the separation of roots.

Finally, it will be shown that in an ordered field which is not real closed, there can exist two maximal Archimedean subfields whcih are not isomorphic. For this purpose, let A be the field of all **real** algebraic numbers ordered in the usual way, and $A(e)$ the field obtained from A by adjoining the base of the natural logarithms, e. Let P be the (real closed) field of all real numbers algebraic over $A(e)$. We now consider the field $G = P(x)$, ordered so that x is infinitely small with respect to P. P and $A(e + x)$ are two Archimedean subfields of G. It is possible to show that both of them are even maximal Archimedean. In spite of this, they are observably not isomorphic, since P is real closed, and $A(e + x)$ is not.

THE MATHEMATICAL SEMINAR, HAMBURG, JUNE 1926

A Characterization of Real Closed Fields*

Emil Artin and Otto Schreier in Hamburg

In the following, our investigation "Algebraische Konstruktion reeller Körper"[1] will be completed at a certain point which had been left unresolved. We recall first the following basic definitions:

A field is called "real" if -1 cannot be represented as a sum of squares.

A field P is called "real closed" if it is real, but no proper algebraic extension of it is real.

Our goal is the proof of the following Theorem

Real closed fields are the same as those fields which have a finite algebraically closed extension, but are not algebraically closed themselves.

One part of this characterization, namely that a real closed field has a finite algebraic extension which is algebraically closed, is contained in Theorem 3 of A.K. The converse is proven for algebraic number fields in R.Z., and for arbitrary fields of characteristic zero it is proven in Theorem 4 of A.K. utilizing the methods of R.Z. For fields of characteristic $p > 0$ new considerations are needed. Therefore, in Sections 1 and 2 we prove several theorems about cyclic extensions of degree p and p^2 for fields of characteristic p. On the basis of these theorems, which perhaps are of interest in themselves, we succeed in Section 3 in proving a main result. For the convenience of the reader, we have given complete proofs of some results for which we could have referred to R.Z.

Section 1. Let K be a field of characteristic p. As is well known, cyclic extensions of degree p cannot be constructed by means of radicals. For the study of cyclic extensions[2] of degree p we must find a substitute for pure equations. It turns out that the equations

$$x^p - x - a = 0, \tag{1}$$

which we will refer to as "normalized" equations, will serve the purpose. Namely, we have

THEOREM 1. *A normalized equation is either cyclic or all its roots are rational. Every cyclic equation extension of degree p can be generated by the roots of a normalized equation.*

---

* *Kennzeichnung der reell algeschlossen Körper*, Hamb. Abh., Bd. 5 (1927), 225–231.

[1] Hamb. Abh., Bd. 5 (1926), pp. 85–99, which we cite in the following as A.K. Also compare this with: E. Artin, *Kennzeichnung des Körpers der reellen algebraischen Zahlen*, Hamb. Abh., Bd. 3 (1924), pp. 319–323, cited hereafter as R.Z.

[2] An extension is called cyclic, if it is separable and Galois with cyclic Galois group.

Let γ be a root of (1). The other roots are then $\gamma+1, \gamma+2, \dots, \gamma+p-1$. Thus, all the irreducible factors of $x^p - x - a$ have the same degree. If γ is not in K, then $x^p - x - a$ is irreducible and γ generates an extension of degree p, which is clearly Galois and so cyclic.

Conversely, suppose $\bar{K}$ is a cyclic extension of K. Suppose $\bar{K} = K(\alpha)$ and let σ be a generator of the Galois group of $\bar{K}$ over K ($\sigma^p = e$). We set , in the usual way, $\sigma^i(\alpha) = \alpha_i$, whereby the subscript is taken modulo p. We choose an exponent k such that

$$s = \sum_{i \mod p} \alpha_i^k \neq 0.$$

This is certainly possible, since otherwise the Vandermonde determinant

$$\det(\alpha_i^k) = \prod_{i<j} (\alpha_j - \alpha_i) \qquad 0 \leq i, j, k < p$$

would be zero since all the rows would sum to zero. In this case, two of the α_i would coincide, which cannot happen.

Now, set

$$\xi = -\frac{1}{s} \sum_{i \mod p} i\alpha_i^k.$$

Then,

$$\sigma(\xi) = -\frac{1}{s} \sum_{i \mod p} i\alpha_{i+1}^k = -\frac{1}{s}\Big(\sum_{i \mod p} (i+1)\alpha_{i+1}^k - s\Big) = \xi + 1.$$

Consequently, ξ is not in K and so must generate $\bar{K}$. One the other hand $\xi^p - \xi$ does lie in K, since $\sigma(\xi^p - \xi) = (\xi+1)^p - (\xi+1) = \xi^p - \xi$. It follows that ξ satisfies a normalized equation over K. This proves Theorem 1.

We want to determine all the generators of $\bar{K}$ over K which satisfy normalized equations. Obviously, for such a generator γ the condition

$$\sigma(\gamma) = \gamma + k \qquad (k = 1, 2, \dots . p-1) \tag{2}$$

is necessary and sufficient. From (2) it follows that $\sigma(\gamma - k\xi) = \gamma + k - k(\xi+1) = \gamma - k\xi$. Thus, $\gamma - k\xi$ is in K. Since these steps can also be reversed, we have proved

Theorem 2. *If ξ is the root of a normalized equation over K, then*

$$\{k\xi + b \mid k = 1, 2, \dots, p-1;\ b \in K\}$$

is the set of all elements in $K(\xi)$ which likewise satisfy normalized equations over K.

Section 2. We now turn our attention to cyclic extensions of degree p^2 over K. Let $K(\eta)$ be a cyclic extension of degree p^2 over K and $K(\xi)$ the unique subfield which is cyclic of degree p over K. By Theorem 1, we may suppose that ξ is the root of a normalized equation in K and that η is the root of a normalized equation over $K(\xi)$.

$$\xi^p - \xi - a = 0, \tag{3}$$

$$\eta^p - \eta - \phi(\xi) = 0. \tag{4}$$

In (4) we may take ϕ to be a polynomial of degree at most $p-1$ with coefficients in K. Let σ be a generator of the Galois group of $K(\eta)$ over K $(\sigma^{p^2} = e)$. The

Galois group of $K(\xi)$ over K is represented by $e, \sigma, \sigma^2, \dots, \sigma^{p-1}$. We may assume $\sigma(\xi) = \xi + 1$. Applying σ to (4) we find that $\sigma(\eta)$ satisfies the normalized equation

$$\sigma(\eta)^p - \sigma(\eta) - \phi(\xi + 1) = 0.$$

By Theorem 2, $\sigma(\eta) = k\eta + \psi(\xi)$, where k is one of the numbers $1, 2, \dots, p - 1$ and $\psi(\xi)$ is a polynomial in ψ of degree at most $p - 1$ with coefficients in K. By iteration we find that for $\nu = 1, 2, \dots$

$$\sigma^\nu(\eta) = k^\nu \eta + k^{\nu-1}\psi(\xi) + k^{\nu-2}\psi(\xi + 1) + \cdots + \psi(\xi + \nu - 1). \tag{5}$$

Now, $\sigma^p(\eta)$ satisfies equation (4).[3] Thus, $\sigma^p(\eta) = \eta + h$ where h is one of the numbers $1, 2, \dots, p - 1$. If we compare this with equation (5) with $\nu = p$, we discover that, because η does not belong to $K(\xi)$, $k^p = 1$ and so $k = 1$. Moreover,

$$h = \psi(\xi) + \psi(\xi + 1) + \cdots + \psi(\xi + p - 1). \tag{6}$$

The right hand side of (6) is easily calculated to be $-c$ where c is the coefficient of ξ^{p-1} in $\psi(\xi)$. To see this, expand $\psi(\xi + k)$ in powers of k and sum over k. Observe that $\sum_{k=0}^{p-1} k^\nu = 0$ for $\nu = 0, 1, \dots, p-2$ and equals -1 when $\nu = p-1$. Thus, $h = -c$. Substitute $-\eta/h$ for η and we arrive at the following: $\sigma(\eta) = \eta + \xi^{p-1} + \psi_1(\xi)$, where $\psi_1(\xi)$ has degree at most $p - 2$, and $\sigma^p(\eta) = \eta - 1$. In order to further normalize $\psi_1(\xi)$ we need the following Lemma.

LEMMA. *If $p(x) = bx^n + \dots$ is a polynomial of degree n with coefficients in K and $n \le p - 2$, then there exists a polynomial $f(x)$ of degree $n + 1$ which satisfies the difference equation $f(x+1) - f(x) = p(x)$. $f(x)$ is determined up to an additive constant and the highest coefficient of $f(x)$ is $b/(n + 1)$.*

For polynomials of degree zero, this is clear. We assume the lemma is true for polynomials of degree less than n, and that $p(x) = bx^n + p_1(x)$ is a polynoimal of degree n with highest coefficent b. Set $f(x) = bx^{n+1}/(n + 1) + f_1(x)$. The polynomial $f(x)$ will satisfy our requirement if $f_1(x)$ is a polynomial of degree at most n satisfying

$$f_1(x + 1) - f_1(x) = p_1(x) + b\left(x^n - \frac{(x + 1)^{n+1} - x^{n+1}}{n + 1}\right).$$

The right hand side of this equation has degree at most $n - 1$ and so $f_1(x)$ exists by our induction assumption. Since the additional result about uniqueness is immediately clear, the Lemma is proven.

We set $\eta' = \eta + f(\xi)$ and want to determine a polynomial $f(\xi)$ of degree at most $p - 1$ so that $\sigma(\eta')$ has the simplest possible form. We have

$$\sigma(\eta') = \sigma(\eta) + f(\xi + 1) = \eta' + \xi^{p-1} + [f(\xi + 1) - f(\xi) + \psi_1(\xi)].$$

By the Lemma, we can choose the polynomial $f(\xi)$ so that the term in brackets vanishes. Once again replacing η' by η we find that η can be chosen to satisfy $\sigma(\eta) = \eta + \xi^{p-1}$. Assuming that η has been so chosen, we have, according to (3) and (4)

$$\begin{aligned}\phi(\xi + 1) &= \sigma(\eta)^p - \sigma(\eta) = (\eta + \xi^{p-1})^p - (\eta + \xi^{p-1}) \\ &= \eta^p + \xi^{p(p-1)} - \eta - \xi^{p-1} = \phi(\xi) + (\xi + a)^{p-1} - \xi^{p-1}.\end{aligned}$$

[3] σ^p generates the Galois group of $K(\eta)$ over $K(\xi)$. *Editor's note*

Thus, $\phi(x)$ satisfies identically the difference equation

$$\phi(x+1) - \phi(x) = (x+a)^{p-1} - x^{p-1}. \tag{7}$$

Therewith we have obtained a complete overview of the cyclic extensions of degree p^2 which contain the field $K(\xi)$.

We now turn our considerations around and show that we can always construct a cyclic extension of degree p^2 of K.[4] So, let $K(\xi)$ be a cyclic extension of K of degree p where ξ is a root of

$$x^p - x - a = 0. \tag{8}$$

Let $\phi(x)$ be a polynomial of degree $p-1$ which satisfies (7). By the last Lemma, this is possible and we know that the first term of $\phi(x)$ is ax^{p-1}.

We now assert:

The polynomial

$$F(y) = \prod_{k=0}^{p-1} (y^p - y - \phi(\xi + k))$$

is irreducible and its roots generate a cyclic extension of degree p^2 over K which contains $K(\xi)$.

We present the proof in several steps.

α) The coefficients of $F(y)$ are symmetric in the roots of $x^p - x - a$ and thus lie in K. Further, since $x^p - x - a$ is irreducible, certainly $a \neq 0$, and so $\phi(\xi)$ is not in K. Consequently, $K(\phi(\xi)) = K(\xi)$. It follows that no two factors of $F(y)$ have a common root. By Section 1, each factor has distinct roots. Thus, $F(y)$ itself has distinct roots.

β) Let η be a root of $y^p - y - \phi(\xi) = 0$. Then, $\phi(\xi)$ is in $K(\eta)$ and, using α), we find that $K(\xi) = K(\phi(\xi))$ is contained in $K(\eta)$. However, η is not contained in $K(\xi)$. For, if we suppose $\eta = b_0\xi^{p-1} + b_1\xi^{p-2} + \cdots + b_{p-1}$ with the b_i in K, then, using the relation $\xi^p = \xi + a$, we find

$$\phi(\xi) = \eta^p - \eta = b_0^p(\xi + a)^{p-1} + b_1^p(\xi + a)^{p-2} + \ldots) - (b_0\xi^{p-1} + b_1\xi^{p-2} + \ldots).$$

Comparing the coefficient of ξ^{p-1} we see that $b_0^p - b_0 = a$ (recall that the leading term of $\phi(x)$ is ax^{p-1}). This implies that $x^p - x - a$ is reducible over K, a contradiction. Thus, our assumption is false and so, by Theorem 1, $K(\eta)$ is cyclic of degree p over $K(\xi)$. It follows that $F(y)$ is irreducible.

γ) If η_k is a root of $y^p - y - \phi(\xi + k)$, the other roots have the form $\eta_k + h$ where $h = 0, 1, \ldots, p-1$. We claim that $\eta_{k+1} = \eta_k + (\xi + k)^{p-1}$ is a root of $y^p - y - \phi(\xi + k + 1)$. To see this, note that

$$\begin{aligned}
&(\eta_k + (\xi + k)^{p-1})^p - (\eta_k + (\xi + k)^{p-1}) \\
&\qquad = \eta_k^p - \eta_k + (\xi + k)^{p(p-1)} - (\xi + k)^{p-1} \\
&\qquad \text{(by (8))} \quad = \eta_k^p - \eta_k + (\xi + a + k)^{p-1} - (\xi + k)^{p-1} \\
&\qquad \text{(by (7))} \quad = \phi(\xi + k) + (\phi(\xi + k + 1) - \phi(\xi + k)) \\
&\qquad = \phi(\xi + k + 1).
\end{aligned}$$

[4] Assuming that K has a cyclic extension of degree p. *Editor's note.*

By repeated application of this conclusion we see that all the roots of $F(y)$ are polynomial combinations of ξ and η and so, by β) of η alone. This shows that $K(\eta)$ is a Galois extension of K of degree p^2.[5]

δ) According to γ), $\eta + \xi^{p-1}$ is a root of $F(y)$. Let σ be the element of the Galois group that takes η to $\eta+\xi^{p-1}$: $\sigma(\eta) = \eta+\xi^{p-1}$. Then, $\sigma(\phi(\xi)) = \sigma(\eta^p-\eta) = (\eta+\xi^{p-1})^p - (\eta+\xi^{p-1}) = \phi(\xi+1)$. From α) we may conclude that $\sigma(\xi) = \xi+1$. It follows that

$$\sigma^\nu(\eta) = \eta + \xi^{p-1} + (\xi+1)^{p-1} + \cdots + (\xi+\nu-1)^{p-1} \quad (\nu = 1, 2, \dots)$$

and, in particular,

$$\sigma^p(\eta) = \eta + \xi^{p-1} + (\xi+1)^{p-1} + \cdots + (\xi+p-1)^{p-1} = \eta - 1.$$

Thus, $\sigma^p \neq e$, which shows that σ must generate the Galois group of $K(\eta)$ over K. This completetes the proof of our assertion.

Our considerations therefore lead us to

THEOREM 3. *Let $K(\xi)$ be a cyclic extension of degree p of K, where ξ is a root of* (8). *Then, K has a cyclic extension of degree p^2 which contains $K(\xi)$. One constructs these and only these fields by adjoining to K a root of $y^p - y = \phi(\xi)$ where $\phi(x)$ is chosen in all possible ways among the solutions to* (7).

Section 3. After these preparations, we are in a position to prove and strengthen the main theorem presented at the beginning.

THEOREM 4. *Let Ω be an algebraically closed field of arbitrary characteristic, and P a proper subfield such that Ω is finite over P. Then, P is real closed and $\Omega = P(i)$, where i is a root of x^2+1.*

We remark to begin with that P is a perfect field. If not, there is an element a of K which is not a pth power. In this case, one can show that $P(\sqrt[p^n]{a})$ is an extension of P of degree p^n. Thus, P would have algebraic extensions of arbitarily high degree, whereas Ω is finite over P.

It follows that Ω is Galois over P. Let q be a prime factor of the degree of Ω over P. The Galois group must have a subgroup H of order q. Let K be the fixed field of H. According to Theorem 3, q cannot be the characteristic of K, since otherwise Ω would have an extension of degree q.

Over K every polynomial is a product of irreducible factors of degree either 1 or q. Since the qth roots of unity satisfy a polynomial of degree $q-1$, they must all lie in K. Thus, we obtain Ω by adjoining the qth root of an element a in K, i.e. $\Omega = K(\sqrt[q]{a})$. We consider the polynomial $f(x) = x^{q^2} - a$. $f(x)$ has a factor $g(x)$ of degree q, reducible or irreducible, with coefficients in K. The constant term of $g(x)$ is of the form $\epsilon\sqrt[q]{a}$, where ϵ is a q^2 root of unity. Therefore, $\epsilon\sqrt[q]{a}$ belongs to K and so $\Omega = K(\epsilon)$. From this one sees that ϵ is a primitive q^2 root of unity.

Now, let R be the prime field contained in K and ν the largest positive integer such that all the q^ν roots of unity are in $R(\epsilon)$.[6] If ζ is a $q^{\nu+1}$ root of unity, it must satisfy a polynomial $h(x)$ of degree q with coefficients in K. In fact, the coefficients

[5] It is a Galois extension because it is the splitting field of $F(y)$ over K, and it is of degree p^2 because $[K(\eta) : K] = [K(\eta) : K(\xi)][K(\xi) : K] = p^2$. In the next section, it is shown that the Galois group of $K(\eta)$ over K is cyclic. *Editor's note.*

[6] In characteristic zero, $\nu = 2$; in characteristic p, ν is determined as follows: let f be the exponent to which p belongs modulo q^2; then, q^ν is the highest power of q dividing $p^f - 1$.

of $h(x)$ must lie in D, the intersection of K and $R(\zeta)$. Since $h(x)$ remains irreducible when considered over D it follows that $R(\zeta)$ has degree q over D. On the other hand, $\zeta^q = \theta$ is a q^ν root of unity and so lies in $R(\epsilon)$. Since all the roots of $x^q - \theta$ are primitive $q^{\nu+1}$ roots of unity, it follows that $R(\zeta)$ has degree q over $R(\epsilon)$. Since ϵ is not in K, it is not in D and so D is not equal to $R(\epsilon)$. This shows that $R(\zeta)$ has degree q over two different subfields. It follows that $R(\zeta)$ is not cyclic over R. This can only happen when the characteristic is zero and $q = 2$.[7]

We now know that Ω has characteristic zero and that a proper subfield over which Ω is finite, cannot contain a primitive 4th root of unity, i.e. i. Since Ω is finite over $P(i)$ it follows that $\Omega = P(i)$.

It remains to show that P is real. Let a and b be non-zero elements in P and consider the polynomial $f(x) = (x^2 - a)^2 + b^2$. The roots of $f(x)$ are $\pm\sqrt{a \pm ib}$ which are not in P. Thus $f(x)$ must be a product of quadratic factors with coefficients in P. The constant terms of these factors must be either $\pm\sqrt{a^2 + b^2}$ or $\pm(a \pm ib)$. Since the quantities $\pm(a \pm ib)$ are not in P, we must have $\sqrt{a^2 + b^2}$ is in P, i.e. $a^2 + b^2 = c^2$, where c is in P. By induction we see that any sum of squares in P is itself a square. Since -1 is not a square in P, it cannot be a sum of squares. So, in fact, P is a real field. Since the only proper algebraic extension of P, namely Ω, is not real, it follows that P is a real closed field.

THE MATHEMATICAL SEMINAR, HAMBURG, JANUARY 1927

[7] If R has characteristic p, then R is finite and every finite extension of R is cyclic. If R has characteristic zero, then R is the field of rational numbers. The Galois group of $R(\zeta)$ over R is the multiplicative group of integers modulo q^3, which is cyclic unless $q = 2$.

THE THEORY OF BRAIDS

By EMIL ARTIN
Princeton University

THE theory of braids shows the interplay of two disciplines of pure mathematics—topology, used in the definition of braids, and the theory of groups, used in their treatment.

The fundamentals of the theory can be understood without too much technical knowledge. It originated from a much older problem in pure mathematics—the classification of knots. Much progress has been achieved in this field; but all the progress seems only to emphasize the extreme difficulty of the problem. Today we are still very far from a complete solution. In view of this fact it is advisable to study objects that are in some fashion similar to knots, yet simple enough so as to make a complete classification possible. Braids are such objects.

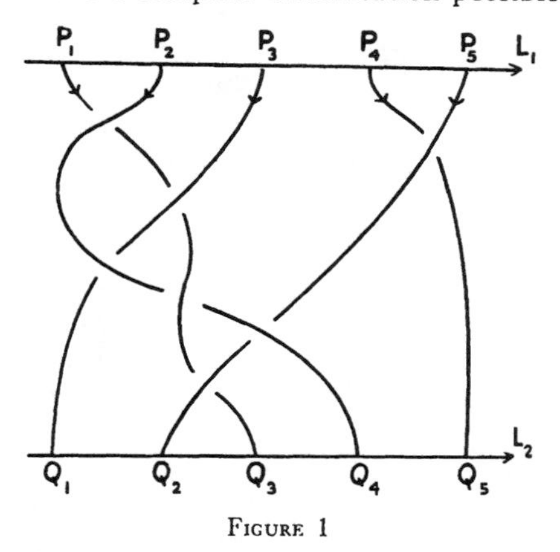

FIGURE 1

In order to develop the theory of braids we first explain what we call a *weaving pattern* of order n (n being an ordinary integral number which is taken to be 5 in Figure 1).

Let L_1 and L_2 be two parallel straight lines in space with given orientation in the same sense (indicated by arrows). If P is a point on L_1, Q a point on L_2, we shall sometimes join P and Q by a curve c. In our drawings we can only indicate the projection of c onto the plane containing L_1 and L_2, since c itself may be a winding curve in space.

The curves c that we shall use will be restricted in their nature by the following condition. If R is a point on the projection of c that moves from P to Q, then its distance from the line L_1 shall always increase. (Therefore a curve moving down a little, then up, and finally down again would be ruled out.) In order to have at our disposal a short name for such curves, let us call them normal curves. We orient them (by arrows) in the sense from P to Q.

Select n points on L_1. Moving along L_1 in the direction indicated by the arrow we shall call the first of the given n points P_1, the next P_2, and the last P_n. In the same way denote by $Q_1, Q_2, \ldots Q_n$, n points on the line L_2. Now we connect each point P_i with one of the points Q_j by a normal curve c_i (c_1 begins at P_1 and ends at some Q_j, which may or

Based on material presented in the Sigma Xi National Lectureships, 1949.
All rights reserved.

may not be Q_1). We only observe the following condition: no two of the curves c_i intersect in space. Consequently no two of the curves c_i end at the same point Q_j.

If we want to indicate this in a drawing, we have to overcome the difficulty that, although the curves do not meet in space, their projections may cross over each other. To indicate that at a certain crossing the curve c_i is below another one, we interrupt its projection slightly (this is the well-known way to indicate such occurrences in technical drawings).

The whole system of straight lines and curves shall be called a weaving pattern.

In order to explain the notion of a braid we start with a given weaving pattern and think of the lines L_1 and L_2 as being made of rigid material, whereas the curves c_i are considered as arbitrarily stretchable, contractible, and flexible. The points P_i and Q_j may also move on their lines provided their ordering is always preserved.

We subject the whole weaving pattern to an arbitrary deformation in space restricted by the following conditions:

(1) L_1 and L_2 stay parallel during the deformation (but otherwise they can be moved freely in space; their distance may change).

(2) No two of the curves c_i intersect each other during the deformation (this means that the material is "impenetrable").

(3) The curves stay normal during the deformation (but otherwise they may be stretched or contracted as the situation demands).

After such a deformation we obtain a weaving pattern that may look quite different from the one we started with. A quite tame-looking pattern may indeed (after the deformation) become hopelessly entangled.

By a braid we mean a weaving pattern together with the permission to deform it according to the previous rules. If we present a weaving pattern, it describes a braid. But infinitely many patterns will describe the same braid, namely all those that can be obtained from the given one by a deformation. The order n of the pattern shall be called the *order of the braid.*

We now have the following fundamental problem. Given two weaving patterns, is it possible to decide whether or not they describe the same braid? In other words, is it possible to decide whether or not a pattern can be deformed into a given other one?

Up to now we have considered braids of all orders n. From now on we assume n to be an arbitrary but fixed integer and restrict ourselves, without saying it explicitly, to braids of that order n.

Let now A and B be two braids. We first explain what we mean by the product AB of A and B. We select definite patterns for A and B. Call L_1, L_2, P_i, Q_i, c_i the lines, points, and curves respectively of A, and $L'_1, L'_2, P'_i, Q'_i, c'_i$ those of B.

We deform B until the plane through L'_1 and L'_2 coincides with the plane through L_1 and L_2, and until the line L_2 coincides (including orientation) with the line L'_1, being careful to have L_1 and L'_2 on different sides of L_2. Finally we deform B until the points Q_i coincide with the points P'_i. This being achieved, we erase the line L_2, obtaining a new composed weaving pattern which shall stand as pattern for the braid AB.

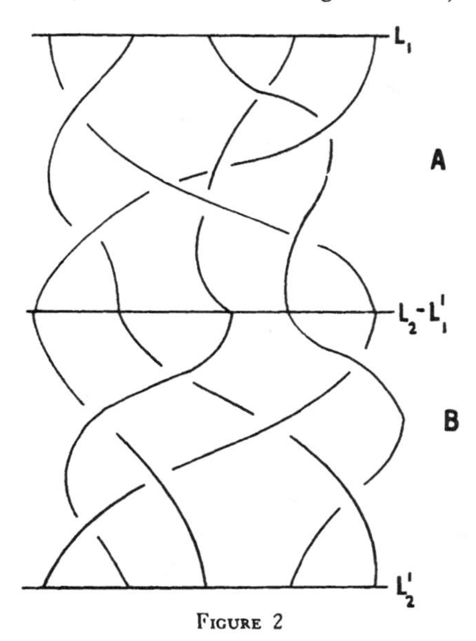

FIGURE 2

Intuitively speaking, this means: AB is obtained by tying the beginning of B to the end of A. Figure 2 explains the process. The reason for calling the result of this process a product lies in the fact that the process has some similarity to the ordinary multiplication of numbers. We first show:

$$(AB)C = A(BC)$$

the so-called associative law of multiplication.

What does $(AB)C$ mean? It means: form first AB and compose this with C. So tie B to A and to the result tie C. What, on the other hand, does $A(BC)$ mean? It asks us first to form BC, that is to tie C to B. The result shall be tied to A. Obviously we obtain the same pattern as $(AB)C$.

But this similarity does not go too far. For instance, the law $AB = BA$ is false in general. Very simple examples already show this. It may hold only accidentally for very special braids. In computations one must, therefore, be careful about the order of terms in a product.

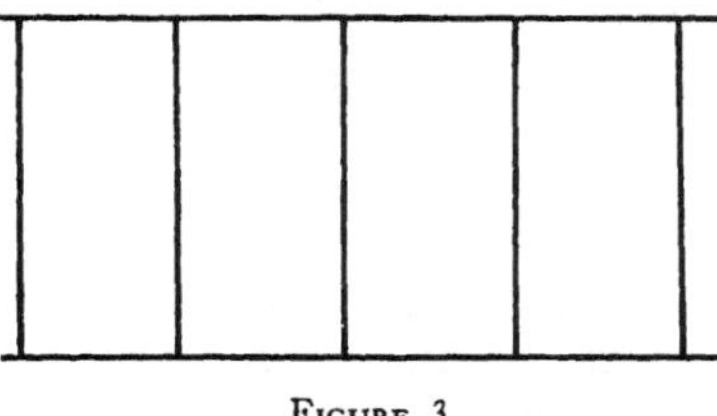

FIGURE 3

Let us denote by I the braid indicated in Figure 3. In its pattern the curves c_i are simply straight lines joining P_i and Q_i without crossings. If we tie I to any braid A, it is almost immediately seen that the resulting braid AI can be changed back to A; indeed the line L_2 is simply replaced by a somewhat lower line. Therefore $A\mathrm{I} = A$ for any braid A; similarly we see $\mathrm{I}A = A$ for any A.

Our braid I has therefore a strong resemblance to the number 1 (since $1.a = a.1$ for any number a). This explains the choice of the name I (roman one).

What does the equation $A = \mathrm{I}$ mean? If A is originally given by some complicated pattern, then $A = \mathrm{I}$ means that by some deformation this pattern can be changed into the pattern of Figure 3. We may say intuitively: $A = \mathrm{I}$ means that A can be combed.

Figure 4 shows the braid A of Figure 1, and tied to it its exact reflexion on the line L_2 which we call A^{-1}. The reader can convince himself that the combined braid AA^{-1} can be disentangled if he starts removing crossings from the middle outwards. In the same way he can see that $A^{-1}A$ can be combed.

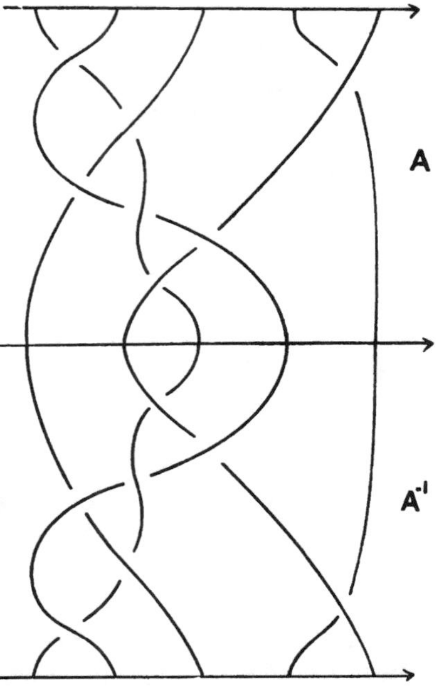

FIGURE 4

There exists therefore to any braid A another braid A^{-1} (its reflexion) such that

$$AA^{-1} = A^{-1}A = \mathrm{I}$$

The symbol A^{-1} is chosen because of an analogy with elementary algebra where a^{-1} stands for the number $\frac{1}{a}$ so that $aa^{-1} = a^{-1}a = 1$ for any non-zero number a.

Reviewing we may say: the braids form a system of objects in which a multiplication is defined. Three properties hold for this multiplication:

(1) The associative law $(AB)C = A(BC)$ is satisfied.

(2) There is a braid called I such that $A\mathrm{I} = \mathrm{I}A = A$ holds for any braid A.

(3) To any braid A another braid A^{-1} can be found such that $AA^{-1} = A^{-1}A = \mathrm{I}$.

If in these three statements we were to replace the word "braid" by the phrase "object of the system," we should obtain the exact definition of what in higher algebra one calls a "group." A group is simply a system of arbitrary objects, together with some kind of multiplication such that our three properties hold. We may say therefore: the system of all braids of order n is a group.

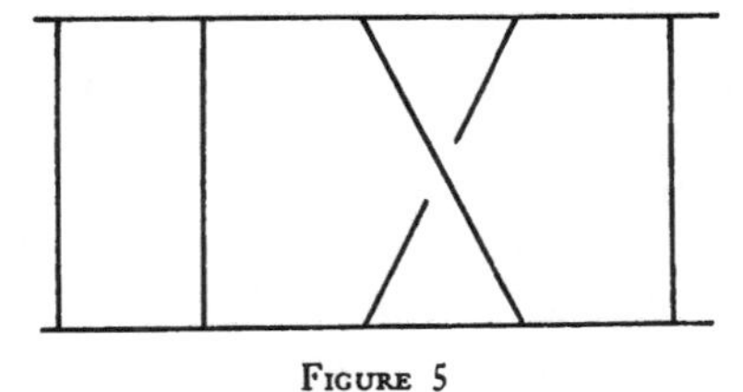
FIGURE 5

The theory of groups has been developed extensively, and its methods may be applied to our problem. Let us look at the special braid indicated in Figure 5. Here the curve c_i goes once over the curve c_{i+1}, whereas all other curves are straight lines connecting P_j and Q_j. We shall call this braid σ_i and obtain in this fashion n–1 braids $\sigma_1, \sigma_2, \ldots \sigma_{n-1}$. ($\sigma_n$ does not exist since it would involve an $n+1$-st curve). The braid where c_i goes *under* c_{i+1} needs no new name. It is the reflexion of σ_i and may therefore be denoted by σ_i^{-1}.

Consider now the pattern of any braid A, for example the braid in Figure 1. In its projection two crossings may occur at exactly the same height. But it is evident that a slight deformation of braid A will produce a pattern where this does not happen.

We cut up our pattern into small horizontal sections, such that only one crossing occurs in each section. Our braid A is obtained from all these sections by tying them together again. Each of these sections is obviously either a braid σ_i or a braid σ_i^{-1} depending on the nature of the crossings. Consequently we can express A as the product of terms each of which is either a σ_i or a σ_i^{-1}.

The braid in Figure 1, for example, is given by:

$$A = \sigma_1^{-1}\,\sigma_4^{-1}\,\sigma_2^{-1}\,\sigma_1\,\sigma_2^{-1}\,\sigma_3\,\sigma_2^{-1}$$

If every element in a group can be expressed as product of some elements σ_i and their inverses, we say that the σ_i are generators of the group. We may therefore state: the $n-1$ elements σ_i are generators of the braid group.

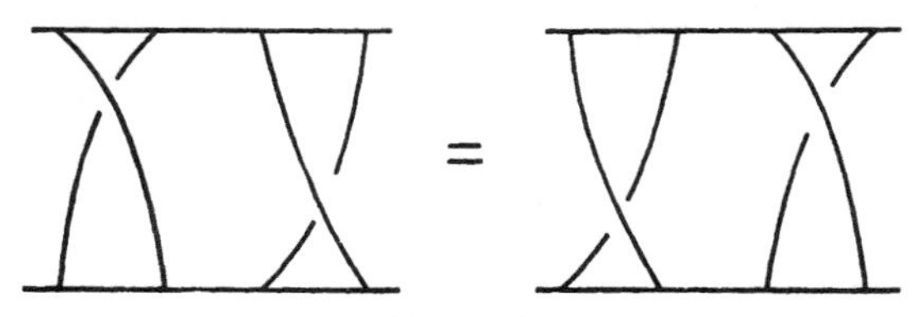

FIGURE 6

We are now in a position to describe any weaving pattern. As an example let us look at the braids in a girl's hair. A close look reveals that such a braid can be described by:

$$A = \sigma_1\,\sigma_2^{-1}\,\sigma_1\,\sigma_2^{-1}\ldots\sigma_1\,\sigma_2^{-1} = (\sigma_1\,\sigma_2^{-1})^k$$

where k is the number of times the elementary weaving pattern is repeated.

Figure 6 shows the equality $\sigma_1\,\sigma_3 = \sigma_3\,\sigma_1$. A similar figure would show $\sigma_i\,\sigma_j = \sigma_j\,\sigma_i$ if j is $i+2$ or more. That $\sigma_1\,\sigma_2$ is different from $\sigma_2\,\sigma_1$ can be seen by a simple sketch; in $\sigma_1\,\sigma_2$ the curve c_1 runs from P_1 to Q_3, whereas in $\sigma_2\,\sigma_1$ it runs from P_1 to Q_2.

But $\sigma_i\,\sigma_{i+1}\,\sigma_i = \sigma_{i+1}\,\sigma_i\,\sigma_{i+1}$. Figure 7 shows it for $i = 1$; the reader readily deforms the two patterns into each other.

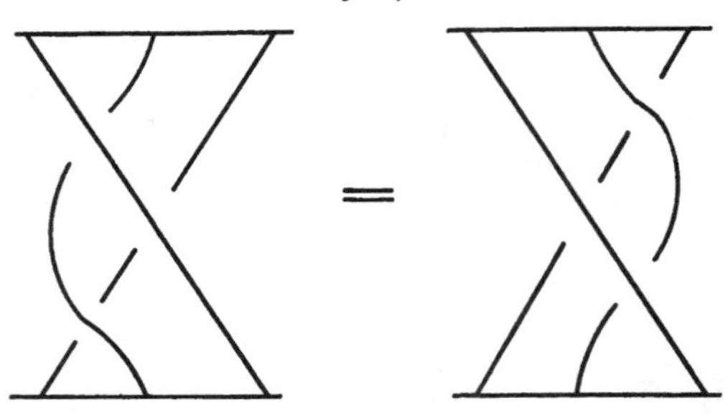

FIGURE 7

We have seen that the group has $n-1$ generators. Actually we can get away with only two: namely σ_1 and the braid

$$a = \sigma_1 \sigma_2 \ldots \sigma_{n-1} \text{ (product)}$$

Let us prove the statement for $n = 5$. We have

$$a \sigma_1 = \sigma_1 \sigma_2 \sigma_3 \sigma_4 \cdot \sigma_1$$

But $\sigma_4 \sigma_1 = \sigma_1 \sigma_4$; therefore $a \sigma_1 = \sigma_1 \sigma_2 \sigma_3 \sigma_1 \sigma_4$.
Now $\sigma_3 \sigma_1 = \sigma_1 \sigma_3$; hence $a \sigma_1 = \sigma_1 \sigma_2 \sigma_1 \cdot \sigma_3 \sigma_4$.
From $\sigma_1 \sigma_2 \sigma_1 = \sigma_2 \sigma_1 \sigma_2$, we obtain $a \sigma_1 = \sigma_2 \sigma_1 \sigma_2 \sigma_3 \sigma_4$.
Therefore $a \sigma_1 = \sigma_2 a$ or $a \sigma_1 a^{-1} = \sigma_2 a a^{-1} = \sigma_2 \cdot I = \sigma_2$.
Hence $\sigma_2 = a \sigma_1 a^{-1}$.

Similarly:

$$a \sigma_2 = \sigma_1 \sigma_2 \sigma_3 \sigma_4 \cdot \sigma_2 = \sigma_1 \cdot \sigma_2 \sigma_3 \sigma_2 \cdot \sigma_4 = \sigma_1 \cdot \sigma_3 \sigma_2 \sigma_3 \cdot \sigma_4 =$$
$$\sigma_3 \cdot \sigma_1 \sigma_2 \sigma_3 \sigma_4 = \sigma_3 a$$

It follows that $a \sigma_2 a^{-1} = \sigma_3$, or $\sigma_3 = a \sigma_2 a^{-1}$.
Substituting our result for σ_2 we obtain

$$\sigma_3 = a a \sigma_1 a^{-1} a^{-1} = a^2 \sigma_1 a^{-2}.$$

Finally:

$$a \sigma_3 = \sigma_1 \sigma_2 \sigma_3 \sigma_4 \sigma_3 = \sigma_1 \sigma_2 \sigma_4 \sigma_3 \sigma_4 = \sigma_4 \sigma_1 \sigma_2 \sigma_3 \sigma_4 = \sigma_4 a$$

Consequently $a \sigma_3 a^{-1} = \sigma_4$. Substituting for σ_3:

$$\sigma_4 = a \sigma_3 a^{-1} = a a^2 \sigma_1 a^{-2} a^{-1} = a^3 \sigma_1 a^{-3}.$$

In one formula:

$$\sigma_i = a^{i-1} \sigma_1 a^{-(i-1)}.$$

Each σ_i can be expressed by a and σ_1 and therefore any braid A can be expressed by a and σ_1.

In the following we shall not make use of this result.

The formulas:

(1) $$\sigma_i \sigma_j = \sigma_j \sigma_i, \text{ if } j \text{ is at least } i+2$$
(2) $$\sigma_i \sigma_{i+1} \sigma_i = \sigma_{i+1} \sigma_i \sigma_{i+1}$$

have the following significance:

Suppose two braids A and B given by patterns. Each pattern may be used to express A and B respectively as a product of terms σ_i or σ_i^{-1}.

If $A = B$, it must in some fashion be possible to change from the expression A to the expression B. It can be shown that this can always be done by a repeated use of either formulas (1) or (2), or of simple algebraic consequences of these formulas. It is this fact one refers to if one says: the braid group has the defining relations (1) and (2). The proof is too long to be reproduced here.

We proceed now to our fundamental problem. Let us first consider a braid A in which the curves c_i connect P_i with Q_i (the Q_i with exactly the same subscript).

Suppose we remove the curve c_1. A certain braid A_1 of order $n-1$ remains. Now we reinsert a curve d_1 between P_1 and Q_1 that is not entangled at all with the other strings (this means that its projection exhibits no crossings at all). This new braid of order n we call B.

Denote now the braid AB^{-1} by C. This braid C has a peculiar property. If the first string of C is removed, then the braid that remains from the A-part of C is A_1, and A_1^{-1} is the part that remains from B^{-1}. (According to our construction, A and B differ only by their first strings.) Therefore removing the first string from C leaves $A_1 A_1^{-1} = I$ —a braid that can be combed. To be sure, C itself cannot necessarily be combed until the first string has been removed.

Suppose now that this combing operation with the last $n-1$ strings of C is performed by force in spite of the presence of the first string. Since the first string is stretchable up to any amount, it may be taken along during this combing operation. At the end the first string will be entangled in a terrible fashion, but the result will look somewhat like Figure 8. A pattern of this type is called 1-pure.

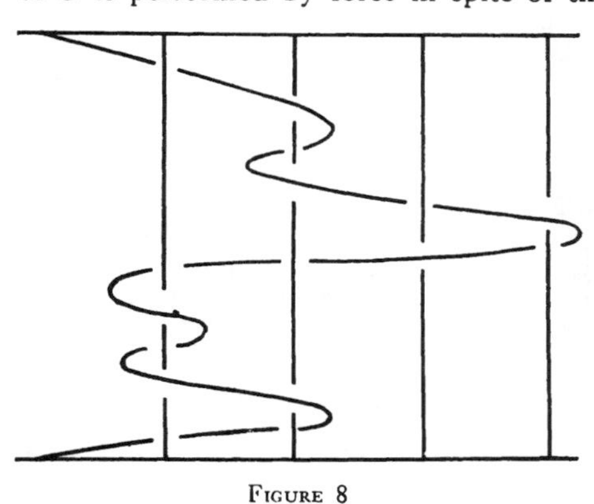

FIGURE 8

Now $AB^{-1} = C$; $AB^{-1}B = CB$; therefore $A = CB$. So A is a product of a 1-pure braid C and another braid B which is obtained from a braid of order $n-1$ by inserting a first string not meeting the others in a projection. The second string of B can be treated in the same way, and so on.

The final result is:

$$A = C_1 C_2 \ldots C_{n-1}$$

where C_i is a braid of the following kind: all strings but the i-th are vertical straight lines, and the i-th is only involved with strings of a higher number. Of course this means that for every braid A a pattern of this special kind can be found.

The solution of our fundamental problem consists in the assertion that a pattern of this type describes the braid uniquely; i.e., that in order to test whether $A = B$ for two braids whose curves c_i connect P_i with Q_i one has only to bring A and B into this form and to see whether exactly the same pattern results. The proof for this fact is very involved and cannot be included here. Nor shall we describe the translation of our geometric procedure into group theoretical language.

It is clear that this procedure contains the solution of the full problem to decide whether $A = B$ for any two braids A and B given by weaving patterns. First $A = B$ means the same as $AB^{-1} = \mathrm{I}$. The braid I connects P_i with Q_i. Should AB^{-1} not do this, then certainly A is not equal to B. In case AB^{-1} connects each P_i with Q_i, the previous method makes it possible to decide whether $AB^{-1} = \mathrm{I}$ or not.

Finally let us mention an unsolved problem of the theory of braids. If we wind a braid once around an axis, close it by identifying P_i and Q_i, and remove the lines L_1 and L_2, we obtain what we call a closed braid. Again we allow all those deformations in the course of which the curves do not cross the axis, nor each other.

The problem of classification of closed braids, at least, can be translated into a group theoretical problem. Let A and B be two open braids. The corresponding closed braids are equal if, and only if, an open braid X can be found such that

$$B = XAX^{-1}$$

A solution to this problem has not yet been found. Since in some ways closed braids resemble knots, such a solution could be applied to the problem of knots. It would also have many applications in pure mathematics.

REFERENCES

1. Artin, E. Theory of braids. *Annals of Mathematics 48*, 1947.
2. Artin, E. Braids and permutations. *Annals of Mathematics 49*, 1948.
3. Bohnenblust, F. The algebraic braid group. *Annals of Mathematics 48*, 1947.
4. Chow, W. L. On the algebraic braid group. *Annals of Mathematics 49*, 1948.

THEORY OF BRAIDS

By E. Artin

(Received May 20, 1946)

A theory of braids leading to a classification was given in my paper "Theorie der Zöpfe" in vol. 4 of the Hamburger Abhandlungen (quoted as Z). Most of the proofs are entirely intuitive. That of the main theorem in §7 is not even convincing. It is possible to correct the proofs. The difficulties that one encounters if one tries to do so come from the fact that projection of the braid, which is an excellent tool for intuitive investigations, is a very clumsy one for rigorous proofs. This has lead me to abandon projections altogether. We shall use the more powerful tool of braid coordinates and obtain thereby farther reaching results of greater generality.

A few words about the initial definitions. The fact that we assume of a braid string that it ends in a straight line is of course unimportant. It could be replaced by limit assumptions or introduction of infinite points. The present definition was selected because it makes some of the discussions easier and may be replaced any time by another one. I also wish to stress the fact that the definition of s-isotopy is of a provisional character only and is replaced later (Definition 3) by a general notion of isotopy.

More than half of the paper is of a geometric nature. In this part we develop some results that may escape an intuitive investigation (Theorem 7 to 10).

We do not prove (as has been done in Z) that the relations (18) (19) are defining relations for the braid group. We refer the reader to a paper by F. Bohnenblust[1] where a proof of this fact and of many of our results is given by purely group theoretical methods.

Later the proofs become more algebraic. With the developed tools we are able to give a unique normal form for every braid[2] (Theorem 17, fig. 4 and remark following Theorem 18). In Theorem 19 we determine the center of the braid group and finally we give a characterisation of braids of braids.

I would like to mention in this introduction a few of the more important of the unsolved problems:

1) Assume that two braids can be deformed into each other by a deformation of the most general nature including self intersection of each string but avoiding intersection of two different strings. Are they isotopic? One would be inclined to doubt it. Theorem 8 solves, however, a special case of this problem.

2) In Definition 3, we introduce a notion of isotopy that is already very general. What conditions must be put on a many to many mapping so that the result of Theorem 9 still holds?

[1] F. Bohnenblust, The algebraical braid group, Ann. Math., vol. 48, (1947), pp. 127–136.

[2] The freedom of the group of k-pure braids has been proved with other methods in: W. Fröhlich, *Über ein spezielles Transformationsproblem bei einer besonderen Klasse von Zöpfen*, Monatshefte für Math. und Physik, vol. 44 (1936), p. 225.

3) Determine all automorphisms of the braid group.

4) With what braids is a given braid commutative?

5) Decide for any two given braids whether they can be transformed into each other by an inner automorphism of the group. Concerning applications of braid theory this is by far the most important problem.

The last three of our questions seem to require an extensive study of the automorphisms of free groups.

We shall have to consider numerous functions of several variables. All of them are meant to be continuous in all the variables involved so that the statement of continuity is always omitted. Only more stringent conditions shall be mentioned.

Let x, y, z be the Cartesian coordinates of a 3-space. By a braid string we mean a curve that has precisely one point of intersection with each plane $z = a$ so that z may be used as parameter. Denoting by X the two dimensional vector (x, y) we may therefore describe the string by a vector function $X = X(z)$. In addition to that we assume the existence of two constants a, b such that $X(z)$ assumes a constant value X^- for all $z \leqq a$ and a constant value X^+ for all $z \geqq b$. X^- and X^+ are called the ends of the string.

By an n-braid we mean a set of n strings $X_i(z)$ $(i = 1, 2, \cdots n)$ without intersections (hence $X_i(z) \neq X_k(z)$ for $i \neq k$) where the numbering of the strings is considered unessential.

Two n-braids $X_i(z)$ and $Y_i(z)$ are called strongly isotopic (s-isotopic) if n vector functions $X_i(z, t)$ can be found with the following properties: They are defined for all z and for all t of a certain interval $c \leqq t \leqq d$, give an n-braid for each special t and are $X_i(z)$ for $t = c$, $Y_i(z)$ for $t = d$. They are constant in z and t if z is large enough and also constant if $-z$ is large enough. We remark at once that the ends remain fixed.

THEOREM 1. *s-isotopy is reflexive, symmetric and transitive.*

This allows us to unite s-isotopic braids into one class. We also obtain without difficulty:

THEOREM 2. *Let $g(z, t)$ be a numerical function defined for all z and all t of a certain t-interval. Assume that it tends with z to ∞ uniformly in t (the sign is unimportant). If $X_i(z)$ is a braid then the new braids $X_i(g(z, t))$ for different values of t are all s-isotopic. (They need not be s-isotopic to $X_i(z)$ itself.)*

COROLLARY 1. *If the numerical function $g(z)$ satisfies $\lim_{z=+\infty} g(z) = +\infty$ and $\lim_{z=-\infty} g(z) = -\infty$, then the braids $X_i(z)$ and $X_i(g(z))$ are s-isotopic.*

PROOF: Put $g(z, t) = (1 - t)\, z + t\, g(z)$ for $0 \leqq t \leqq 1$.

COROLLARY 2. *$X_i(z)$ is s-isotopic to any z-translation $X_i(z + t)$.*

COROLLARY 3. *The braids $X_i(\,|\,z\,|\, + t)$ are s-isotopic among themselves for different values of t. For large positive values of t all braid strings are constant $= X_i^+$. For large negative values of t the part of the braid above the xy-plane looks like a z-translation of $X_i(z)$, the part below this plane like its reflection on the xy-plane.*

We mention a few theorems of which little use shall be made in this paper, but that are useful if braids are to be studied by means of their projection on the yz-plane from the positive x-direction. By lexicographical arrangement of vectors we mean, as usual, their arrangement according to the size of their y-coordinate and, in case the y-coordinates are equal, according to their x-coordinate.

Two braids are said to have the same z-pattern of projection if the lexicographical arrangement of the vectors $X_i(z)$ is for each value of z the same as that of the vectors $Y_i(z)$.

LEMMA. *Two braids with the same z-pattern of projection are s-isotopic.*

PROOF: Put $X_i(z, t) = (1 - t)X_i(z) + t\, Y_i(z)$ for $0 \leqq t \leqq 1$. Assume we could find values z, t and $i \neq k$ such that $X_i(z, t) = X_k(z, t)$. This means that $(1 - t)(X_i(z) - X_k(z)) + t(Y_i(z) - Y_k(z)) = 0$. Let $X_i(z) - X_k(z) = (a, b)$ and $Y_i(z) - Y_k(z) = (a', b')$. The y-coordinate of our equation shows that b and b' can not have the same sign hence $b = b' = 0$. Now the x-coordinate of the equation leads to $a = a' = 0$. But then $X_i(z)$ would not be a braid.

Two braids $X_i(z)$ and $Y_i(z)$ are said to have the same pattern of projection if a monotonically increasing function $g(z)$ with infinite limits exists such that $X_i(z)$ and $Y_i(g(z))$ have the same z-pattern of projection.

THEOREM 3. *Two braids with the same pattern of projection are s-isotopic.*

PROOF: Use the lemma and Corollary 1 of Theorem 2.

THEOREM 4. *Let d be less than half the minimal distance between two of the strings of $X_i(z)$. Let $Y_i(z)$ be a braid with the same ends as $X_i(z)$ and assume that equally numbered strings of the two braids have at each z level a distance less than d. Then the braids are s-isotopic.*

The proof is done with the same device as in the lemma and is trivial.

DEFINITION: Two braids $X_i(z)$ and $Y_i(z)$ are called composable if:

1) They have the same number of strings.

2) After a suitable change in the numbering of the strings we have $Y_i^+ = X_i^-$. So the upper ends of $Y_i(z)$ must fit the lower ends of $X_i(z)$.

If (after a suitable translation) we join these ends we obtain a new braid which is said to be composed of $X_i(z)$ and $Y_i(z)$ (in this order). The formal definition would be:

Select an a and a b such that $X_i(z + a)$ is constant for negative z and $Y_i(z + b)$ is constant for positive z. Put $Z_i(z) = X_i(z + a)$ for positive z and $= Y_i(z + b)$ for negative z. This braid and any translation of it is called the composed braid. We still have a great freedom in the selection of a, b and the translation. All these braids are however s-isotopic.

If we replace both braids by others that are s-isotopic to them we obtain braids that are s-isotopic to $Z_i(z)$. The proof follows from the fact that after suitable selection of a and b the necessary deformation can be carried out independently in both sections of the composed braid. This leads to:

DEFINITION 1. Two classes of braids A and B are called *composable* if the braids in these classes are composable. The resulting braids form a class denoted by AB. It is to be remarked that if A is composable with B then B may not be composable with A. And even if the classes are composable from both sides then the commutative law may not hold. But the associative law does. If A and B as well as AB and C are composable then B and C as well as A and BC are composable and we have $A(BC) = (AB)C$.

THEOREM 5. *The classes of n-braids form a groupoid under our composition.*

PROOF: The postulates that we must verify are:

1) The kind of associative law we just have mentioned.

2) The existence of two classes U and U' for each given class A (dependent on A) such that $AU = U'A = A$. U is obviously the class containing a constant braid with the same lower ends as A and U' the similar class connected with the upper ends of A.

3) The existence of a class A^{-1} such that $A^{-1}A = U$ and $AA^{-1} = U'$. Corollary 3 shows that the reflection of a braid on the x, y plane gives such a class.

4) If A and B are given classes there exists a class C such that AC as well as CB can be formed. This just means to construct an example of a braid with given ends.

If U is one of the unit classes call G_U the set of all A that have U as left as well as right unit. They form a group. If V is another unit and C a class such that $UC = CV = C$ then $G_V = C^{-1}G_UC$ and the transformation thus indicated is an isomorphism. The knowledge of one of these groups reveals the structure of all and as a matter of fact the structure of the whole groupoid. The braids in such a group are those whose upper ends are only a permutation of the lower ends.

Next we prove that s-isotopy can be extended to the whole space:

THEOREM 6. *Let $X_i(z, t)$ be the n functions describing an s-isotopy. Then we can find a function $F(X, z, t)$ defined for all X and z and all necessary t, whose value is a vector, and that has the following properties:*

1) *For any fixed z it is a deformation of the plane. That means that it is a one to one correspondence of the plane if t also is fixed and it is identity for $t = a$ if that is the beginning of the t-interval.*

2) *Should for any special value of z the original functions $X_i(z, t)$ be independent of t, then $F(X, z, t) = X$ for that z and all X and t.*

3) *If a point (X, z) of the 3-space has a sufficiently large distance from the origin then $F(X, z, t) = X$ for all t.*

4) *$F(X_i(z, a), z, t) = X_i(z, t)$. So the deformation of the space moves the braid-strings precisely as the s-isotopy does.*

PROOF: 1) Select an $r > 0$ such that $| X_i(z, t) - X_k(z, t) | < 3r$ for all $i \neq k$ and all z and t. We first construct an auxiliary function $G(X, P_\nu, Q_\nu)$ of X and $2n$ points P_ν, Q_ν ($\nu = 1, 2, \cdots n$) of the plane. The points P_ν are restricted by the condition that their mutual distance shall always be greater than $3r$, the points Q_ν by the condition that Q_i lies in the interior of a circle C_i of radius r around P_i. The value of G shall be X if X is outside of all the circles or on the periphery of one. For $X = P_i$ the value shall be Q_i. If X is in the interior of C_i but different from P_i draw a radius through X and call R its intersection with C_i. Define the functionvalue as that point on the straight line segment

$R\ Q_i$ that bisects it in the same ratio as X bisects $R\ P_i$. Our function is continuous in all the variables, is a one to one correspondence of the plane for fixed P_ν and Q_ν and reduces to identity if $P_\nu = Q_\nu$ for all ν.

2) Divide the t interval into a finite number of parts $t_i \leqq t \leqq t_{i+1}$ such that the variation of every $X_i(z, t)$ in that interval is less than r for fixed z.

In $a = t_0 \leqq t \leqq t_1$ we define $F(X, z, t) = G(X, X_\nu(z, a), X_\nu(z, t))$. It has all the necessary properties. Assume that we succeeded to define $F(X, z, t)$ for all t of $t_0 \leqq t \leqq t_m$ and to check on the required properties. For $t_m \leqq t \leqq t_{m+1}$ we define:

$$F(X, z, t) = G(F(X, z, t_m), X_\nu(z, t_m), X_\nu(z, t)).$$

For $t = t_m$, we get the old value so it is a continuous continuation. The properties 1, 2, 3 follow immediately. For $X = X_i(z, a)$ we get $F(X, z, t_m) = X_i(z, t_m)$ hence $F(X, z, t) = X_i(z, t)$.

The extension of an s-isotopy to the whole space is not the only use of Theorem 6. We also use it to introduce new coordinates for the points of the space called braid coordinates. They are much more flexible in dealing with braids and the principal tool in the proofs of most of the following theorems.

Let $X_i(z)$ be a given braid, constant for $z \leqq a$ and for $z \geqq b$. Consider the braids $X_i(z + t)$ for $0 \leqq t \leqq b - a$. They are all isotopic and let $F(X, z, t)$ be the extension of this isotopy to the whole space. Then we have:

$$F(X_i(z), z, t) = X_i(z + t), \qquad 0 \leqq t \leqq b - a. \tag{1}$$

With each point (x, z) of the space we associate now a 2 dimensional vector $Y = Y(X, z)$ in the following way:

For $z \leqq a$ let $Y = X$.

For $a \leqq z \leqq b$ let Y be the unique solution of $F(Y, a, z - a) = X$. For $z = a$ we have $F(X, a, 0) = X$, therefore $X = Y$ again.

For $z \geqq b$ put $Y(X, z) = Y(X, b)$.

If z is fixed then the mapping $Y = Y(X, z)$ is a one to one correspondence of the plane that is certainly identity, outside a large circle whose radius does not depend on z. It is identity for all X if $z \leqq a$. For $z \geqq b$ it is in general not identity but at least is the same mapping for all $z \geqq b$.

$F(X_i(a), a, z - a) = X_i(a + (z - a)) = X_i(z)$ if $a \leqq z \leqq b$ because of (1). This shows that the Y for the point $(X_i(z), z)$ of the ith string is $X_i(a) = X_i^-$. The same is true for $z \leqq a$ and for $z \geqq b$ for trivial reasons.

We associate now with the point (X, z) the corresponding combination $\{Y, z\}$ and call it the braid coordinates of that point. They equal the ordinary coordinates for all $z \leqq a$ and also for all large $|X|$. All points on the ith string have the simple braid coordinates $\{X_i^-, z\}$.

Another way to look upon the braid coordinates is this: Interpret them as ordinary coordinates of a point of a 3-space. Then our 3-space is mapped by a one to one correspondence onto this new one and the braid strings are mapped onto vertical lines with the same lower ends.

Let now u be any real number. The mapping $\{Y, z\} \to \{Y, z + u\}$ in our old space is a one to one correspondence that has all the essential features of a translation and shall therefore be called translation by u along the braid. Each single string remains fixed as a whole. For large $|X|$ and also if $|z|$ is large in comparison to $|u|$ it is an ordinary translation. For the other points the z-coordinate does at least behave in the ordinary way.

We can of course also find the inverse function $X = H(Y, z)$ that describes the passage from Y back to X. Let now $X(z)$ be any other braidstring (it may intersect the strings of our braid). Apply to it a translation by u along our braid. The braid coordinates of $(X(z), z)$ are $\{Y, z\}$ where $Y = Y(X(z), z)$. The translation moves it into $\{Y, z + u\}$. The ordinary coordinates are $(X', z + u)$ where $X' = H(Y(X(z), z), z + u)$. Looking at the new string as a whole, we may replace z by $z - u$ and obtain the new braidstring

$$X(z, u) = H(Y(X(z - u), z - u), z)$$

such that $(X(z, u), z)$ are the points of the new braidstring. Letting now the u change, we see that we have before us an s-isotopic change where the points move only in horizontal planes. Should the original string not intersect the braidstrings, then its translation does not either and the braid formed out of the old braid by adding our new string undergoes an s-isotopy under translation.

We use this in the following way: Let $X_i(z)$ be an n-braid and replace its nth string by any other string $X'_n(z)$ with the same ends. The new string may intersect $X_n(z)$ but shall not intersect the other strings of our braid. This gives a new n-braid and we apply now to our old braid a translation by a large u along our new braid. What happens? The $n - 1$ first strings remain fixed. If u is sufficiently large then a very low portion of the nth string will now be in the main part of the braid. In that very low portion the string $X_n(z) = X'_n(z)$. The string $X'_n(z)$ does not change under our translation. This shows that in the main portion of our braid $X_n(z)$ has moved into the position $X'_n(z)$. This is of course compensated by the fact that the nth string is now entangled in the other strings above the main portion of the braid. But above the main section of the braid, the first $n - 1$ strings are very simple, namely parallel lines. Remember finally that we have shown in the preceding paragraph that the translation can also be considered as a horizontal motion, as an s-isotopy.

THEOREM 7. *It is possible to apply to a braid an s-isotopy moving one string only whereby this string may be brought into any other position provided this is compensated by a motion of the string above the main section of the braid where the $n - 1$ other strings are parallel.*

This suggests the definition:

DEFINITION 2. A braid with the same upper and lower ends is called *i-pure*, if all but the ith string are constant. A class is called i-pure if it contains an i-pure representative.

We see: if B and B' are braids or classes of braids having $n - 1$ strings in common then $B = AB'$ where A is i-pure.

Another useful notion connected with braid coordinates is that of projection along the braid. Consider the plane $z = z_0$. The mapping $\{Y, z\} \to \{Y, z_0\}$ is this projection. It carries the braidstrings into their intersection with the plane and a point not on the braid into a point different from these intersections.

Let now R_B be the complementary set to B in the 3-space. We introduce the usual notion of homotopy of paths in R_B. Two paths $a = \{Y(t), z(t)\}$, $b = \{Y'(t), z'(t)\}$, $0 \leqq t \leqq 1$ are called homotopic in $R_B : a \sim_B b$ if a function of two variables t and s (interval 0, 1) $\{Y(t, s), z(t, s)\}$ can be found, that is constant for $t = 0$ and $t = 1$, gives the first path for $s = 0$ and the second for $s = 1$. All points of the deformation have to belong to R_B which means simply that the function Y avoids the values X_i^-.

The composition of homotopy classes is introduced in the usual way and leads to a groupoid. If we select a point P in R_B and consider the homotopy classes of those paths that have beginning as well as endpoint at P, we obtain the Poincaré group of R_B. Let z_0 be the z coordinate of P and $a = \{Y(t), z(t)\}$ any element of this group. The projection $a' = \{Y(t), z_0\}$ is homotopic to a as the function $\{Y(t), z_0 + s(z(t) - z_0)\}$ shows. If two paths in this plane are B-homotropic, say by the function $\{Y(t, s), z(t, s)\}$, then the function $\{Y(t, s), z_0\}$ shows that the paths are already homotopic in the plane. The Poincaré group is therefore the same as that of a plane punctured in the n points $X_i(z_0)$. So it is a free group with n generators.

We must now carefully describe the generators we want to use. The plane will be either in the region of z where the braidstrings assume the constant values X_i^- and shall then be called a lower plane or in the region of the X_i^+ when we call it an upper plane. Take an upper plane and draw in it a ray that does not meet any of the upper or lower ends. We intend to take the point P on that ray sufficiently far away. Each of the points X_i^+ shall be connected with the beginning point Q of the ray by a broken line without self intersection such that two of the lines and also the ray have only the endpoint Q in common. By l_i we denote the connection thus established between the beginning point X_i^+ and the point P on the ray. By $l_i(\epsilon)$ we mean the same path but starting with the parameter value ϵ. An orientation of the plane is selected. By $c_i(\epsilon)$ we mean a curve with the winding number 1 around X_i^+ starting and ending at the beginning point of $l_i(\epsilon)$ that stays within a small neighborhood of X_i^+. It is well known that the paths (for small ϵ)

$$t_i = l_i(\epsilon)^{-1} c_i(\epsilon) l_i(\epsilon) \tag{2}$$

are free generators of the Poincaré group of the punctured plane. This pattern of paths is then transferred to all other upper planes by projection (in the ordinary sense) including the point P. We use the same names for the paths in all the upper planes.

In a lower plane we first transfer the ray, the orientation and the point P to it by ordinary projection. Since we now have to take care of the lower ends, new paths are selected denoted by l'_i, $l'_i(\epsilon)$, $c'_i(\epsilon)$. The corresponding generators shall be called t'_i.

What we intend to do with this setup is roughly this: If two braids are s-isotopic they must by necessity have the same ends. It therefore suffices to consider braids with given ends. For all these braids we use the same pattern of paths in the lower and upper planes. If B is a given braid then we project by braid projection the generators t'_i into the upper plane. We obtain paths $\bar{t}_i$ that are now generators in the upper plane. Therefore they can be expressed in terms of the t_i. It turns out that these expressions are a complete set of invariants for the isotopy classes.

As a first indication that the study of the homotopy classes must give a solution of our problem let us consider the following special case:

Let C be an $(n - 1)$-braid. Form two n-braids by inserting in C an nth string in two ways but both time with the same ends. Select a z_0 such that the two strings are equal for $z \leqq z_0$ and call P_0 the point on the nth strings at that z_0-level. In a similar fashion select z_1 and P_1 for the upper end. Consider now the two pieces of the nth strings between P_0 and P_1. If they are homotopic relative to C we may call the two n^{th} strings C-homotopic. Then the following rather surprising theorem holds:

THEOREM 8. *Let B and B' be obtained from C by insertion of an nth string with given ends. If the two strings are C-homotopic then B and B' are s-isotopic. Every homotopy class can be realized by a braid. The converse is also true but will be proved later.*

PROOF: 1) We use braid coordinates of C and express the two inserted strings in terms of these coordinates: $Y_n(z)$, $Y'_n(z)$. The fact that they do not intersect the strings of C means just that the functions avoid the values X_i^-. By assumption there exists a function $\{Y(t, s), z(t, s)\}$ defined in $z_0 \leqq t \leqq z_1$, $0 \leqq s \leqq 1$ describing the homotopy of the strings. So $Y(t, s)$ will avoid the values X_i^-, have fixed beginning and end points for all s and will be $Y_n(t)$ for $s = 0$ and $Y'_n(t)$ for $s = 1$. The method consists now in forgetting about the function $z(t, s)$ altogether and to define a function $Y_n(z, s)$ as equal to $Y(z, s)$ for $z_0 \leqq z \leqq z_1$ and equal to X_i^- for all other z. The function avoids X_i^- and reduces to the given strings for $s = 0$ and $s = 1$. So it gives the required s-isotopy.

2) If $\{Y(t), z(t)\}$ is any curve defined in $z_0 \leqq z \leqq z_1$ that avoids the strings of C and joins P_0 and P_1, put as before $Y_n(z) = Y(z)$ in that interval and $= X_n^-$ for all other z. The function $\{Y(t), s \cdot z(t) + (1 - s)t\}$ shows that this n^{th} string is homotopic to the given curve.

REMARK. The s-isotopy of Theorem 8 moves the nth string only.

Let us now return to our upper plane. Join one of the points X_i^+ to P by a curve h that avoids the braid with exception of its beginning point. Define $h(\epsilon)$ as before and let $d(\epsilon)$ be a curve analogous to $c_i(\epsilon)$. Consider the element $t = h(\epsilon)^{-1}d(\epsilon)\, h(\epsilon)$. If we join the beginning point of $l_i(\epsilon)$ to that of $h(\epsilon)$ by a path e that stays in a small neighborhood of X_i^+ then $e^{-1}c_i(\epsilon)e$ is homotopic to

$d(\epsilon)$. The path $S' = l_i(\epsilon)^{-1}e\, h(\epsilon)$ is a certain element of the group and may be expressed in terms of the t_ν. The element

$$S'^{-1}t_iS' = h(\epsilon)^{-1}e^{-1}l_i(\epsilon)l_i(\epsilon)^{-1}c_i(\epsilon)l_i(\epsilon)l_i(\epsilon)^{-1}e\, h\,(\epsilon)$$

is homotopic to t. Assume now that a similar expression $t = S^{-1}t_iS$ is known from some other source and may even not be in a reduced form. We first perform the possible cancellations in S only; a further simplification of the expression $S^{-1}t_iS$ is then possible only if S begins with a power of t_i. Then S^{-1} ends with the reciprocal of that power and we see that this term may indeed be dropped. This shows that S is uniquely determined but for a power of t_i and we have therefore $S' = t_i^rS$ or

$$l_i(\epsilon)^{-1}e\, h(\epsilon) = t_i^rS. \tag{3}$$

Let us now reinsert in our plane the one point X_i^+ and consider homotopies in that new plane, punctured in $n - 1$ points only. This homotopy shall be denoted by $\sim_i$. It amounts to put $t_i = 1$ in all previous expressions. The resulting element shall still be denoted by S. We obtain:

$$l_i(\epsilon)^{-1}e\, h(\epsilon) \sim_i S.$$

The path $l_il_i(\epsilon)^{-1}e\, h(\epsilon)h^{-1}$ is i-homotopic to a closed curve starting at X_i^+ and remaining in a small neighborhood of that point. It is therefore i-homotopic to this point X_i^+. This proves the formula

$$\bar{l}_i^{-1}h \sim_i S. \tag{4}$$

Let now B be a braid with the given ends. If we apply braid projection to a generator t_i' of a lower plane onto an upper plane and assume that the point P has been selected sufficiently far out, then the image $\bar{t}_i$ will be an element of the Poincaré group for P. If l_i is the projection of l_i' we obtain equations of the form:

$$\bar{t}_i = S_i^{-1}t_iS_i \tag{5}$$

$$l_i(\epsilon)^{-1}e_i\overline{l_i(\epsilon)} \sim t_i^rS_i \tag{6}$$

$$\bar{l}_i^{-1}l_i \sim_i S_i. \tag{7}$$

It is to be remarked that the properties of the braid coordinates show that the form of the equation (5) does not depend on the precise location of the upper and the lower plane. It is also clear that it does not depend on ϵ provided it is only small enough. The position of P plays also no role in it provided that it is far enough out. The equations (6) and (7) change of course their meaning and the exponent r may depend on ϵ and e_i.

We may look upon this process in yet another way. Call g the straight line segment that connects P with its projection in the lower plane and put $\tau_i = gt_i'g^{-1}$. They are elements of the Poincaré group for P and as a matter of fact a set of generators. If we subject them to braid projection they will go over again into the $\bar{l}_i$. But being elements with the same beginning and end point τ_i is, as we have seen before, homotopic to $\bar{l}_i$. So equation (5) becomes

$$\tau_i \sim S_1^{-1} t_i S_i . \tag{8}$$

We are now ready for a generalized notion of isotopy:

DEFINITION 3. A braid is called *isotopic to another braid* if the space can be mapped into itself in such a way that points on the first braid but no other points of the space are mapped onto points of the second braid. In addition to this we assume that the mapping is identity outside of a certain sphere. Inside that sphere the mapping must of course be continuous but need not be one to one.

Consider now two isotopic braids B and C. Locate the lower and the upper plane outside the sphere and select P so that g is also outside this sphere. The two elements τ_i and its B-projection $\bar{l}_i$ are B-homotopic. The surface connecting the two paths does not meet B. Its image under our mapping will therefore avoid C. This shows that the images of our paths are C-homotopic. But our paths remain fixed. So τ_i and $\bar{l}_i$ are C-homotopic. τ_i on the other hand is C-homotopic to its C-projection. This C-projection is therefore C-homotopic to $\bar{l}_i$ and this proves:

THEOREM 9. *If B is isotopic to C then the exexpressions in formula* (5) *are the same for B and C.*

The i-homotopy of formula (7) may be interpreted as homotopy with respect to the braid resulting from a cancellation of the ith string. Denote by Σ_i the piece of the ith string that starts at the upper plane and ends at the lower plane. The path $l_1^{-1}\ \Sigma_i l_i' g^{-1}$ is a closed path starting at P and as such i-homotopic to its projection onto the upper plane. But this projection is obviously the left side of (7). Computing Σ_i out of the resulting homotopy we get:

$$\Sigma_i \sim_i l_i S_i g l_i'^{-1} \tag{9}$$

and this shows that S_i determines the homotopy class of the ith string.

(9) Interprets the ith string but it would not completely explain S_i since it is only an i-homotopy. Let Σ_i' be any path connecting the beginning points of $l_i(\epsilon)$ and $l_i'(\epsilon)$ that stays in the immediate vicinity of the nth string without intersecting it. Its projection is then a curve that may be used as e_i. Now the projection of $l_i(\epsilon)^{-1}\Sigma_i' l_i'(\epsilon) g$ is the left side of (6):

$$t_i^r S_i \sim l_i(\epsilon)^{-1}\Sigma_i' l_i'(\epsilon) g \tag{10}$$

which provides the full geometric meaning of S_i. The converse of Theorem 8 is:

THEOREM 10. *Let B and C be two braids with the same ends and with the same first $n - 1$ strings. Assume either that B and C are s-isotopic or that they are isotopic or that the expressions* (5) *are the same for both braids. Then the nth strings are n-homotopic and there exists an s-isotopy moving the nth string only.*

PROOF: If they are s-isotopic then they are isotopic because of Theorem 6. If they are isotopic then the expressions (5) are the same because of Theorem 9.

If (5) is the same even for $i = n$ only then the nth strings are n-homotopic because of formula (9). The remark to Theorem 8 shows the rest of the contention.

THEOREM 11. *If two braids have the same ends and if the expressions* (5) *are the same for both braids then they are s-isotopic.*

PROOF: For $n = 1$ our theorem is trivial since the two strings are 1-homotopic. Let it be proved for braids with $n - 1$ strings. If B and C are two n-braids for which the assumption holds, let B' and C' be the braids resulting from cancellation of the nth string. The expressions (5) for B' and C' are obtained by putting $t_n = 1$. So (5) is also the same for B' and C'. Therefore B' and C' are s-isotopic. Extend the s-isotopy to the whole space and apply this mapping to the braid B. It carries B into an s-isotopic braid D having again the same expressions (5). D and C have now the first $n - 1$ strings in common so that Theorem 10 shows that they are s-isotopic. This completes the proof.

THEOREM 12. *Isotopy and s-isotopy imply each other.*

The proof follows from Theorems 6, 9 and 11.

THEOREM 13. *The expressions* (5) *do not depend on the special braid-coordinates used. They depend even only on the class and give together with the ends a full system of invariants of the class that determines the class completely.*

The proof is now obvious. It is to be remarked, however, that the expressions depend on the selection of generators t_i and t'_i. We must now develop methods that allow the actual computations of these invariants and reveal the structure of our groupoid.

To do so we have first to change our notation slightly. Select in a plane n points X_i, a ray and paths l_i. Up to this point the numbering was considered unessential. Now we get a natural arrangement of our points by starting with our ray and going around Q in the positive sense of rotation (in a neighborhood of Q). The first path that we meet shall be called l_1, the next l_2 and so on. The points X_i are now numbered precisely as the paths leading to them. The very same pattern is now used for the upper planes as well as for the lower ones. The points X_i are now used as lower and upper ends of braids. We restrict ourselves to the investigation of braids whose lower ends are a subset of the X_i and whose upper ends are another subset. These subsets may or may not be the whole set, no restriction being put upon them. If B is such a braid and X_i, X_j the lower respectively upper end of one of its strings we write:

$$j = B(i).$$

Thus B maps a certain subset of the numbers 1, 2, $\cdots$ n onto another subset. The numbering of the generators t_i and t'_i so far was connected with the numbering of the string. Now we change that and attach to them the subscript of the point around which they run. We also drop the accent on the t'_i and write uniformly t_i for all the generators in the different planes. This leads to the following situation:

We have a group F before us with the n free generators t_i. For the Poincaré group of a braid with the reference point in an upper or a lower plane, not all the generators are used; the Poincaré group of such a plane is therefore considered a subgroup of F generated by a subset of the t_i. Braid projection of a lower unto an upper plane will provide us with an isomorphic mapping of the group in the lower plane onto the group in the upper one. If T is an element of the group in the lower plane then its image under braid projection shall be denoted by $B(T)$. In this new notation (5) takes on the form:

$$B(t_i) = S_j^{-1} t_j S_j \quad \text{where} \quad j = B(i), \tag{11}$$

and where the numbering of all generators has been changed according to our new convention. (11) alone already gives us the isomorphic mapping and in this form contains also the information about the upper and lower ends of the strings of the braid. In case all the points are used for the lower ends, it will be an automorphism of F. Otherwise it maps a subgroup of F onto another subgroup.

Let A and B be two composable braids and form the composed braid in such a way that in AB the part B corresponds to negative, the part A to positive z. Returning for a moment to the interpretation of our projection in terms of the generators τ_i which allow to express projection in terms of homotopies we see:

$$AB(T) = A(B(T)).$$

We project namely a lower plane of AB first onto the plane $z = 0$ and the result onto an upper plane. Making use of the fact that (11) completely determines the class we see:

THEOREM 14. *The groupoid of braid classes whose lower and upper ends are subsets of the X_i is isomorphic to the groupoid of mappings in F indicated by* (11).

Consequently we express the braid class B in form of a substitution

$$B = \begin{pmatrix} t_i \\ S_j^{-1} t_j S_i \end{pmatrix} \tag{12}$$

where t_i runs of course only through certain of the generators. It is convenient to consider also more general substitutions

$$B = \begin{pmatrix} t_i \\ T_i \end{pmatrix}$$

in the free group F where certain t_i are mapped onto power products regardless of whether the substitution is derived from a braid or not. If the substitution is derived from a braid class then we say briefly that it is a braid. Also in this general case we denote by $B(T)$ the result of applying the substitution B onto the power product T.

The braid substitutions have one special property that we must derive. Draw a huge circle in a lower plane starting at the reference point of the Poincaré group and running around the braid. It is well known from the theory of the homotopies in a punctured plane that this element of the Poincaré group is homotopic to the product of the generators t_i (of course only those that we need for our braid) taken in the natural arrangement of the subscripts according to their size. Braid projection onto an upper plane carries the circle into a similar circle starting at P. This proves:

THEOREM 15. *If B is a braid then*

$$B\left(\prod t_i\right) = \prod t_i \tag{13}$$

both products taken in the natural arrangement of their subscripts.

Select a subscript $i < n$ and put $X_\nu(z) = X_\nu = \text{const for } \nu \neq i, i+1$. We connect now X_i and X_{i+1} by a broken line starting at X_i and running then parallel to l_i until it comes near the ray; then it runs parallel to l_{i+1} until it comes close to X_{i+1} with which it is then connected. If $X(t)$ $(0 \leqq t \leqq 1)$ is the parametric representation of this line we put

$$X_i(z) = \begin{cases} X_i & \text{for } z \leqq 0 \\ X(z) & \text{for } 0 \leqq z \leqq 1 \\ X_{i+1} & \text{for } z \geqq 1. \end{cases}$$

Then we draw a similar parallel curve between X_{i+1} and X_i running farther out and not intersecting the previous one but at the ends. The string $X_{i+1}(z)$ is explained in a similar fashion than $X_i(z)$ but it has X_{i+1} as lower and X_i as upper end. That this braid carries t_ν into itself if $\nu \neq i, i+1$ is seen by using ordinary projection which shows that τ_ν is homotopic to t_ν. t_i is mapped into a transform of t_{i+1}. To find the transformer we go back to (10). As parallel curve we use one that will under ordinary projection become the parts of $l_i(\epsilon)$ and $l_{i+1}(\epsilon)$ up to the ray. Consider now the right side of (10). Instead of l_i we have to write l_{i+1} in our new notation. The path projects by ordinary projection still into a homotopic path. But this homotopic path is now obviously 1. So t_i is carried into t_{i+1}. The image of t_{i+1} can now be found by a simpler method, namely, by Theorem 15. Since the product of the t_i must remain fixed we find by a simple computation that t_{i+1} is mapped into $t_{i+1}^{-1}t_it_{i+1}$. The class of this particular braid shall be called σ_i and the corresponding substitution is:

$$\sigma_i = \begin{pmatrix} t_i, & t_{i+1} \\ t_{i+1}, & t_{i+1}^{-1}t_i t_{i+1} \end{pmatrix} \tag{14}$$

with the understanding that the generators that are not mentioned are left unchanged. For σ_i^{-1} we have to compute the inverse substitution and an easy computation gives:

$$\sigma_i^{-1} = \begin{pmatrix} t_i, & t_{i+1} \\ t_i t_{i+1} t_i^{-1}, & t_i \end{pmatrix}. \tag{15}$$

To check whether a given braid is σ_i it suffices, however, to check the following properties: dropping the ith and $i+1^{\text{st}}$ string we must obtain a unit. In it the $i+1^{\text{st}}$ string must correspond to the unit homotopy. After reinserting it the ith string must have unit homotopy (always using the simpler formula (9) rather then (10)). According to our theory these checks already determine the class.

From now on the nature of our proofs will be mostly algebraic.

Consider a rather general substitution B that maps each generator t_i onto a transform $Q_i^{-1}t_kQ_i = T_i$ (in reduced form) of some generator t_k. When is B a braid? The answer is given by the condition of Theorem 15, namely that

$$T_1T_2 \cdots T_n = t_1t_2 \cdots t_n . \tag{16}$$

The necessity is obvious. To prove the sufficiency we assume each Q_i written as a product of terms t_ν^ϵ, $\epsilon = \pm 1$. The number of terms shall be called the length of Q_i and the sum of all these lengths the length of B. If the length of B is 0 then (16) can hold only if each $T_i = t_i$ or $B = 1$ which is the unit braid. So we may assume our contention proved for all braids with smaller length than B. (16) can hold only if some cancellations take place on the left side. Since each T_i is already reduced these cancellations must take place between adjacent factors of the left side. Two cases are conceivable:

1) In a cancellation between two neighbors the middle terms are never affected. Carrying them out in (16) there will be a residue R_i left from each T_i and this residue must contain the middle term of T_i. We obtain:

$$R_1R_2 \cdots R_n = t_1t_2 \cdots t_n$$

and no further cancellation is possible. This proves $R_i = t_i =$ middle term. The terms on the left side of the middle term of T_1 never could be cancelled at all since no factor is on their left. So $Q_1 = 1$. Now there is no further chance for Q_2^{-1} to be cancelled, so it must be 1 too. This shows $T_i = t_i$ so that this case is settled.

2) Or else there are two neighbors T_i and $T_{i+1} = Q_{i+1}^{-1}t_sQ_{i+1}$ such that one or both of the middle terms are reached in a cancellation. They cannot be affected at the same time since their positive exponent prevents it. Now two alternatives are forced upon us:

a) t_k is affected first. Consider the product $T_iT_{i+1}T_i^{-1}$. Because of the special form of the T_r a cancellation is now possible on both sides of T_{i+1}. Carry it out, term by term, on both sides until the middle term of T_i and T_i^{-1} is reached and stop the cancellation at that point even if it is possible to go on. More than half of T_i and T_i^{-1} will have been absorbed, the middle term of T_{i+1} will not yet be reached and T_{i+1} will have lost as many factors as the other two. A few remnants from these factors will remain but they will be shorter than the loss. It is of course very easy to write this down formally. What is important is, that the length of this product is shorter than that of T_{i+1}. Consider now the substitution $B\sigma_i^{-1}$. We find:

$$B\sigma_i^{-1} = \begin{pmatrix} t_\nu, & t_i, & t_{i+1} \\ T_\nu, & T_iT_{i+1}T_i^{-1}, & T_i \end{pmatrix} \qquad \nu \neq i, i+1.$$

The product of the second line is $T_1T_2 \cdots T_n = t_1t_2 \cdots t_n$. It has still the general form of B but is shorter. So this substitution A is a braid. Since $B = A\,\sigma_i$, we see that B is a braid.

b) t_s is affected first. Then we find that $T_{i+1}^{-1}T_iT_{i+1}$ is shorter than T_i so that

$$B\sigma_i = \begin{pmatrix} t_\nu, & t_i, & t_{i+1} \\ T_\nu, & T_{i+1}, & T_{i+1}^{-1}T_iT_{i+1} \end{pmatrix} \qquad \nu \neq i, i+1$$

is a braid. This proves that $B = A\sigma_i^{-1}$ is a braid.

Our proof also shows that B can be expressed as a powerproduct of the σ_i.

The proof would also have worked if the subscripts i ran through a subset of all indices only. No condition need be put on the t_k and Q_i. (15) has to be replaced by the condition that B leaves $\prod t_i$ (in the natural order) invariant. The braids σ_i have of course to be replaced by the corresponding braids for this subset of ends.

The most general case would be finally this. A subset of the t_i and mappings of the previous kind are given. B carries $\prod t_i$ into $\prod t_j$ where the t_j form another subset also in the natural order. To reduce this case to the previous one let B_1 be a braid having the X_i as lower ends and the X_j as upper ones. The substitution $B\,B_1^{-1}$ maps $\prod t_j$ onto itself and is therefore a braid. So B is a braid. Let (i) and (j) be subsets of the indices both equal in number. Our result shows that

$$B_{(i)(j)} = \begin{pmatrix} t_i \\ t_j \end{pmatrix} \text{ is a braid.} \quad (i, j \text{ natural arrangement}). \tag{17}$$

Our results may be expressed in the theorem:

THEOREM 16. *A substitution is a braid if, and only if, it has the general form of a transformation and if it satisfies the condition of Theorem* 15. *The full group of n-braids has the σ_i as generators. A general braid can be expressed as a product of a braid of the form* (17) *followed by generators like the σ_i but concerning the lower ends of the braid only.*

A simple computation of substitutions shows that the following relations hold between the σ_i:

$$\sigma_i\sigma_k = \sigma_k\sigma_i \qquad \text{if } |i - k| \geqq 2 \tag{18}$$

$$\sigma_i\sigma_{i+1}\sigma_i = \sigma_{i+1}\sigma_i\sigma_{i+1}. \tag{19}$$

In Z. I have shown that these relations form a full set of defining relations for the group. The method is geometric and can easily be made rigorous by means of the tools developed in this paper. However a more interesting proof shall be given in a paper by F. Bohnenblust which is essentially algebraic and leads deeper into the theory of the group. I shall therefore omit a proof here especially since no use shall be made of this fact in this paper. All we shall use is that these relations hold.

A simple operation is that of removing a string. What does it mean for the substitution? Let A be the braid and remove the string with X_r as lower and

X_m as upper end. That means that we have to cancel the column referring to the image of t_r. In addition to that we have to substitute everywhere in the second row $t_m = 1$ since that describes the shrinkage in the homotopy generators.

The inverse problem in a somewhat simplified form would be:

Given a braid in the form (12), assume r does not occur among the i and m not among the j. Form a new braid A by inserting one string with X_r and X_m as respective lower and upper ends. The simplification will consist in the special position which the new string is in, we shall namely get $A(t_r) = t_m$. To do it, enlarge the meaning of B by prescribing of a new substitution C that it shall have the same effect on the t_i as B and map t_r onto t_m. This substitution will in general not be a braid since the condition concerning the product will be violated. Define two new substitutions α and β by their effect on the t_i respectively t_j and t_r and t_r resp t_m:

$$
(20) \qquad \alpha(t_i) = \begin{cases} t_i & \text{if} \quad i < r \\ t_r^{-1} t_i t_r & \text{if} \quad i > r, \end{cases} \qquad \alpha(t_r) = t_r;
$$

$$
\beta(t_j) = \begin{cases} t_j & \text{if} \quad j < m \\ t_m t_j t_m^{-1} & \text{if} \quad j > m \end{cases}, \qquad \beta(t_m) = t_m .
$$

Then $A = \beta C \alpha$ is the desired braid. We first prove that it is a braid by showing that A carries the product

$$
\prod_{i<r} t_i \cdot t_r \cdot \prod_{i>r} t_i \quad \text{into} \quad \prod_{j<m} t_j \cdot t_m \cdot \prod_{j>m} t_j .
$$

Indeed α carries it into

$$
\prod_{i<r} t_i \cdot t_r \cdot t_r^{-1} \prod_{i>r} t_i \cdot t_r = \prod t_i \cdot t_r
$$

C transforms it into

$$
\prod t_j \cdot t_m = \prod_{j<m} t_j \cdot \prod_{j>m} t_j \cdot t_m
$$

and β into

$$
\prod_{j<m} t_j \cdot t_m \prod_{j>m} t_j \cdot t_m^{-1} \cdot t_m = \prod_{j<m} t_j \cdot t_m \cdot \prod_{j>m} t_j .
$$

It is seen immediately that cancellation of the rth string leads back to B. So A is the desired braid. Obviously $A(t_r) = t_m$.

We may now combine both operations. If A is a braid, we may first cancel the rth string and then reinsert it with the same ends so that it maps t_r onto t_m. The new braid shall be denoted by $A^{(r)}$. Theorem 7 shows that A may be obtained from $A^{(r)}$ by multiplying it from the left by a uniquely determined m-pure braid (m and not r-pure because of our change of notation). That $A^{(r)}$ is uniquely determined by A follows from Theorem 8 since $A^{(r)}(t_r) = t_m$ describes the homotopy class of the rth string. We may write:

$$
A = U_m A^{(r)}, \quad U_m \ m\text{-pure}.
$$

We know how to compute $A^{(r)}$ from A and shall see a little later how U_m can be computed.

The elements S_j in their dependency on the braid shall be denoted by $S_j(A)$. In order to have well defined elements before us, we must still make an agreement about the arbitrary power of t_j that still may be added as a left factor. We choose it in such a way that the sum of the exponents of t_j in S_j is 0.

Let now A and B be two composable braids. B maps t_i into $S_j(B)^{-1}t_jS_j(B)$. A maps this into $A(S_j(B))^{-1}\cdot S_k(A)^{-1}\cdot t_k\cdot S_k(A)\cdot A(S_j(B))$, where $k = A(j)$. Since the transformer is determined to within a power of t_k we get to within such a power

$$S_k(AB) = S_k(A)A(S_j(B)), \qquad k = A(j). \tag{21}$$

On the left side and in the first factor on the right the sum is 0; in $S_j(B)$ the sum of the exponents of t_j is 0. A maps it into a power product where the sum of the exponents of t_k is 0. So (21) is correct as it stands.

A rather elementary invariant can be derived from (21). Calling $H_j(A)$ the sum of all exponents in $S_j(A)$ we obtain:

$$H_k(AB) = H_k(A) + H_j(B) \qquad k = A(j). \tag{22}$$

Defining now the "twining number" $T(A)$ as the sum of all $H_k(A)$ we get:

$$T(AB) = T(A) + T(B). \tag{23}$$

Since $T(\sigma_i^\epsilon) = \epsilon$ for $\epsilon = \pm 1$ this invariant can in case of the full group of n-braids also be explained as the sum of all exponents of the σ_i in any expression of A by the σ_i. This allows us to determine the factor commutator group without making use of the fact that the system of relations (18), (19) is complete. Making all σ_i commutative (19) gives the equality of all the σ_i. In the factor commutator group a braid A shrinks therefore to $\sigma_1^{T(A)}$. So this group is infinite cyclic and $T(A)$ gives the position of A in it.

The homology class of a string is obtained from $S_j(A)$ by the substitution of $t_j = 1$. The result may be denoted by $\bar{S}_j(A)$. In order to obtain a formula similar to (21) we must see what effect that substitution has on the second term on the right of (21). A maps t_j onto a transform of t_k. After substituting $t_k = 1$ all terms coming from t_j will disappear. Hence we may substitute $t_j = 1$ in $S_j(B)$. In addition to that, we must also substitute $t_k = 1$ in the result wherever it appears from the rest of the substitution. The same effect is achieved if the braid A is replaced by one where the jth string has been dropped. Let us denote this braid by A_{-j}. Then we have:

$$\bar{S}_k(AB) = \bar{S}_k(A)\cdot A_{-j}(\bar{S}_j(B)), \qquad k = A(j). \tag{24}$$

Consider the special case that A is k-pure. Then $k = j$, A_{-j} a unit. Hence:

$$\bar{S}_k(AB) = \bar{S}_k(A)\cdot\bar{S}_k(B); \text{ if } A \text{ is } k\text{-pure}. \tag{25}$$

A still more special case is obtained when both A and B are k-pure. Theorem 8 tells that the homotopy class of the kth string together with the ends determines any k-pure braid completely. It also shows that every homotopy class is possible. (25) means therefore that $\bar{S}_k(A)$ gives an isomorphic mapping of the group of k-pure braids with given ends onto the free group of the generators t_i with $i \neq k$. If we denote by A_{ik} the k-pure braid that is mapped onto t_i then the A_{ik} are the generators of our group and the mapping $\bar{S}_k(A)$ means just a replacement of each A_{ik} by t_i. These braids satisfy $A_{ik} = A_{ki}$. We prove this by giving at the same time the full substitution of A_{ik}^{ϵ} for $i < k$ and ϵ any integer.

$$(26)\qquad A_{ik}^{\epsilon} = A_{ki}^{\epsilon} = \begin{pmatrix} t_r, & t_i, & t_r, & t_k \\ t_r, & t_i C_{ik}, & C_{ik}^{-1} t_r C_{ik}, & C_{ik}^{-1} t_k \\ r < i \text{ or } > k, & & i \leqq r \leqq k, & \end{pmatrix}$$

$$C_{ik} = (t_i^{-1} t_k^{-1})^{\epsilon} \cdot (t_i t_k)^{\epsilon}.$$

Writing out the critical terms we see that all are transformations of generators. The product property holds so they are braids. $\bar{S}_i(A_{ik}^{\epsilon}) = t_k^{\epsilon}$, $\bar{S}_k(A_{ik}^{\epsilon}) = t_i^{\epsilon}$ and the braid reduces to a unit if we drop either the ith or the kth string. This proves all our contentions.

It is convenient to introduce also the inverse mapping F_k to $\bar{S}_k$. It maps t_i onto A_{ik}. Let it also have a meaning for t_k whose image shall be 1.

Let A be a braid and assume $k = A(j)$. We can write $A = U \cdot A^{(j)}$ where U is k-pure. Making use of (25) we obtain $\bar{S}_k(A) = \bar{S}_k(U)\bar{S}_k(A^{(j)}) = \bar{S}_k(U)$ since $A^{(j)}$ maps by definition t_j onto t_k, whence $\bar{S}_k(A^{(j)}) = 1$. Now applying F_k gives $U = F_k(\bar{S}_k(A))$ or:

$$(27)\qquad A = F_k(\bar{S}_k(A)) \cdot A^{(j)}, \qquad A(j) = k.$$

This is the algebraic form of Theorem 7.

(27) also solves the general question: given a braid B; insert a new string with the given homotopy class $\bar{S}_k(A)$. We have learned to form a braid with the homotopy class 1, it is the braid $A^{(j)}$. If we substitute this and the given $\bar{S}_k(A)$ in (27) we obtain A expressed by the A_{ik} and $A^{(j)}$. Use now (26).

Another application of (24) is this: let A be j-pure and assume of B that $B(t_j) = t_k$ hence $B^{-1}(t_k) = t_j$. Then $\bar{S}_k(B) = \bar{S}_j(B^{-1}) = 1$. This leads to $\bar{S}_k(BAB^{-1}) = B_{-j}(\bar{S}_j(A))$. BAB^{-1} is k-pure so we can apply F_k. This leads to:

$$BAB^{-1} = F_k(B_{-j}(\bar{S}_j(A))).$$

To get a still more general formula let B be now any braid. We first replace in the previous formula B by $B^{(j)}$. On the right side $B_{-j}^{(j)}$ appears. It is followed by the mapping F_k which anyhow maps t_k onto 1. So this braid may be replaced by B itself. Use now (27) on B. We obtain

$$(28)\qquad BAB^{-1} = F_k(\bar{S}_k(B)) \cdot F_k(B(\bar{S}_j(A))) \cdot (F_k(\bar{S}_k(B)))^{-1}, \quad B(j) = k.$$

B is here completely general so that we have before us the general transformation formula for a j-pure braid A. If A is given as a power product of the A_{ij} then $\bar{S}_j$ is a very trivial mapping and so is F_k. The right side is directly expressed as a power product of the A_{ik}. It is a very powerful formula that allows us to write down transformation formulas whose direct computation would be very painful.

As an application let $A = A_{ij}$ and $B = \sigma_r$. We give only the result of the computation which is now very easy:

$$(29)\qquad \sigma_r A_{ij} \sigma_r^{-1}$$

$$= \begin{cases} A_{ij} & \text{if } r \neq i-1, i, j-1, j \\ A_{i+1,j} & \text{if } r = i \\ A_{j,j-1} A_{i,j-1} A_{j,j-1}^{-1} & \text{if } r = j-1 \end{cases} \text{ but } i \neq j-1; \qquad A_{j,j-1} \text{ for } i = j-1$$

$$\begin{cases} A_{ij}^{-1} A_{i-1,j} A_{ij} & \text{if } r = i-1 \\ A_{i,j+1} & \text{if } r = j \end{cases} \text{ but } i \neq j+1; \qquad A_{j,j+1} \text{ for } i = j+1.$$

It is to be remarked that the symmetry $A_{ij} = A_{ji}$ gives other expressions for the same transforms. We note the very special cases $r = i, j$. Since a very simple computation gives $A_{i,i+1} = \sigma_i^2$ we obtain for $i < j$ the following explicit expressions of the A_{ij} in terms of the generators σ_r:

$$(30\qquad A_{ij} = \sigma_i^{-1} \sigma_{i+1}^{-1} \cdots \sigma_{j-2}^{-1} \sigma_{j-1}^2 \sigma_{j-2} \cdots \sigma_i$$
$$= \sigma_{j-1} \sigma_{j-2} \cdots \sigma_{i+1} \sigma_i^2 \sigma_{i+1}^{-1} \cdots \sigma_{j-2}^{-1} \sigma_{j-1}^{-1}.$$

As a second application we study the structure of the group I of n-braids with identity permutation. We fist prove:

LEMMA: *If A is an element of I that maps t_i onto itself for $i \leqq j$ then the S_k for $k > j$ do not depend on these t_i.*

PROOF: Dropping in the substitution A the first j columns will give a substitution that still satisfies the product condition, so it is an $(n - j)$-braid. This proves the lemma.

Now we use (27) for $j = k = 1$. On $A^{(1)}$ we use it for $k = 2$ and so on. Making use each time of the lemma we obtain a unique expression of A:

THEOREM 17. *The A_{ik} are generators of the group I. Every element can be expressed uniquely in the form:*

$$(31)\qquad A = U_1 U_2 \cdots U_{n-1}$$

where each U_j is a uniquely determined power product of the A_{ij} using only those with $i > j$. (Of course $A_{ij} = A_{ji}$).

The simple geometric meaning of this normal form shall be given later when we interpret our results in terms of projection.

What are the defining relations between these generators? Obviously those that permit the change of an arbitrary power product of them into the normal form. We must therefore find rules for interchanging factors of U_j with factors of U_i. For this purpose we derive all transformation rules for the expression $A_{rs}^{\epsilon} A_{ik} A_{rs}^{-\epsilon}$ ($\epsilon = \pm 1$). We use (28) with i instead of k for $B = A_{rs}^{\epsilon}$, $A = A_{ik}$. A simple computation yields:

THEOREM 18. *The braid $A_{rs}^{\epsilon} A_{ik} A_{rs}^{-\epsilon}$ ($\epsilon = \pm 1$) is i-pure. The following rules give its expression as i-pure braid:*

1) *If $i = r$ or s then it has already the desired form. Since the i-pure braids form a free group no other expression can be expected.*

2) *If all indices are different and if the pairs r, s and i, k do not separate each other we simply get A_{ik}.*

In all other cases we change, if necessary, first the names r, s in such a way that the arrangement i, r, s as compared with the natural arrangement of these three numbers is a permutation with the same sign as ϵ.

3) *If $k = r$ we get: $A_{is}^{-\epsilon} A_{ir} A_{is}^{\epsilon}$.*

4) *If $k = s$ we obtain:*

$$A_{is}^{-\epsilon} A_{ir}^{-\epsilon} \cdot A_{is} \cdot A_{ir}^{\epsilon} A_{is}^{\epsilon}.$$

5) *If finally the subscripts are all different and if the pairs r, s and i, k separate each other the result is:*

$$A_{is}^{-\epsilon} A_{ir}^{-\epsilon} A_{is}^{\epsilon} A_{ir}^{\epsilon} \cdot A_{ik} \cdot A_{ir}^{-\epsilon} A_{is}^{-\epsilon} A_{ir}^{\epsilon} A_{is}^{\epsilon}.$$

As defining relations the ones where i is the smallest index of all are sufficient. It also suffices to have only $\epsilon = +1$ since $\epsilon = -1$ is just only the inverse automorphism.

For braids whose permutation is not identity, a normal form is also easily obtained. Select to each of the $n!$ permutations π a braid B_π with this permutation. Any braid can then be written as a product like that in Theorem 17 followed by a B_π. This form is again unique.

The operation $A^{(j)}$ obviously satisfies:

$$(AB)^{(j)} = A^{(k)} B^{(j)}, \qquad k = B(j). \tag{32}$$

For the group I, it is therefore a homomorphic mapping and it suffices to know the result for the generators A_{ik}. $A_{ik}^{(r)} = 1$ for $r = i$ or k since A_{ik} is i- and k-pure. It is A_{ik} if $r < i$ or $> k$ as the substitution shows. If r is between i and k we apply (27) and Theorem 18. The result is:

$$A_{ik}^{(r)} = \begin{cases} 1 \quad \text{for} \quad r = i \text{ or } r = k \\ A_{ik} \quad \text{if} \quad r \text{ not between } i \text{ and } k \\ A_{ir} A_{ik} A_{ir}^{-1} = A_{kr}^{-1} A_{ki} A_{kr} \quad \text{if} \quad i < r < k. \end{cases} \tag{33}$$

We return now to the general group of n-braids. Let A be a braid that leaves t_j fixed if either $j < i$ or $j > k$. For the same reason as that in the previous lemma, we find that A maps $t_i, t_{i+1} \cdots t_k$ onto expressions depending on these variables alone. A can therefore be expressed in terms of the generators σ_i,

$\sigma_{i+1}, \cdots \sigma_{k-1}$. These braids form a subgroup denoted by G_{ik}. They behave as if they were braids of $k - i + 1$ strings (from i to k) only.

Put $c_{ik} = t_i t_{i+1} \cdots t_k$ and consider the following substitution C_{ik}:

$$C_{ik} = \begin{pmatrix} t_r & t_s \\ t_r & c_{ik}^{-1} t_s c_{ik} \\ r < i \text{ or } > k & i \leqq s \leqq k \end{pmatrix}. \tag{34}$$

It is a braid of G_{ik} since the product property holds. It is obviously commutative with any element of G_{ik} hence in the center of this group. If we drop the kth string in C_{ik} and $C_{i,k-1}$ or the ith in C_{ik} and $C_{i+1,k}$ we always obtain identical braids. $C_{i,k-1}$ leaves t_k and $C_{i+1,k}$ leaves t_i fixed. Hence:

$$C_{ik}^{(k)} = C_{i,k-1}, \qquad C_{ik}^{(i)} = C_{i+1,k} \qquad (C_{ii} \text{ means } 1). \tag{35}$$

Formula (27) gives:

$$C_{ik} = A_{ik}A_{i+1,k} \cdots A_{k-1,k}C_{i,k-1} = A_{i,i+1}A_{i,i+2} \cdots A_{ik}C_{i+1,k}. \tag{36}$$

This gives us the explicit expressions by the A_{ik}.

C_{ik} is closely related to a braid D_{ik} of G_{ik} which maps $t_i, t_{i+1} \cdots t_{k-1}$ onto t_{i+1}, $t_{i+2}, \cdots t_k$ but t_k onto $c_{ik}^{-1} t_1 c_{ik}$. We see that

$$C_{ik} = (D_{ik})^{k-i+1}. \tag{37}$$

The product $\sigma_i \sigma_{i+1} \cdots \sigma_{k-1}$ has the same effect on $t_i, t_{i+1}, \cdots t_{k-1}$ as D_{ik}. Because of the product property this suffices to establish equality. Hence:

$$C_{ik} = (\sigma_i \sigma_{i+1} \cdots \sigma_{k-1})^{k-i+1}. \tag{38}$$

THEOREM 19. *If $n = 2$ all braids are commutative. If $n \geqq 3$ let k be one of the numbers $\leqq n$. If B is commutative with every k-pure braid, then B is a power of $C_{1,n}$. This also determines of course the centers of the whole group, of I, G_{ik} and $I_{ik} = I \cap G_{ik}$.*

PROOF: If A is r-pure and $B(r) = s$, then BAB^{-1} is s-pure. Put $A = A_{ik}$. It is pure only for i and k. Since $BA_{ik}B^{-1} = A_{ik}$ we see that B can at most interchange i and k. If $n \geqq 3$ then i may be replaced by another index which shows that k remains fixed. Consequently i remains fixed too so B has identity as permutation.

We now make use of (28) where $j = k$. The braid B in the middle term on the right side came originally from B_{-k} which we introduce again. For A we take any k-pure braid so that the left side is A again. If we then apply the operation $\bar{S}_k$ to both sides we get:

$$\bar{S}_k(A) = \bar{S}_k(B) \cdot B_{-k}(\bar{S}_k(A)) \cdot (\bar{S}_k(B))^{-1}.$$

In this formula $\bar{S}_k(A)$ may be any power product T of the t_i with $i \neq k$. This shows:

$$B_{-k}(T) = d^{-1}Td, \text{ where } d = \bar{S}_k(B).$$

For $T = T_0 = \prod_{i \neq k} t_i$ we have on the other hand $B_{-k}(T_0) = T_0$. So T_0 is commutative with d. Since T_0 occurs in a generator system of the free group d is a power of T_0 say T_0^r. B_{-k} transforms all T with T_0^r. The same transformation is produced by $(C_{1,n})_{-k}^r$; B_{-k} is therefore this braid. Put now $C = B\, C_{1,n}^{-r}$. We find $C_{-k} = 1$ so C is k-pure. But C is still commutative with all k-pure braids. They form a free group with at least 2 generators whence $C = 1$ or $B = C_{1,n}^r$. This proves the contention.

For our next question we need a certain result about automorphisms of free groups. Let F be a free group with the generators t_i. Divide the subscripts into two classes p and q and in some other way into the classes g and h. We assume that there are at least two t_q and two t_h. Let x be a power product of the t_h that appears among a generator system of F and y a similar power product of the t_q. Since we assumed that there are at least two t_h, x will not be commutative with every t_h. Define now the automorphisms C and D by:

$$C(t_g) = t_g, \quad C(t_h) = x^{-a} t_h x^a; \quad D(t_p) = t_p, \quad D(t_q) = y^{-b} t_q y^b, \tag{39}$$

where a and b are positive integers. We ask for all automorphisms A that satisfy:

$$D\, A = A\, C. \tag{40}$$

C leaves all t_g as well as x invariant. Their image T under A will satisfy (because of (40)):

$$D(T) = T. \tag{41}$$

The power products $T = A(t_h)$ give $D(T) = AC(t_h) = A(x^{-a} t_h x^a)$ hence:

$$D(T) = z^{-a} T\, z^a \text{ where } z = A(x). \tag{42}$$

The equations (42) (41) exhaust (40) and only the condition that A is an automorphism will have to be taken care of.

Denote by the letter P any power product of the t_p alone, by Q one of the t_q and by R one of the t_p and y. Any T can be written in the form:

$$T = P_1 Q_1 P_2 Q_2 \cdots \tag{43}$$

where P_1 may be absent. Then:

$$D(T) = P_1 y^{-b} Q_1 y^b P_2 \cdots \tag{44}$$

Assume T satisfies (41). Each Q_i must be commutative with y^b. But y occurs in a generator system of F so Q_i is a power of y. Hence T is an R. We get

$$A(t_g) = R_g, \qquad A(x) = z = R_0. \tag{45}$$

This takes care of (41). Assume now that T satisfies (42). We get:

$$P_1 y^{-b} Q_1 y^b P_2 y^{-b} Q_2 \cdots = z^{-a} P_1 Q_1 P_2 Q_2 \cdots z^a. \tag{46}$$

If every Q is a power of y then T is an R and $D(T) = T$ must be commutative with z^a. But z is the image of a generator so is itself a generator. So T is a power of z. Since one t_h at least will not be commutative with x, its image T will not be commutative with z. So this case does not always happen.

Assume now that Q_i is the first of the Q in (46) that is not a power of y. Then the whole segments on the left of this factor in (46) must be equal since z on the right side is also an R. We obtain: (the earlier Q are powers of y)

$$P_1Q_1P_2 \cdots P_i y^{-b} = z^{-a}P_1Q_1 \cdots P_i \qquad \text{or}$$
$$z^a = R^{-1}y^bR.$$

Since y and z are generators this is only possible if $a = b$. $R^{-1}yR$ is also a generator and we get:

$$z = R^{-1}yR. \tag{47}$$

With this R put now $T = R^{-1}T_0R$. Because of $D(R) = R$ (42) gives:

$$D(T_0) = y^{-b}T_0y^b. \tag{48}$$

Writing now T_0 in the form (43) we get:

$$P_1y^{-b}Q_1y^bP_2 \cdots = y^{-b}P_1Q_1P_2 \cdots y^b.$$

The right side shows that P_1 must be absent. But also the presence of P_2 leads to a contradiction. T_0 is therefore a Q. Our results so far are:

$$A(t_g) = R_g, \; a = b, \quad A(x) = R^{-1}yR, \quad A(t_h) = R^{-1}Q_hR. \tag{49}$$

A maps the group generated by the t_g and x into the group of the t_p and y. A^{-1} satisfies $A^{-1}D^{-1} = C^{-1}A^{-1}$ where the roles of g and p are interchanged. It maps therefore the group of the t_p and y into the group of the t_g and x. The mappings are therefore one to one and this shows that the number of subscripts g is the same as that of the subscripts p.

We go now back to the braids and ask what automorphisms satisfy:

$$C_{ik}^aA = AC_{rs}^a, \qquad a > 0. \tag{50}$$

Our conditions are satisfied. The p are $< i$ or $> k$, the q satisfy $i \leqq q \leqq k$, the g are $< r$ or $> s$, the h satisfying $r \leqq h \leqq s$. $x = c_{rs}$ and $y = c_{ik}$ are generators. We must have $k - i = s - r$.

Assume a little more about the automorphism A namely that it maps each t_j onto a transform of another. Then:

$$A(t_g) = R_p^{-1}t_pR_p, \quad A(t_h) = R^{-1}Q_q^{-1}t_qQ_qR, \quad A(c_{rs}) = R^{-1}c_{ik}R \tag{51}$$

where each Q_q is a power product of the t_q, R_p and R power products of the t_p and of c_{ik}.

Split A into two substitutions:

$$B(t_g) = R_p^{-1}t_pR_p, \qquad B(t_{r+j}) = R^{-1}t_{i+j}R, \qquad 0 \leqq j \leqq k - i = s - r. \tag{52}$$

$$E(t_p) = t_p, \quad E(t_{i+j}) = Q_q^{-1}t_qQ_q \text{ where } r + j \text{ corresponds to } q. \tag{53}$$

A maps c_{rs} on one hand onto $R^{-1}c_{ik}R$; computed directly onto $\prod R^{-1}Q_q^{-1}t_qQ_qR$ whence $\prod Q_q^{-1}t_qQ_q = c_{ik}$. Now we see that E leaves c_{ik} fixed and therefore

$$A = E\,B. \tag{54}$$

We see at the same time that E is always a braid. The condition that A be an automorphism is therefore that B is one and that E is a braid. Should A be also a braid then B is one and conversely.

In case A is a braid the geometric significance of (52), (53) and (54) is obvious. B behaves as if the ith up to k^{th} string were just only one strand (only the product of the t_h and of the t_q plays a role). A is obtained by weaving the pattern E into the ith up to the kth string, then, considering this partial braid as one string only, interweaving it with the other strings according to pattern B.

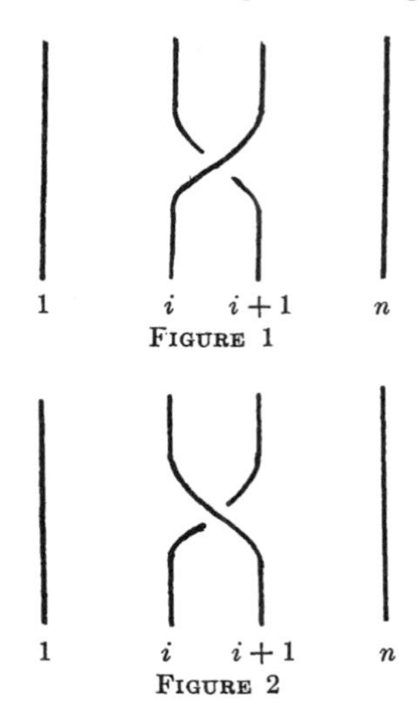

FIGURE 1

FIGURE 2

The question whether a given braid A can be considered as braid of braids amounts to checking relations of the type (50). It suffices of course to consider $a = 1$. Since we can decide whether or not they hold, this question is decided too.

THEOREM 20. *The number n is a group invariant of the group of braids with n strings.*

PROOF: Theorem 19 shows that $C_{1,n}$ is a generator of the center and therefore together with its reciprocal characterized by an inner property of the group. The number $T(C_{1,n}) = n(n - 1)$ is therefore also an invariant since it gives the position in the factor commutator group.

The structure of the group does not depend on the position of the ends. We may therefore put the ends in the special points with the coordinates $x = 0$, $y = i$, $(i = 1, 2, \cdots n)$. As ray for the Poincaré group we may select $x = 1$,

$y \leqq 0$ and as paths l_i the straight line segments form the beginning point of the ray to the ends. It is advisable to use as orientation of the plane the nega-

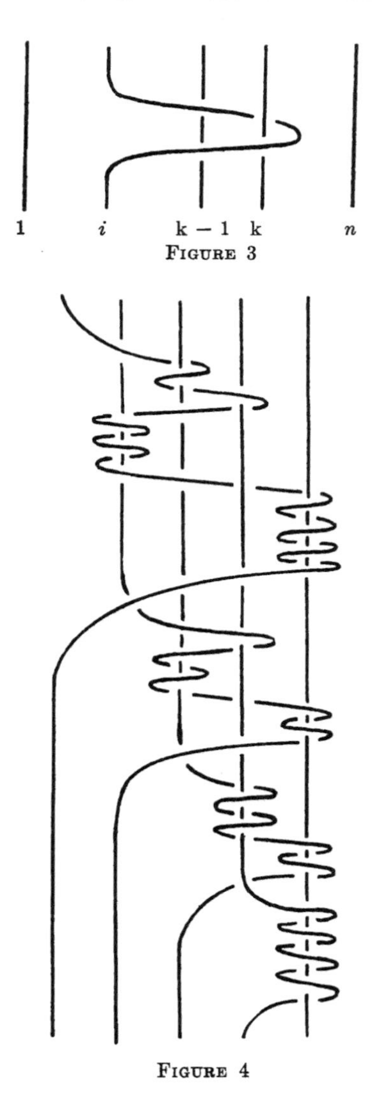

FIGURE 3

FIGURE 4

tive one, so the sense of rotation from the positive y-axis to the positive x-axis. This fixes all the necessary data and we are now in a position to interpret our results in the projection from the positive x-direction onto the yz-plane.

Theorem 4 shows us that any braid is isotopic to another one whose strings are broken lines. It leaves us so much freedom that we can assume in addition that the projection is free from any triple points and that no two double points occur at the same z-level.

Figures 1 and 2 show the generators σ_i and σ_i^{-1} in their projection. Corollary 3 to Theorem 2 shows indeed that they are reciprocal. The braid in Fig. 1 maps t_ν obviously onto itself if $\nu \neq i, i + 1$. It maps t_i onto t_{i+1} and, because of the product property, that is sufficient to establish its identity. Theorem 3 teaches how to read off from the projection of a braid its expression in terms of the generators σ_i.

Formula (38) shows that C_{ik} is simply the full twist of all the strings from the ith to the kth and that gives the geometric meaning of Theorem 19.

Formula (30) gives now the projection of the generator A_{ij}. It is indicated in Fig. 3.

The geometric meaning of the normal form of Theorem 17 and that mentioned after Theorem 18 is now also clear. Every braid is isotopic to another one whose pattern of projection is especially simple and is indicated in Fig. 4 for a special case. This pattern is unique. The braid in Fig. 4 has the expression:

$$A = A_{13}^{2}A_{14}^{-1}A_{12}^{-3}A_{14}A_{15}^{4}\cdot A_{24}^{-1}A_{23}^{2}A_{24}A_{25}^{-2}\cdot A_{34}^{3}A_{35}^{-2}\cdot A_{34}^{-1}\cdot A_{45}^{-4}.$$

Although it has been proved that every braid can be deformed into a similar normal form the writer is convinced that any attempt to carry this out on a living person would only lead to violent protests and discrimination against mathematics. He would therefore discourage such an experiment.

INDIANA UNIVERSITY

On the Theory of Complex Functions

The following pages are intended to exhibit some of the advantages obtained by a more extensive use of topological methods and notions in courses on complex variables. These methods simplify the proofs and are more flexible in their application.

We shall need as preparation the following simple properties of closed and open sets:

1) The complementary set $\overline{D}$ of an open set D is closed.

2) A continuous image of a closed and bounded set is itself closed and bounded.

3) A function that is continuous on a closed and bounded set is uniformly continuous on that set.

4) By distance of two sets S_1 and S_2 we mean the greatest lower bound ρ of the distances between any point z_1 of S_1 and any point z_2 of S_2. If S_1 and S_2 are closed and one of them is bounded we can find a special pair of points z_1 and z_2 with precisely the distance ρ. It follows that if in addition the two sets are disjoint, we have $\rho > 0$.

The proofs are so well known that we omit them here.

By arc we mean a continuous image $z(t)$ of the interval $0 \leq t \leq 1$. It is a closed and bounded set. We consider it as oriented by the orientation of the interval. Now let ζ be a point not on the arc A. We first try to find a *continuous* function $\varphi(t)$ whose value is one of the possible values of the argument of $z(t) - \zeta$. Such a $\varphi(t)$ is easily constructed if our arc is contained in a circle that does not contain ζ because it is then possible to define the argument of $z - \zeta$ as a continuous function of z in the whole circle. All we have to do, therefore, in the general case is to subdivide our arc into a finite number of parts of the previous kind. To do so, let $\rho > 0$ be the distance of ζ and A; because of the uniform continuity of $z(t)$ we can find a subdivision of A into a finite number of parts such that each part is contained in a circle of diameter ρ.

Let us assume our point ζ is not fixed but moves on a closed set S whose distance from A is $\rho > 0$. Any subdivision of A with this ρ will then work for all the points ζ at once.

The so-constructed $\varphi(t)$ is uniquely determined but for a multiple of 2π. This follows easily from the meaning of $\varphi(t)$ and its continuity.

What we really want to construct is the uniquely determined value $\varphi(1) - \varphi(0) = V(A, \zeta)$. We call it the *variation of argument* of A with respect to ζ. It is easy to show that it depends continuously on ζ and that it satisfies the equation $V(A, \zeta) = V(B, \zeta) + V(C, \zeta)$ if the arc A is subdivided into the two arcs B and C.

Returning to our closed set S and any subdivision of A into parts of diameter $< \rho$, let us connect each two consecutive endpoints of these parts by a straight line segment. We obtain thus an inscribed polygon A' that is also disjoint from S. Each of the line segments has the same variation of argument with respect to any ζ of S as the corresponding part of A. This proves $V(A', \zeta) = V(A, \zeta)$ for all ζ of S.

Theorem 1. Let S be a closed set disjoint from the arc A. If A' is an inscribed polygon that belongs to a sufficiently fine subdivision of A then $V(A', \zeta) = V(A, \zeta)$ for all ζ of S.

It is convenient to use not only arcs but also chains of arcs as paths of integration. By a chain C we mean a formal sum $\sum A_\nu$ of a finite number of arcs A_ν, each arc being oriented. One and the same arc can enter in this sum repeatedly and with either of its orientations. If ζ is not on C we generalize the variation of argument $V(C, \zeta)$ to chains by the definition

$$V(C, \zeta) = \sum_\nu V(A_\nu, \zeta).$$

Obviously this definition is additive in C.

If we disregard multiples of 2π then $V(C, \zeta) \equiv \sum_\nu (\alpha_\nu - \beta_\nu) \pmod{2\pi}$ where α_ν and β_ν are the arguments of the vectors from ζ to the endpoint and to the beginning point of A_ν. We remark, however, that it is just this neglected multiple of 2π that we wanted to define by the previous discussion.

A chain C is called closed if each point is beginning point of just as many of the arcs A_ν as it is endpoint. $V(C, \zeta)$ is then a multiple of 2π; therefore we frequently use the *winding number* $W(C, \zeta) = \frac{1}{2\pi} V(C, \zeta)$ instead of $V(C, \zeta)$. Its value is an integer; being continuous in ζ it is constant on any connected and open set D that is disjoint from the arcs of C.

If all the arcs A_ν of a chain $C = \sum A_\nu$ are rectifiable and if $f(z)$ is integrable on each A_ν we may introduce the integral of $f(z)$ on the chain C by the definition:

$$\int_C f(z)\, dz = \sum_\nu \int_{A_\nu} f(z)\, dz.$$

We are now in a position to state and to prove the most general form of the theorem of Cauchy:

Theorem 2. Let $f(z)$ be analytic in the open set D, and let C be a closed chain in D that satisfies the following condition:

The winding number $W(C, \zeta) = 0$ for every ζ of the complementary set $\overline{D}$ of D.

Then
$$\int_C f(z)\, dz = 0.$$

Proof: (A) Let C be a triangle. $W(C, \zeta) = \pm 1$ if ζ is in the interior of the triangle. Our assumption about C means, therefore, that the triangle C and its interior belong to D. The proof in this case is well known and need not be repeated since the reader can find it in most of the books on complex variables.

(B) Let C be a polygonal closed chain where each A_ν is a segment of a straight line L_ν. We assume that all the straight lines L_ν have been drawn. Each of them decomposes the plane into two convex sets, namely, two halfplanes. The intersection of a finite number of convex sets is either empty or itself convex. It thus follows that our straight lines L_ν decompose the plane into a finite number of convex sets each of them bounded by segments of the L_ν. Each convex set is either bounded and therefore an ordinary convex polygon, or else extends to infinity. In case it is bounded we select one of its vertices and draw all the diagonals from it. In this way we obtain a decomposition of the plane into triangles and into convex sets extending to infinity.

$W(C, \zeta)$ is constant in the interior of each of these parts of the plane. A point ζ_0 at the boundary of such a part either belongs to C, so that $W(C, \zeta_0)$ is undefined, or else leads to a value of $W(C, \zeta_0)$ equal to that in the interior because of the continuity of $W(C, \zeta)$.

Now let ζ be very large. Then $W(C, \zeta)$ is very small and consequently 0. This shows that $W(C, \zeta) = 0$ in each part that extends to infinity.

Next consider a triangle Δ with $W(C, \zeta) \neq 0$ for the interior of Δ. Since $W(C, \zeta) = 0$ for all ζ of $\overline{D}$, all the points of the interior of Δ belong to D. Those on the boundary of Δ also belong to D because they are either on C which is in D, or else again $W(C, \zeta)$ is $\neq 0$ for them. Thus, for such a triangle we get
$$\int_\Delta f(z)\, dz = 0.$$

Now let $\Delta_1, \Delta_2, \ldots, \Delta_n$ be all the triangles for which $W(C, \zeta) = w_i \neq 0$ if ζ is in Δ_i where $W(C, \zeta) = 0$ if ζ is in any other triangle. We assume Δ_i oriented in such a way that $W(\Delta_i, \zeta) = +1$ in the interior of Δ_i. Consider the new chain:
$$C' = C - w_1\Delta_1 - w_2\Delta_2 \cdots - w_n\Delta_n.$$

We contend: $W(C', \zeta) = 0$ for any ζ not on C'. Indeed,

a) If ζ is on the boundary of one of the parts but not on C' we can shift it a little so that it falls in the interior of a part.

b) If ζ is in Δ_i then $W(C, \zeta) = w_i$, $W(\Delta_i, \zeta) = 1$, $W(\Delta_k, \zeta) = 0$ for $k \neq i$. Hence, $W(C', \zeta) = W(C, \zeta) - w_i W(\Delta_i, \zeta) = 0$.

c) If ζ is in any other part then $W(C, \zeta) = 0$, $W(\Delta_j, \zeta) = 0$, so $W(C', \zeta) = 0$.

Now $\int_{C'} f(z)\, dz = \int_C f(z)\, dz - \sum_\nu w_\nu \int_{\Delta_\nu} f(z)\, dz = \int_C f(z)\, dz$ since $\int_{\Delta_\nu} f(z)\, dz = 0$. This reduces the proof to the case of the chain C'. We first break up each arc of C' into largest line segments Λ such that the interior of each Λ does not contain any vertex of C'. Assume now that C' contains Λ r times in one and s times in opposite orientation so that we have $C' = r\Lambda - s\Lambda + E$ where E is a chain that does not contain Λ any more. Then $0 = V(C', \zeta) = (r - s)V(\Lambda, \zeta) + V(E, \zeta)$ or $V(E, \zeta) = (s - r)V(\Lambda, \zeta)$ for all ζ not on C'. $V(E, \zeta)$ is defined and continuous even on Λ but $V(\Lambda, \zeta)$ is close to π on one side of Λ and close to $-\pi$ on the other. Hence, $r = s$. Therefore, C' contains each line segment equally often in both orientations, a fact that makes $\int_{C'} f(z)\, dz = 0$ obvious.

(C) In the general case we inscribe a sufficiently fine polygon A'_ν in each arc A_ν and replace $C = \sum A_\nu$ by $C' = \sum A'_\nu$. According to Theorem 1 we have $W(C', \zeta) = W(C, \zeta) = 0$ for all ζ of $\overline{D}$ if only the subdivision of each arc was fine enough. Then $\int_{C'} f(z)\, dz = 0$. If we could only show that for all sufficiently fine C' we have:

$$\left| \int_C f(z)\, dz - \int_{C'} f(z)\, dz \right| \leq \epsilon$$

our theorem would be proved. It obviously suffices to show the following lemma:

Let $f(z)$ be continuous in the open set D containing an arc A. If $\epsilon > 0$ is given we shall have for all sufficiently fine inscribed polygons A' the inequality:

$$\left| \int_A f(z)\, dz - \int_{A'} f(z)\, dz \right| \leq \epsilon.$$

Though the proof is well known, it may be inserted here for the convenience of the reader.

Let $z_0, z_1, \ldots, z_n$ be the vertices of the inscribed polygon A' and call $g(z, A')$ the following discontinuous function on A': $g(z, A)$ has on the line segment from $z_{\nu-1}$ to z_ν (excluding $z_{\nu-1}$) the constant value $f(z_\nu)$. Then

$$\int_{A'} g(z, A')\, dz = \sum_{\nu=1}^{n} \int_{z_{\nu-1}}^{z_\nu} f(z_\nu)\, dz = \sum_{\nu=1}^{n} (z_\nu - z_{\nu-1}) f(z_\nu)$$

is obviously a Riemann-sum for the integral $\int_A f(z)\,dz$ and is therefore close to $\int_A f(z)\,dz$ for all sufficiently fine polygons A'. If we prove on the other hand that $\int_{A'} g(z, A')\,dz$ also comes close to $\int_{A'} f(z)\,dz$ we have all that is needed.

Since the length of A' is bounded by the length of A, it suffices to show:

$$\operatorname*{Max}_{z \text{ on } A'} |f(z) - g(z, A')| \leq \epsilon \quad \text{or} \quad \operatorname*{Max}_{\substack{z \text{ on } \nu\text{-th} \\ \text{segment}}} |f(z) - f(z_\nu)| \leq \epsilon$$

provided A' is sufficiently fine.

To prove this we merely have to imbed our arc A into a closed set B that is part of D and that on the other hand contains any sufficiently fine polygon A'. The function $f(z)$ would indeed be uniformly continuous on B and the subdivision of A is found immediately.

Obviously, the set B of all points that have a distance $\leq \frac{1}{2}\rho$ from A where ρ is the distance of A and $\overline{D}$, has all the required properties.

A chain C that satisfies the conditions of our theorem shall be called homologous 0 in the open set D. If D is part of the open set D' and if C is homologous 0 in D' it need not be homologous 0 in D; we would still have to investigate whether $W(C, \zeta) = 0$ for all points of D' that are not in D. The following special case will prove important:

Assume that D can be obtained from D' by omitting the finite number of points $z_1, z_2, \ldots, z_n$ of D'. Assume that C is homologous 0 in D'. We have then to compute $W(C, z_i) = w_i$. Only if the w_i turn out to be 0 will C be homologous 0 in D. Now suppose that the w_i are not necessarily 0. We construct around each point z_i a circle C_i of so small a radius ρ that its interior with exception of z_i belongs to D. The new chain

$$C' = C - w_1C_1 - w_2C_2 - \cdots - w_nC_n$$

is then homologous 0 in D and hence $\int_{C'} f(z)\,dz = 0$ or

$$\int_C f(z)\,dz = \sum_{\nu=1}^{n} w_\nu \int_{C_\nu} f(z)\,dz. \tag{1}$$

As an example, put $f(z) = \dfrac{1}{z}\cdot$ D' is then the whole plane and $z_1 = 0$. ρ may be taken as 1. We get from (1)

$$\int_C \frac{dz}{z} = W(C, 0) \cdot \int_{C_1} \frac{dz}{z}$$

where C_1 is the unit circle. The integral on C_1 may be computed by a Riemann-sum: $\sum_{\nu=1}^{n} (z_\nu - z_{\nu-1}) \dfrac{1}{\xi_\nu}$, using as intermediate point ξ_ν the point on the unit circle halfway between $z_{\nu-1}$ and z_ν. The geometric significance of the points shows $(z_\nu - z_{\nu-1}) \dfrac{1}{\xi_\nu} = i|z_\nu - z_{\nu-1}|$ so that our Riemann-sum is $i \cdot 1$ where 1 is the length of the inscribed polygon. The integral on C_1 is therefore $2\pi i$ and we obtain:

$$\int_C \frac{dz}{z} = 2\pi i W(C, 0).$$

We now use the following definition:

Definition. If $f(z)$ is analytic in a certain neighborhood of the point z_0, with possible exception of z_0 itself, but if $f(z)$ remains bounded in that neighborhood, we call z_0 an R-point of $f(z)$.

Let us now assume that all the points z_ν in (1) are R-points of $f(z)$ and that M_i is the bound for $f(z)$ in the neighborhood of z_i. For small values of ρ we obtain the estimate

$$\left|\int_{C_i} f(z)\, dz\right| \leq 2\pi\rho M_i. \tag{2}$$

Now ρ can be taken as small as we like in (2). This shows that $\int_C f(z)\, dz = 0$, yielding the following generalization of Theorem 2.

Theorem 3. Let D be an open set and assume $f(z)$ analytic in D with exception of a finite number of R-points. Assume the closed chain C homologous 0 in D. Then $\int_C f(z)\, dz = 0$.

We next prove the integral formula of Cauchy:

Theorem 4. Make the same assumptions about $f(z)$, D and C as in the previous theorem. If z is a point not on C, where $f(z)$ is analytic, then

$$2\pi i W(C, z) f(z) = \int_C \frac{f(t)\, dt}{t - z}.$$

For a fixed z, consider the following function of t:

$$g(t) = \frac{f(t) - f(z)}{t - z}.$$

$g(t)$ is analytic in D with exception of the R-points of $f(t)$ which are also R-points of $g(t)$, and with exception of $t = z$. Since $g(t)$ approaches the limit $f'(z)$ as t approaches z, the function $g(t)$ has z as an R-point.

This shows:

$$\int_C g(t)\, dt = 0$$

or

$$\int_C \frac{f(t)}{t-z}\, dt = \int \frac{f(z)}{t-z}\, dt = f(z) \int \frac{dt}{t-z} = 2\pi i W(C, z) f(z).$$

The proof of the following lemma is well known:

Lemma. Let C be any closed or open chain and let $\varphi(t)$ be a function defined and continuous on C. The function

$$F(z) = \int_C \frac{\varphi(t)}{(t-z)^n}\, dt$$

is then defined for all z not on C and has the derivative:

$$F'(z) = \int_C \frac{n\varphi(t)}{(t-z)^{n+1}}\, dt.$$

Proof:

$$\frac{F(z+h) - F(z)}{h} - \int_C \frac{n\varphi(t)}{(t-z)^{n+1}}\, dt$$

$$= \int_C \frac{(t-z)^{n+1} - (t-z)(t-z-h)^n - nh(t-z-h)^n}{h \cdot (t-z-h)^n (t-z)^{n+1}} \varphi(t)\, dt.$$

If we expand the polynomial in the numerator and collect the terms free from h and those of the first power of h, we see that they cancel. Therefore, the numerator has the form $h^2 \cdot P(t, z, h)$ where P is a polynomial in the three variables. Our integral is therefore:

$$h \int_C \frac{P(t, z, h)}{(t-z-h)^n (t-z)^{n+1}}\, dt.$$

There is now no difficulty in getting from it an estimate of the form $|h| \cdot M$ where M is a certain bound. This proves the lemma.

Now suppose z_0 to be an R-point of $f(z)$ and take C as a small circle around z_0. We find then for all points in the interior of C and $\neq z_0$:

$$f(z) = \frac{1}{2\pi i} \int_C \frac{f(t)}{t-z}\, dt.$$

According to our lemma, the right side of this formula is a function that is analytic also for $z = z_0$ so that we get:

Theorem 5. If z_0 is an R-point of $f(z)$ we can complete the definition of $f(z)$ at z_0 in such a fashion that the completed function is analytic at z_0. Or, an R-point is a removable singularity of $f(z)$.

This theorem makes superfluous the mentioning of R-points in the preceding theorems.

Another application of our lemma is the fact that an analytic function has an infinity of derivatives, and also the generalized formulas of Cauchy:

$$2\pi i W(C, z) f^{(n)}(z) = n! \int_C \frac{f(t)}{(t - z)^{n+1}}\, dt.$$

We turn now to a discussion of the zeros of a function.

If $f(z)$ is analytic at z_0 and $f(z_0) = 0$, then z_0 is called a zero of $f(z)$. The quotient $\varphi(z) = \dfrac{f(z)}{z - z_0} = \dfrac{f(z) - f(z_0)}{z - z_0}$ is analytic in D except at $z = z_0$. Since $\lim_{z \to z_0} \varphi(z) = f'(z_0)$ the point z_0 is an R-point of $\varphi(z)$. This shows:

Theorem 6. If z_0 is a zero of $f(z)$ we can find a function $\varphi(z)$ analytic in D (especially at z_0 itself) such that $f(z) = (z - z_0)\varphi(z)$. In other words, $f(z)$ is "divisible" by $z - z_0$.

We must now decide whether an analytic function can be divisible by an arbitrarily high power of $z - z_0$, that is, if we can find for every n a function $\varphi_n(z)$ analytic in D such that

$$f(z) = (z - z_0)^n \varphi_n(z).$$

Such a point might be called a zero of infinite order.

In this case we draw a circle $|z - z_0| \leq r$ that belongs completely to D and call its periphery C. Then

$$\varphi_n(z) = \frac{1}{2\pi i} \int_C \frac{\varphi_n(t)}{t - z}\, dt$$

for any point z in the interior of C. This leads to

$$f(z) = \frac{(z - z_0)^n}{2\pi i} \int_C \frac{f(t)\, dt}{(t - z)(t - z_0)^n}$$

and to the estimate

$$|f(z)| \leq \left(\frac{|z - z_0|}{r}\right)^n \frac{Mr}{\delta} \qquad \text{for} \qquad |z - z_0| < r$$

where $M = \operatorname{Max} |f(z)|$ on C and $\delta =$ distance of z and C. If we keep z fixed we get $f(z) = 0$ for $n \to \infty$. Hence $f(z) = 0$ in any circle around z_0 that belongs completely to D and obviously any z in such a circle is now also a zero of infinite order. Indeed if z_1 is within this circle and if we define

$$\psi_n(z) = \begin{cases} 0 \text{ in our circle} \\ \dfrac{f(z)}{(z - z_1)^n} \text{ outside} \end{cases}$$

then $\psi_n(z)$ is analytic in D and $f(z) = (z - z_1)^n\psi_n(z)$. Call z_2 any point in D that can be connected with z_0 by an arc A in D. Let ρ be the distance of A and $\overline{D}$ and subdivide A into arcs of diameter $\leq \frac{\rho}{2}\cdot$ Then any point on the first part is a zero of infinite order; therefore, any point on the second part, and so on. Consequently, $f(z_2) = 0$.

This is now the point where the usual restriction of analytic functions becomes understandable. We assume that D is not only an open set but is also connected. Then we find

Theorem 7. The only analytic function with a zero of infinite order is the constant 0.

If we exclude this exceptional case of the constant 0 there will always be a maximal n such that $f(z) = (z - z_0)^n\varphi_n(z)$ and $\varphi_n(z)$ analytic at z_0. Then $\varphi_n(z_0) \neq 0$ or else $\varphi_n(z)$ in turn would be divisible by $z - z_0$ and we could therefore increase our n. Because of its continuity $\varphi_n(z)$ is $\neq 0$ not only at z_0 but also in a certain neighborhood of it. Within this neighborhood $f(z) = (z - z_0)^n\varphi_n(z)$ vanishes only at z_0. The point z_0 is called a zero of order n and we have

Theorem 8. Every zero of $f(z)$ is isolated and of finite order unless $f(z)$ is identically 0.

This is the well known theorem about the uniqueness of analytic continuation.

We derive finally the classification of isolated singularities by Riemann and Weierstrass.

z_0 is called an isolated singularity of $f(z)$ if the function is analytic in a certain neighborhood of z_0 with exception of z_0 itself.

Now assume the existence of a complex number a and of a certain neighborhood of z_0 such that $f(z)$ does not come arbitrarily close to a in that neighborhood, in other words, that $|f(z) - a| \geq \eta$ for a certain $\eta > 0$.

The function $\varphi(z) = \dfrac{1}{f(z) - a}$ is then regular in this neighborhood except for z_0 itself. Since $|\varphi(z)| \leq \dfrac{1}{\eta}$ the point z_0 is an R-point of $\varphi(z)$, and

$\varphi(z)$ may be considered analytic at z_0. Now:

$$f(z) = a + \frac{1}{\varphi(z)} \qquad \text{and} \qquad \frac{1}{f(z)} = \frac{\varphi(z)}{a\varphi(z) + 1} = \psi(z).$$

In case $\varphi(z_0) \neq 0$ the first formula shows that $f(z)$ can be defined at z_0 in such a way that it is analytic there.

Should $\varphi(z_0) = 0$, then $\psi(z)$ is analytic at z_0 and $\psi(z_0) = 0$. Assume $\psi(z) = (z - z_0)^n \chi(z)$ with $\chi(z_0) \neq 0$ and we get:

$$f(z) = (z - z_0)^{-n} \cdot \Phi(z)$$

with an analytic $\Phi(z)$ and $\Phi(z_0) \neq 0$. This is the case of a pole of order n.

Excluding the case of a regular $f(z)$ and the case of a pole, z_0 is an essential singularity. $f(z)$ must then come arbitrarily close to any complex number a in any neighborhood of z_0.

A proof of the Krein-Milman Theorem*

Let **V** be a vector space over the reals of finite or infinite dimension. The intersection, a homomorphic image and the inverse of a homomorphic image of convex sets is convex.

Definition 1: A convex subset K' of a convex set K is called a face of K if the following is true: $x \in K'$; $y_1, y_2 \in K$; $x = ty_1 + (1 - t)y_2$ with $0 < t < 1$ (note the inequality) implies that $y_1, y_2 \in K'$.

Examples: The empty set and K itself are faces of K.

Definition 2: If the empty set and K itself are the only faces of K we call K ultimate.

Examples: **V**, a linear subspace of **V**, a point of **V**.

Definition 3: An ultimate face of **V** that is a point is called an extremal point of K.

Lemma 1: Any intersection of faces of K is a face of K.

Proof: Let $K' = \bigcap_{\alpha} K_\alpha$, each K_α a face of K. Suppose $x \in K'$, $y_1, y_2 \in K$ $x = ty_1 + (1 - t)y_2$, $0 < t < 1$. Since $x \in K_\alpha$ we have $y_1, y_2 \in K_\alpha$ so $y_1, y_2 \in K'$.

Lemma 2: If K_1 is a face of K and K_2 a face of K_1, then K_2 is a face of K.

Proof: $x \in K_2$, $y_1, y_2 \in K$, $x = ty_1 + (1 - t)y_2$, $0 < t < 1$ implies, since $x \in K$, that $y_1, y_2 \in K_1$. Now $x \in K_2$ implies $y_1, y_2 \in K_2$.

Lemma 3: Let ϕ be a homomorphic map of V into some space $\bar{V}$, $\bar{K}$ the image of K, $\bar{K}'$ a face of $\bar{K}$. Then $K' = \phi^{-1}(\bar{K}') \cap K$ is a face of K and is certainly $\neq K$ if $\bar{K}' \neq \bar{K}$.

Proof: The convexity of K' follows from the initial remarks. Let $x \in K'$, $y_1, y_2 \in K$, $x = ty_1 + (1 - t)y_2$, $0 < t < 1$. Then $\phi(x) = t \cdot \phi(y_1) + (1 - t)\phi(y_2)$ and $\phi(x) \in \bar{K}'$, $\phi(y_1)$, $\phi(y_2) \in \bar{K}$. Hence $\phi(y_1)$, $\phi(y_2) = \bar{K}'$ or $y_1, y_2 \in \phi^{-1}(\bar{K}')$. Since y_1, y_2 are in K we get $y_1, y_2 \in K'$.

---

*) A letter from Artin to M. Zorn, *Picayune Sentinel*, University of Indiana (1950).

The Krein-Milman Theorem

We do not formulate it in the dual, but in the space V. Let $\hat{V}$ be any linear space of functionals (maps into real numbers) of V ($\hat{V}$ need not be the dual) such that for any $x \neq 0$ there is a $\varphi \in \hat{V}$ with $\varphi(x) \neq 0$; so $\hat{V}$ is big enough to distinguish the vectors of $\mathbf{V}$. In $\mathbf{V}$ we introduce the weak topology induced by $\hat{V}$: A neighborhood of 0 in $\mathbf{V}$ is described by an $\epsilon > 0$ and a finite number $\varphi_1, \varphi_2, \ldots, \varphi_n$ of elements of $\hat{V}$ and consists of all x such that $|\varphi_\nu(x)| < \epsilon$, $\nu = 1, 2, \ldots, n$.

Let $\mathfrak{F}$ be the family of *convex* and *compact* subsets of $\mathbf{V}$. It satisfies the "finite intersection property": If each $K_\alpha \in \mathfrak{F}$ and a finite number of K_α has a non empty intersection then $\bigcap_\alpha K_\alpha$ is non empty and an element of $\mathfrak{F}$.

The word face is from now on used only for compact faces (so $\in \mathfrak{F}$).

Let K be a non empty element of $\mathfrak{F}$. Consider the set of all *non empty* faces of K (K itself is in this set). We order them partially by inclusion property and look in the descending direction.

The unmentionable lemma shows: On every non empty face of K there is at least one non empty ultimate face of K.

Each $\varphi \in \hat{V}$ gives a continuous map of V into the real line and $\varphi(K)$ is compact on the real line, hence $-\varphi(K)$ is convex—a closed and bounded interval. If a is real, $\varphi^{-1}(a)$ is a closed and convex set of V, so $\varphi^{-1}(a) \cap K$ is in $\mathfrak{F}$. If a is an endpoint of $\varphi(K)$, then a is a face of $\varphi(K)$ (the only case where we really use the inequality $0 < t < 1$), hence (Lemma 3) $\varphi^{-1}(a) \cap K$ is a face of K; suppose K is ultimate and non empty, then $\varphi^{-1}(a) \cap K = K$ hence $\varphi(K) = a$. So $\varphi(K)$ is a point for all $\varphi \in \hat{V}$. Since $\hat{V}$ distinguishes all points of V it follows that K is a point. This shows:

Lemma 4. On every non empty face of K there are extremal points of K. Next we show:

Lemma 5. Let $K \in \mathfrak{F}$ and $x_0 \notin K$. There exists a $\varphi \in \hat{V}$ such that $\varphi(x_0)$ is *not* in the interval $\varphi(K)$.

Proof: Since K is closed, there is a neighborhood of x_0, described by ϵ, $\varphi_1, \varphi_2, \ldots, \varphi_n$ and consisting of all x such that $|\varphi_\nu(x) - \varphi_\nu(x_0)| < \epsilon$, $\nu = 1, 2, \ldots, n$ which is *not* in K.

Let R_n be an n-space of n-vectors $(\xi_1, \xi_2, \ldots, \xi_n)$ and map $\mathbf{V}$ into R_n by $\phi(x) = (\varphi_1(x), \varphi_2(x), \ldots, \varphi_n(x))$. $\phi(K)$ is convex and compact in R_n, call it $\bar{K}$. A point of $\bar{K}$ has the form $(\varphi_1(x), \ldots, \varphi_n(x))$ with $x \in K$, so that for at least one ν we must have $|\varphi_\nu(x) - \varphi_\nu(x_0)| \geq \epsilon$. This shows that the neighborhood $|\xi_\nu - \varphi_\nu(x_0)| < \epsilon$, $\nu = 1, \ldots, n$ of $\phi(x_0)$ in R_n is not in $\bar{K}$.

Let now $c_1\xi_1 + c_2\xi_2 + \cdots + c_n\xi_n = b$ be the equation of a hyperplane in R_n that separates $\bar{K}$ and $\phi(x_0)$; (this finite dimensional fact is easily proved by putting a metric on the R_n, finding a point $y_0 \in \bar{K}$ closest to x_0

and constructing a perpendicular plane). Put

$$\varphi(x) = c_1\varphi_1(x) + c_2\varphi_2(x)$$

Then $\varphi \in \hat{V}$ and $\varphi(K)$ is an interval not containing $\varphi(x_0)$.

Theorem. Let S be the set of extremal points of K and K^* the smallest element of $\mathfrak{F}$ that contains S. Then $K = K^*$.

Proof: $K^* \subset K$ is trivial. Assume $x_0 \in K$ but $x_0 \notin K^*$. Let φ be in $\hat{V}$ such that $\varphi(x_0)$ is not in $\varphi(K^*)$. $\varphi(K)$ is then larger than $\varphi(K^*)$. Let c be an endpoint of $\varphi(K)$ not in $\varphi(K^*)$. $\varphi^{-1}(c) \cap K$ is a face of K that has no point in common with K^*. On this face there is according to Lemma 4 an element of S. This contradicts $S \subset K^*$. Hence $K = K^*$.

THE INFLUENCE OF J. H. M. WEDDERBURN ON THE DEVELOPMENT OF MODERN ALGEBRA

It is obvious that the title of this paper is presumptuous. Nobody can give in a short article a really exhaustive account of the influence of Wedderburn on the development of modern algebra. It is too big an undertaking and would require years of preparation. In order to present at least a modest account of this influence it is necessary to restrict oneself rather severely. To this effect we shall discuss only the two most celebrated articles of Wedderburn and try to see them in the light of the subsequent development of algebra. But even this would be too great a task. If we would have to mention all the consequences and applications of his theorems we could easily fill a whole volume. Consequently we shall discuss only the attempts the mathematicians made to come to a gradual understanding of the *meaning* of his theorems and be satisfied just to mention a few applications.

For the understanding of the significance that Wedderburn's paper *On hypercomplex numbers* (Proc. London Math. Soc. (2) vol. 6, p. 77) had for the development of modern algebra, it is imperative to look at the ideas his predecessors had on the subject.

The most striking fact is the difference in attitude between American and European authors. From the very beginning the abstract point of view is dominant in American publications whereas for European mathematicians a system of hypercomplex numbers was by nature an extension of either the real or the complex field. While the Europeans obtained very advanced results in the classification of their special cases with methods that were not well adapted to generalization, the Americans achieved an abstract formulation of the problem, developed a very suitable terminology, and discovered the germs of the modern methods.

On the American side one has first of all to consider the very early paper by B. Peirce, *Linear associative algebras* (1870, published in Amer. J. Math. vol. 4 after his death). In it he states explicitly that mathematics should be an *abstract* logical scheme, the absence of a special interpretation of its symbols making it more useful in that the same logical scheme will in general reflect many diverse physical situations. Although it is true that he was actually able to introduce and treat only the general linear associative algebra over the complex field, yet he clearly had in mind much more, and it is his attitude which leads to the modern postulational method. In his treatment of algebras he gives a rational proof of the existence of an idempotent and employs the well known Peirce decomposition of an algebra relative to an idempotent. His results were put in a more readable form by H. E. Hawkes, *On hypercomplex number systems* (Trans. Amer. Math. Soc. vol. 3 (1902)). A correct definition of an associative algebra over an arbitrary field seems to be given for the first time by L. E. Dickson, *Definitions of a linear associative algebra by independent postulates* (Trans. Amer. Math. Soc. vol. 4 (1903)).

At that time several European mathematicians, Molien, Cartan, and Frobenius, had already arrived (always for the special cases of the real or complex field) at many of the results of the modern theory. The notions radical, semisimple, simple had been found and the decomposition of a semisimple algebra into simple components proved. Cartan derived the structure of the simple algebras but apparently without recognizing the possibility of stating the result in the very simple form Wedderburn discovered. It has to be borne in mind that all these authors had the complex field at their disposal and were therefore never hestitant to use roots of algebraic equations. This fact made a direct generalization of their results to arbitrary fields very difficult.

Wedderburn succeeded in a synthesis of these two lines of investigation. He extended the proof of all the structural theorems found by the European mathematicians for the special cases of the real and complex field to the case of an arbitrary field. By the effective use of a calculus of complexes (analogous to that which had been used in the treatment of finite groups) combined with the Peirce decomposition relative to an idempotent, he was able to prove his theorems within the given field and in a simpler way. He was the first to find the real significance and meaning of the structure of a simple algebra. We mean by this the gem of the whole paper, his celebrated:

"THEOREM 22—*Any simple algebra can be expressed as the direct product of a primitive algebra and a simple matric algebra.*"

In his terminology primitive algebra means the same thing as what we now call division algebra.

This extraordinary result has excited the fantasy of every algebraist and still does so in our day. Very great efforts have been directed toward a deeper understanding of its meaning.

In the first period following his discovery the work consisted mainly in a polishing up of his proofs. But the fundamental ideas of all these later proofs are already contained in his memoir.

In the meantime a great change in the attitude of the algebraists had taken place. The European school had discovered the great advantage of the abstract point of view which had been emphasized so early in the American school. The algebraists began to analyze Wedderburn's methods and tried to find an even more abstract background.

The essential point in the definition of an algebra is that it is a vector space of finite dimension over a field. This fact allows us to conclude that ascending and descending chains of subalgebras will terminate. After the great success that Emmy Noether had in her ideal theory in rings with ascending chain condition, it seemed reasonable to expect that in rings where the ascending and the descending chain condition holds for left ideals one should obtain results similar to those of Wedderburn. As one of the papers written from this point of view we mention E. Artin, *Zur Theorie der hyperkomplexen Zahlen* (Abh. Math. Sem. Hamburgischen Univ. vol. 5 (1926)). In 1939 C. Hopkins showed (*Rings with minimal condition for left ideals*, Ann. of Math. vol. 40) that the descending chain condition suffices.

Independently of Wedderburn's paper, the representation theory of groups had been developed under the leadership of Frobenius, Burnside, and I. Schur. These mathematicians had been very well aware of the connection with algebras, a connection given by the notion of a group ring. But little use was made of the theory of algebras.

It was Emmy Noether who made the decisive step. It consisted in replacing the notion of a matrix by the notion for which the matrix stood in the first place, namely, a linear transformation of a vector space.

Emmy Noether introduced the notion of a representation space—a vector space upon which the elements of the algebra operate as linear transformations, the composition of the linear transformations reflecting the multiplication in the algebra. By doing so she enables us to use our geometric intuition. Her point of view stresses the essential fact about a simple algebra, namely, that it has only one type of irreducible space and that it is faithfully represented by its operation on this space. Wedderburn's statement that the simple algebra is a total matrix algebra over a quasifield is now more understandable. It simply means that all transformations of this space which are linear with respect to a certain quasifield are produced by the algebra. This treatment of algebras may be found in van der Waerden's *Moderne Algebra*.

Recently it has been discovered that this last described treatment of simple algebras is capable of generalization to a far wider class of rings.

One considers a ring R and an additive group V with R as left operator domain—V playing the role of the representation space and called R-space for short. Chevalley and Jacobson proved a direct generalization of Wedderburn's theorem if two simple axioms are satisfied: That the ring R is faithfully represented by its action on V and that V is irreducible (this means that 0 and V are the only R-subspaces of V). In these terms the proof is essentially simple and geometrical, no idempotents being required, and no finiteness assumption on R.

In homage to J. H. M. Wedderburn we present in fuller detail this modern proof of his theorem.

Let R be a ring, V an R-space satisfying the two axioms stated above.[1] We shall show that V is naturally a vector space over a certain quasifield D and that practically all D-linear transformations of V are produced by elements of R.

To construct the quasifield is easy. Let D be the set of all homomorphisms of V into itself. D is a ring from first principles. Since the kernel of a nonzero element of D is an R-subspace of V which is different from V, this kernel is zero, and the element is an isomorphism. Since the image of V under this isomorphism is an R-subspace of V which is not zero, it is all of V, and we have an isomorphism of V onto V. Such a map has an inverse and we see that D is a quasifield. We have obtained in a natural, invariant manner the quasifield which Wedderburn obtained only in a noninvariant way as subring of R.

We denote the typical element of D by d and write these elements on the right of V so that our space V becomes now a right vector space over the quasifield D.

If W is a D-subspace of V, then the set of all elements of R which annihilate W is a left ideal of R which we shall call $W^{\#}$. If L is a left ideal of R, then the set of all elements of V annihilated by L is a D-subspace of V which we shall denote by $L^{\flat}$.

We can now state and prove the fundamental lemma:

$$(L \cap (\xi \cdot D)^{\#})^{\flat} = L^{\flat} + \xi \cdot D \tag{1}$$

for any left ideal L of R and any element ξ of V.

PROOF. The right-hand side is trivially contained in the left. If $L\xi = 0$, the equation becomes $L^{\flat} = L^{\flat}$. It remains only to prove that the left-hand side is contained in the right under the assumption that $L\xi \neq 0$. Since $L\xi$ is a subspace of V, and V is irreducible, $L\xi = V$: every element of V can be expressed in the form $l\xi$ where $l \in V$. Let η be an

[1] I follow a presentation given by Mr. J. T. Tate.

element of the left side of (1). It is annihilated by $L \cap (\xi D)^{\#}$, hence by every $l \in L$ which is in $(\xi \cdot D)^{\#}$. η is consequently annihilated by every $l \in L$ for which $l\xi = 0$. Let now ζ be any element of V; if we write it in the form $\zeta = l\xi$ and map it onto the element $l\eta$ of V we have before us a well defined map. If indeed $\zeta = l\xi = l\xi_1$ then $(l - l_1)\xi = 0$ hence $(l - l_1)\eta = 0$ or $l\eta = l_1\eta$. That this map $l\xi \to l\eta$ is a homomorphism of V into V follows from the fact that L is a left ideal; as such it is a certain element d of D and satisfies $(l\xi)d = l\eta$ for all $l \in L$. The element $\eta - \xi d$ is therefore annihilated by L and is consequently an element of $L^{\flat}$. This shows

$$\eta \in L^{\flat} + \xi \cdot D,$$

which is what we were trying to prove. It is of course the construction of the element d which is the heart of this method.

Let W be any D-subspace of V. If we substitute $L = W^{\#}$ in (1), the left side becomes

$$(W^{\flat} \cap (\xi D)^{\#})^{\flat}.$$

Since obviously $(A + B)^{\#} = A^{\#} \cap B^{\#}$ for any two D-subspaces A and B of V we obtain from (1)

$$(W + \xi D)^{\#\flat} = W^{\#\flat} + \xi D. \tag{2}$$

This we can use to argue in the following manner.

(3) If $W = W^{\#\flat}$ *then* $(W + \xi D) = (W + \xi D)^{\#\flat}$.

Combining a repeated application of (3) with

$$0^{\#\flat} = 0 \tag{4}$$

we obtain the

THEOREM.

$$W_0^{\#\flat} = W_0 \tag{5}$$

for any finite-dimensional $W_0 = \xi_1 D + \xi_2 D + \cdots + \xi_r D$.

The only gap in the argument was the proof of (4): $0^{\#\flat} =$ (trivially) $R^{\flat} =$ an R-subspace of V (which is not all of V) $= 0$, again using the irreducibility of V.

Now let $\xi_1, \xi_2, \cdots, \xi_r$ be a finite number of elements of V, linearly independent over the quasifield D. Let $W = \xi_1 D + \cdots + \xi_{r-1} D$. Since $\xi_r \notin W = W^{\#\flat}$ it follows that $W^{\#} \cdot \xi_r \neq 0$. Therefore, as usual, by the irreducibility of V we have $W^{\#} \cdot \xi_r = V$. Consequently there exists an element of $W^{\#}$ which annihilates $\xi_1, \xi_2, \cdots, \xi_{r-1}$ and sends ξ_r into any element of V. Combining such elements together we find an element of R which sends the vectors ξ_i independently into any set of r elements of V. Viewing the ξ_i as a basis for a finite-dimensional D subspace W_0 of V we see that any given D-linear map of W_0 into V can be produced by an element of the ring R. This is what we meant by the statement that "practically all" D-linear maps of V were produced by elements of R.

To specialize this result we must add the axiom that R satisfies the descending chain condition on left ideals (this is obviously true if R is Wedderburn's simple algebra). An ascending chain $W_1 \subset W_2 \subset \cdots$ of finite-dimensional D-subspaces of V leads to a descending chain $W_1^{\#} \supset W_2^{\#} \cdots$ of left ideals of R because of the statement $W^{\#\flat} = W$. Therefore V must satisfy the ascending chain condition on finite-dimensional D-subspaces. This is possible only if V itself is finite-dimensional over D. In this case our previous result shows that *every* D-linear map of V is produced by an element of R, and we have therefore obtained Wedderburn's theorem in geometric form.

As we have stated at the beginning it is not our intention to discuss the many applications Wedderburn's theorem has found, for instance, the investigations on division algebras by Wedderburn, Dickson, and others. They lead finally to a complete description of all simple algebras over an algebraic number-field by A. Albert, R. Brauer, H. Hasse, and Emmy Noether, or the theory of modular representations of algebras and groups by R. Brauer.

Let us now consider the theorem of Wedderburn concerning finite fields (*A theorem on finite algebras*, Trans. Amer. Math. Soc. vol. 6 (1905)) and its influence on the development of modern algebra. One sees immediately that the characteristic of such a field K is a prime $p > 0$ and that the number of elements of K is a power p^r of p.

In 1903 E. H. Moore had determined all commutative fields of this type. The result was that to a given number p^r of elements there exists (apart from isomorphisms) only one field, namely, the Galois field of degree r and characteristic p. The proof for this fact was simplified considerably by Steinitz. It is his proof one finds in modern books on algebra.

In 1905 Wedderburn found the complete answer to our question in a paper entitled *A theorem on finite algebras*, where he proves that every field with a finite number of elements is automatically commutative (under multiplication) and therefore a Galois field.

Wedderburn introduces the center C of K and also the normalizer N_α of any element α of K. It is obvious that C and N_α are subfields and that $C \subset N_\alpha$ for each α. Denoting by q the number of elements of

C we find q^{n_α} resp. q^n for the number of elements in N_α resp. K where n_α and n are the degrees of N_α resp. K over C. Since K is an extension of N_α, the degree n_α divides n.

Wedderburn then considers the multiplicative group of K of order q^n-1. He divides it into classes of conjugate elements and obtains an identity of the form:

$$q^n - 1 = (q - 1) + \sum_{n_\alpha | n, n_\alpha < n} \frac{q^n - 1}{q^{n_\alpha} - 1} \tag{6}$$

where he unites the classes with only one element in the term $q-1$ and where the sum runs over certain divisors n_α of n, the same divisors possibly several times.

In §4 of his paper he shows the impossibility of (6) for $n>1$, making use of divisibility properties of numbers of the form a^n-b^n which are hard to establish. In §5 he gives another arrangement of this proof, again making use of these divisibility properties. A third proof by Dickson is based on similar ideas.

This result of Wedderburn has fascinated most algebraists to a very high degree and several attempts were made to simplify the proofs. Artin (Abh. Math. Sem. Hamburgischen Univ. vol. 5) gave a proof that did not make use of (6) and the divisibility properties but the proof is somewhat lengthy.

The first really simple proof of our theorem was given by E. Witt, *Über die Kommutativität endlicher Schiefkörper* in 1931 (Abh. Math. Sem. Hamburgischen Univ. vol. 8). Witt starts from (6) and makes the following simple remark:

If $\phi_n(x)$ is the nth cyclotomic polynomial, then each term in the sum on the right of (6) and also q^n-1 are obviously divisible by $\phi_n(q)$. Consequently $\phi_n(q) \mid q-1$. Since $\phi_n(q) = \prod(q-\epsilon)$ where ϵ runs through the primitive nth roots of unity, we have $\phi_n(q) > q-1$ if $n>1$, and this shows the impossibility of $n>1$.

In 1933 a paper by C. C. Tsen, *Divisionalgebren über Funktionenkörpern* (Nachr. Ges. Wiss. Göttingen (1933)) shed a new light on the whole question. Tsen did not investigate finite fields, but he worked with algebraic fields F of transcendency degree 1 with an algebraically closed field of constants. He proved that there does not exist any non-commutative extension field of finite degree. The method of his proof yielded really a much stronger theorem, namely:

If $N(x_1, x_2, \cdots, x_n)=0$ is an algebraic equation in F without constant term and if the total degree d is smaller than the number n of unknowns x_i, there exists a nontrivial solution in F.

If one knows this theorem for a given field F then F cannot have any noncommutative extension field E of finite degree. To see this let $\xi = x_1\omega_1 + \cdots + x_n\omega_n$ be the generic element of E ($\omega_1, \cdots$ a basis) and let $N(x_1, x_2, \cdots, x_n)$ be the reduced norm in E/F. N is a homogeneous form of $x_1, x_2, \cdots, x_n$ of a degree d which is less than n if E is noncommutative. The theorem would give the existence of a $\xi \neq 0$ whose norm is 0, which is a contradiction.

It occurred immediately to the mathematicians that possibly a Galois' field F would have the same property, so that Wedderburn's theorem would appear as a consequence of a much more general theorem on Galois fields.

In 1935 C. Chevalley (*Demonstration d'une hypothèse de M. Artin*, Abh. Math. Sem. Hamburgischen Univ. vol. 11) proved this conjecture.

Wedderburn's theorem is therefore the special case of a more general Diophantine property of fields and thus has opened an entirely new line of research.

EMIL ARTIN

Titles in This Series